**ALSO BY CHARLES FISHMAN**

The Big Thirst

The Wal-Mart Effect

# ONE GIANT LEAP

## THE IMPOSSIBLE MISSION THAT FLEW US TO THE MOON

CHARLES FISHMAN

SIMON & SCHUSTER
New York London Toronto Sydney New Delhi

Simon & Schuster
1230 Avenue of the Americas
New York, NY 10020



First Simon & Schuster hardcover edition June 2019



Interior design by Lewelin Polanco

Manufactured in the United States of America

1 3 5 7 9 10 8 6 4 2

Library of Congress Cataloging-in-Publication Data is available.

ISBN 978-1-5011-0629-3
ISBN 978-1-5011-0631-6 (ebook)

PHOTO CREDITS

*Insert One*—1, 2: Cecil Stoughton, White House Photographs, John F. Kennedy Presidential Library; 3, 4, 5, 6, 7: The Charles Stark Draper Laboratory; 8: NASA; 9: NASA/The Tindall Family; 10: The Tindall Family; 11, 12, 13: Northrop Grumman Corporation; 14, 15, 16, 17: NASA; 18: NASA, John Hancock/Apollo Lunar Surface Journal

*Insert Two*—1, 2, 3, 4, 5, 6, 7, 8, 9: NASA; 10: NASA, Erik van Meijgaarden/Apollo Lunar Surface Journal; 11: Getty Images; 12, 13, 14: NASA

**TO NICOLAS**

*Who always reaches for the Moon,*
*and whose footprints are on these pages,*
*and on my heart*

# CONTENTS

# PREFACE
## The Mystery of Moondust

Boy this thing sure flies nice.

> **Pete Conrad, Apollo 12 commander**
> *at the controls of lunar module* Intrepid, *preparing to fly to a pinpoint landing on the Moon*[1]

The Moon has a smell.

It has no air, but it has a smell.

Each pair of Apollo astronauts to land on the Moon tramped lots of Moondust back into the lunar module—it was deep gray, fine-grained and extremely clingy—and when they unsnapped their helmets, they immediately noticed the smell.

"We were aware of a new scent in the air of the cabin," said Neil Armstrong, the first man to set foot on the Moon, "that clearly came from all the lunar material that had accumulated on and in our clothes." To Armstrong, it was "the scent of wet ashes." To his Apollo 11 crewmate Buzz Aldrin, it was "the smell in the air after a firecracker has gone off."

All the astronauts who walked on the Moon noticed it, and many of them commented on it to Mission Control. Harrison Schmitt, the geologist who flew on Apollo 17, the last lunar landing, said after his second Moon walk, "Smells like someone's been firing a carbine in here." Almost unaccountably, no one had warned lunar module pilot Jim Irwin about the dust. When he took off his helmet inside the cramped lunar module cabin, he said, "There's a funny smell in here." His Apollo 15 crewmate

Dave Scott said: "Yeah, I think that's the lunar dirt smell. Never smelled lunar dirt before, but we got most of it right here with us."[2]

Moondust was a mystery that the National Aeronautics and Space Administration had, in fact, thought about. Cornell University astrophysicist Thomas Gold warned NASA that the dust had been isolated from oxygen for so long that it might well be highly chemically reactive. If too much dust was carried inside the lunar module's cabin, the moment the astronauts repressurized it with air and the dust came into contact with oxygen, it might start burning, or even cause an explosion. (Gold, who correctly predicted early on that the Moon's surface would be covered with powdery dust, also had warned NASA that the dust might be so deep that the lunar module and the astronauts themselves could sink irretrievably into it.)[3]

Among the thousands of things they were keeping in mind while flying to the Moon, Armstrong and Aldrin had been briefed about the very small possibility that the lunar dust could ignite. It was, said Aldrin, "the worry of a few. A late-July fireworks display on the Moon was not something advisable."

Armstrong and Aldrin did their own test. They took a small sample of lunar dirt that Armstrong had scooped into a lunar sample bag and put in a pocket of his spacesuit right as he stepped onto the Moon—a contingency sample in case, for some reason, the astronauts had to leave suddenly without collecting rocks. Back inside the lunar module the astronauts opened the bag and spread the lunar soil out on top of the ascent engine. As they repressurized the cabin, they watched to see if the dirt started to smolder. "If it did, we'd stop pressurization, open the hatch and toss it out," explained Aldrin. "But nothing happened."[4]

The Moondust turned out to be so clingy and so irritating that on the one night that Armstrong and Aldrin spent in the lunar module on the surface of the Moon, they slept in their helmets and gloves, in part to avoid breathing the dust floating around inside the cabin.[5]

NASA had anticipated the dust, and the danger. The smell was a surprise.

By the time the Moon rocks and dust got back to Earth—a total of 842 pounds from six lunar landings—the smell was gone. Scientists think the rocks and dirt were chemically reactive, as Gold theorized, but that the air and moisture the rocks were exposed to in their sample boxes on the way to Earth "pacified" them, releasing whatever smell there was to be released.

Scientists who have studied the rocks and dirt and handled them and sniffed them say they have no odor at all. And no one has quite figured out what caused it, or why it was so like spent gunpowder, which is chemically nothing like Moon rock. "Very distinctive smell," said Apollo 12 commander Pete Conrad. "I'll never forget. And I've never smelled it again since then."[6]

---

In 1999, as the century was ending, the Pulitzer Prize–winning historian Arthur Schlesinger Jr. was among a group asked what the most significant human achievement of the 20th century was. In ranking the events, Schlesinger said, "I put DNA and penicillin and the computer and the microchip in the first 10 because they've transformed civilization." But in 500 years, if the United States of America still exists, most of its history will have faded to invisibility. "Pearl Harbor will be as remote as the War of the Roses," said Schlesinger. "The one thing for which this century will be remembered 500 years from now was: This was the century when we began the exploration of space." He picked the first Moon landing, Apollo 11, as the most significant event of the 20th century.[7]

The leap from one small planet to its even smaller nearby Moon may well look modest when space travel has transformed the solar system—a trip no more ambitious than the way we think of a flight from Dallas to New York City today. But it is hard to argue with Schlesinger's larger observation: in the chronicle of humanity, the first missions by people from Earth through space to another planetary body are unlikely ever to be lost to history, to memory, or to storytelling.

The leap to the Moon in the 1960s was an astonishing accomplishment. But why? What made it astonishing? We've lost track not just of the details; we've lost track of the plot itself. What exactly was the hard part?

The answer is simple: when President John Kennedy declared in 1961 that the United States would go to the Moon, he was committing the nation to do something we couldn't do. We didn't have the tools, the equipment—we didn't have the rockets or the launchpads, the spacesuits or the computers or the zero-gravity food—to go to the Moon.

And it isn't just that we didn't have what we would need; we didn't even know what we would need. We didn't have a list; no one in the world had a list. Indeed, our unpreparedness for the task goes a level deeper: we didn't even know how to fly to the Moon. We didn't know what course to fly to get there from here. And, as the small example of lunar dirt shows, we didn't know what we would find when we got there. Physicians worried that people wouldn't be able to think in zero gravity. Mathematicians worried that we wouldn't be able to work out the math to rendezvous two spacecraft in orbit—to bring them together in space, docking them in flight both perfectly and safely. And that serious planetary scientist from Cornell worried that the lunar module would land on the Moon and sink up to its landing struts in powdery lunar dirt, trapping the space travelers.

Every one of those challenges was tackled and mastered between May 1961 and July 1969. The astronauts, the nation, flew to the Moon because hundreds of thousands of scientists and engineers, managers and factory workers unraveled a series of puzzles, a series of mysteries, often without knowing whether the puzzle had a good solution.

In retrospect, the results are both bold and bemusing. The Apollo spacecraft ended up with what was, for its time, the smallest, fastest, and most nimble and most reliable computer in a single package anywhere in the world. That computer navigated through space and helped the astronauts operate the ship. But the astronauts also traveled to the Moon with paper star charts so they could use a sextant to take star sightings—like the explorers of the 1700s from the deck of a ship—and cross-check their computer's navigation. The guts of the computer were stitched together by women using wire instead of thread. In fact, an arresting amount of work across Apollo was done by hand: the heat shield was applied to the spaceship by hand with a fancy caulking gun; the parachutes were sewn by hand, and also folded by hand. The only three staff members in the country who were trained and licensed to fold and pack the Apollo parachutes were considered so indispensable that NASA officials forbade them to ever ride in the same car, to avoid their all being injured in a single accident.[8]

The astronauts went to the Moon, and their skill and courage is undeniable, and also well-chronicled. But the astronauts aren't the ones who made it possible to go to the Moon.

The race to the Moon in the 1960s was also a real race, motivated by the Cold War and sustained by politics. It's been only 50 years—not 500—and yet that part of the story too has faded.

One of the ribbons of magic running through the Apollo missions is that an all-out effort born from bitter rivalry ended up uniting the world in awe and joy and appreciation in a way it had never been united before and has never been united since.

The mission to land astronauts on the Moon is all the more compelling because it was part of a tumultuous decade of transformation, tragedy, and division in the United States. Civil rights protesters, led by the Reverend Ralph Abernathy, marched on Cape Kennedy on the eve of the launch of Apollo 11.

In that way, the story of Apollo holds echoes and lessons for our own era. A nation determined to accomplish something big and worthwhile can do it, even when the goal seems beyond reach, even when the nation is divided over other things. Kennedy said of the Apollo mission that it was hard—that we were going to the Moon precisely because doing so was hard—and that it would "serve to organize and measure the best of our energies and skills." And to measure the breadth of our spirit as well.[9]

Putting spaceships and astronauts on the Moon, and bringing them back again, required surmounting 10,000 challenges. That extraordinary accomplishment was done by ordinary people, each, as Neil Armstrong said, taking one small step. Theirs is a story with unexpected surprises at every turn, like the moment when Armstrong, safely back inside the lunar module, took off his space helmet, took a breath, and discovered that the Moon has a smell.

# 1

# Tranquility Base & the World We All Live In

In ancient days, men looked at stars and saw their heroes in the constellations. In modern times, we do much the same, but our heroes are epic men of flesh and blood.

> **William Safire**
> *speechwriter to President Richard Nixon, text of an undelivered speech*[1]

For the first Moon walk ever, Sonny Reihm was inside NASA's Mission Control building, watching every move on the big screen. Reihm was a supervisor for the most important Moon technology after the lunar module itself: the spacesuits, the helmets, the Moon walk boots. And as Neil Armstrong and Buzz Aldrin got comfortable bouncing around on the Moon and got to work, Reihm got more and more uncomfortable.

The spacesuits themselves were fine. They were the work of Playtex, the folks who brought America the "Cross Your Heart Bra" in the mid-1960s. Playtex had sold the skill of its industrial division to NASA in part with the cheeky observation that the company had a lot of expertise developing clothing that had to be flexible and also form-fitting.[2]

It was when the cavorting started on the Moon that Reihm got butterflies in his stomach. Aldrin had spent half an hour bumping around in his spacesuit, with his big round helmet, when all of a sudden, here he came bounding from foot to foot like a kid at a playground, right at the video camera he and Armstrong had set up at the far side of their landing site.

Aldrin was romping straight at the world, growing larger and larger, and he was talking about how he'd discovered that you have to watch yourself when you start bouncing around exactly like he was bouncing around, because you couldn't quite trust your sense of balance in Moon gravity; you might get going too fast, lose your footing, and end up on your belly, skidding along the rocky lunar ground.

"You do have to be rather careful to keep track of where your center of mass is," Aldrin said, as if his fellow Earthlings might soon find this Moon walk advice useful. "Sometimes, it takes about two or three paces to make sure you've got your feet underneath you."[3]

Reihm should have been having the most glorious moment of his career. He had joined the industrial division of Playtex, ILC Dover, in 1960 at age twenty-two, and by the time of the Moon landing, before he turned thirty, he had become the Apollo project manager. His team's blazing white suits were taking men on their first walk on another world. They were a triumph of technology and imagination, not to mention politics and persistence. The spacesuits were completely self-contained spacecraft, with room for just one. They had been tested and tweaked and custom-tailored. But what happened on Earth really didn't matter, did it—that's what Reihm was thinking. There was only one test that mattered, and Aldrin was conducting it right there, right now, in full view of the whole world, on the airless Moon, with unabashed enthusiasm.

If Aldrin should trip and land hard on a Moon rock, well, a tear in the suit wouldn't be a seamstress's problem. It would be a disaster. The suit would deflate instantly, catastrophically, and the astronaut would die, on TV, in front of the world. That's what Reihm was thinking about.

The TV camera, set up on a tripod, would have a perfect view. Aldrin ran left, planted his left leg, then cut to the right like an NFL running back dodging tacklers. He did kangaroo hops right past the American flag, but announced that this wasn't a good way of moving around. "Your forward mobility is not quite as good as it is in the more conventional one foot after another," he said. Then he disappeared from the camera's view.

By this time Reihm could barely contain his fretfulness. "That silly bastard is out there running all over the place," he thought.[4]

Seconds ticked by. The Moon base was quiet. Armstrong was working by the lunar module, his back to the camera. Suddenly Aldrin came dashing in from the left, straight across the landing site, Moon dirt flying from his boots. His narration back to Mission Control was calm, but his speed was anything but. He was doing a Moon run: "As far as saying what a sustained pace might be, I think the one that I'm using now would get rather tiring after several hundred feet."[5]

Reihm was in a technical support room adjacent to Mission Control, with a group of spacesuit staff, standing by in case anything went wrong. Even though everything was going perfectly, and even though the whole point of the spacesuits was to explore the Moon, Reihm couldn't wait for it to end. Why in the world was Aldrin acting crazy on the Moon, of all places?

Reihm's worries weren't unique to him. Eleanor Foraker had supervised the women who sewed the spacesuits, each suit painstakingly stitched by hand. When the jumping around started, she started thinking about the pressure garment, one of the inner layers of the spacesuit that sealed the astronaut against the vacuum of space. What if all that hopping and tugging caused a leak?

Joe Kosmo was one of the spacesuit designers on the NASA side. He was at home, watching with his family, thinking exactly the same thing Reihm was: "This is great. I hope he doesn't fall over."[6]

Reihm knew, of course, that the astronauts were just out there "euphorically enjoying what they were doing." If the world was excited about the Moon landing, imagine being the two guys who got to do it. In fact, according to the flight plan, right after the landing, Armstrong and Aldrin were scheduled for a five-hour nap. They told Mission Control they wanted to ditch the nap, suit up, and get outside. They hadn't flown all the way to the Moon in order to sleep.[7]

And there really wasn't anything to worry about. There was nothing delicate about the spacesuits. Just the opposite. They were marvels: 21 layers of nested fabric, strong enough to stop a micrometeorite, but still flexible enough for Aldrin's kangaroo hops and quick cuts. Aldrin and Armstrong moved across the Moon with enviable light-footedness.

Still, watching Aldrin dash around, Reihm could "think of nothing but, Please go back up that ladder and get back into the safety of that lunar module. When [they] went back up that ladder and shut that door, it was the happiest moment of my life. It wasn't until quite a while later that I reveled over the accomplishment."[8]

Reihm wasn't thrilled by the Moon walk that he and his colleagues had worked for years to make possible; he was thrilled by its being over, by Armstrong and Aldrin going back inside, sealing the hatch, and re-pressurizing their cabin.

That anxiety—not just of one man but of a trio on the same team—seems such an unexpected reaction to the climax of the space program's dash through the 1960s. For Sonny Reihm and Eleanor Foraker and Joe Kosmo, the moment of maximum triumph was also the moment of maximum risk: they knew the thousand things that might go wrong.

Reihm's anxiety, in fact, is a kind of time machine.

We know how the story ended: every Moon mission was a success. Even Apollo 13, which was a catastrophe, was a triumph. Every spacesuit worked perfectly. Astronauts did trip and fall—they skipped, bunny-hopped, skidded to their knees, did pushups to stand upright, jumped too high, and fell over backward. As crews got more experience and more confidence, they would trot at high speed across the Moon's surface—carefree—in that distinctive one-sixth-gravity locomotion. Once we got to the Moon, nothing much went wrong, not with the spacesuits or anything else.[9]

But the rockets and spaceships that flew the astronauts to the Moon were far and away the most complex machines ever created. The vast system of support assembled to manage those spaceships was far and away the most elaborate support in history for an expedition. There were so many ways things could go wrong that, for that first mission, President Nixon's staff had a speech ready in case the astronauts died during the mission. That speech was written before they even blasted off.

Reihm's anxiety takes us back to the moment when every spaceflight was dangerous and daring. Over and over, NASA pushed the limits of its experience and tested the reliability of what it had created—21 layers of spacesuit fabric—against the unforgiving forces of spaceflight physics. His anxiety takes us back to a moment when the standing and reputation of the United States around the world hung on the soaring ambition of its space program and on the success of those space missions.

Reihm's anxiety is a time machine because it puts us back in the moments before anyone knew how the story would come out. And it's a reminder of the mostly unsung men and women who made it possible for Armstrong and Aldrin to leave those distinctive bootprints at Tranquility Base.

---

Today the race to the Moon seems touched by magic. The Moon landing has ascended to the realm of American mythology. In our imaginations, it's a snippet of crackly audio, a calm and slightly hesitant Neil Armstrong stepping from the ladder onto the surface of the Moon, saying, "That's one small step for man, one giant leap for mankind." It's the video clip of the Saturn V roaring off the launchpad in Florida, with almost inhuman power, smoke and fire streaming behind it. It's a brilliant color picture of an astronaut standing on the Moon, saluting the American flag. It's a phrase: "If we can put a man on the Moon, why can't we . . . ?"

It is such a landmark accomplishment that the decade-long journey has been concentrated into a single event, as if on a summer day in 1969, three men climbed into a rocket, flew to the Moon, pulled on their spacesuits, took one small step, planted the American flag, and then came home. How they got there, how many times they went, even why they went—the myth has polished all that away.

The Moon landing was 50 years ago, but the event itself has an immediacy in our minds—a singular brilliant destination, a well-scrubbed cast of astronauts, a well-ordered place called Mission Control staffed with people of calm competence, a series of astonishing accomplishments that managed to get more routine as they became more astonishing.

America reached the Moon without conquering it or capturing it. We landed, and the world came along with us. But the magic, of course, was the result of an incredible effort—an effort unlike any that had been seen before. Three times as many people worked on Apollo as on the Manhattan Project to create the atomic bomb. In 1961, the year Kennedy formally announced Apollo, NASA spent $1 million on the program for the year. Five years later NASA was spending $1 million every three hours on Apollo, 24 hours a day.[10]

On that day, May 25, 1961, when Kennedy asked Congress to send Americans to the Moon before the 1960s were over, NASA had no rockets to launch astronauts to the Moon, no computer portable enough to guide a spaceship to the Moon, no spacesuits to wear on the way, no spaceship to land astronauts on the surface (let alone a Moon car to let them drive around and explore), no network of tracking stations to talk to the astronauts en route. On the day of Kennedy's speech, no human being had ever opened a hatch in space and gone outside; no two manned spaceships had ever been in space together or ever tried to rendezvous with each other. No one had any real idea what the surface of the Moon was like and what kind of landing craft it would support, because no craft of any kind had landed safely on the Moon and reported back. As Kennedy gave that speech, there was an argument—at MIT no less—about whether engineers could do the math required, could do the navigation required, and do it fast enough, to fly to the Moon and back.

"When [Kennedy] asked us to do that in 1961, it was impossible," said Chris Kraft, the NASA engineer who created Mission Control. "We made it possible. We, the United States, made it possible."[11]

And just eight years later, the spacesuit designers were worried that the astronauts were being too exuberant in their first Moon walk. With perspective, that's an understandable worry, and also a worry with a certain charm.

The big myth about the race to the Moon contains many small myths. One is that, during this golden age of space exploration, Americans enthusiastically supported NASA and the space program, that Americans wanted to go to the Moon. In fact two American presidents in a row hauled the space program all the way to the Moon with not even half of Americans saying they thought it was worthwhile. The sixties were a wildly tumultuous decade, and while Apollo sometimes seemed to exist in its own bubble of intensity and focus, in a place somehow separate from the Vietnam War and the urban riots and the assassinations, in fact Americans constantly questioned why we were going to the Moon when we couldn't handle our problems on Earth.

As early as 1964, when asked if America should "go all out to beat the Russians in a manned flight to the moon," only 26 percent of Americans said yes.[12] Public support for Apollo actually faded as the 1960s went along, despite the saturation coverage of astronauts headed to the Moon. During Christmas 1968, NASA sent three astronauts in an Apollo capsule all the way to the Moon, where they orbited just 70 miles over the surface, and on Christmas Eve, in a live, primetime TV broadcast, they showed pictures of the Moon's surface out their spaceship windows. Then the three astronauts, Bill Anders, Jim Lovell, and Frank Borman, read the first 10 verses of Genesis to what was then the largest TV audience in history. From orbit, Anders took one of the most famous pictures of all time, the photo of the Earth floating in space above the Moon, the first full-color photo of Earth from space, later titled *Earthrise*, a single image credited with helping inspire the modern environmental movement.[13]

At the end of a chaotic and catastrophic year, with the assassinations of Martin Luther King Jr. and Robert Kennedy; the riots that followed in 168 U.S. cities, including Washington, D.C.; the war protests and campus protests; the rioting around the 1968 Democratic National Convention in Chicago; the election of Richard Nixon as the president to replace Lyndon Johnson; that moment of arresting unity as Apollo 8 orbited the Moon on Christmas Eve seemed to briefly redeem an irredeemable year. Out of everything that happened that year, *Time* magazine chose as "Man of the Year" for 1968 that Apollo 8 crew, Anders, Lovell, and Borman, their triumphant voyage "a particularly welcome gift after a year of disruption and despond."[14]

Their trip meant, among other things, that the United States had made it to the Moon first. Americans had won "the race." There would be no "Red Moon." It also meant that the landing Kennedy envisioned would almost certainly happen, as promised, before the end of the 1960s. Apollo 8 was a worldwide triumph for the United States and the prelude to an even greater one. It was thrilling. It provided a sense of satisfaction and pride, even catharsis, for a country that was losing confidence in its ability to do anything, from run its universities to wage war to protect its leaders. The *Time* "Men of the Year" story said that in 1968, America's "self-confidence sank to a nadir" because of a growing sense "that American society was afflicted with some profound malaise of spirit and will." The excitement and anticipation for the actual Moon landing should have been extraordinary.[15]

In fact it was anything but universal. Four weeks after Apollo 8's telecast from lunar orbit, the Harris Poll conducted a survey of Americans about the mission. Asked if they favored landing a man on the Moon, only 39 percent said yes—even as the Moon landings were about to happen. Asked if they thought the space program was worth the $4 billion a year it was costing, 55 percent of Americans said no. That year, 1968, the war in Vietnam had cost $19.3 billion, more than the total cost of Apollo to that point, and had taken the lives of 16,899 U.S. soldiers—almost 50 dead every single day—by far the worst single year of the war for the U.S. military. Americans were delighted to be flying to the Moon, but they were not preoccupied by it.[16]

Another myth about the race to the Moon in the sixties—perhaps the core myth of the whole enterprise—is that in the end, Apollo was a kind of cosmic disappointment, that in terms of space exploration, it led nowhere.

Way back in the 1960s we stretched 240,000 miles to the Moon and back—not just to touch it and return; we flew electric cars to the Moon and spent hours driving around in them. Now, more than 50 years into the Space Age, all we do in space—all any astronauts do—is circle rather monotonously around Earth. The International Space Station orbits at about 240 miles up. And as America hit the 50th anniversary of that first Moon landing, the nation didn't have a single rocket and spaceship of its own for launching astronauts into space. The country that flew people all the way to the Moon had, 50 years later, no way at all to fly its own astronauts into space. The Space Shuttle flew its last mission in 2011. In 2018 none of the private companies jockeying to provide civilian space transportation had yet succeeded. The only access the U.S. had to space was by buying seats to the Space Station on the Russian Soyuz spacecraft, a modestly modernized version of a Soviet spaceship that has been flying to orbit since 1968. The United States was actually paying the nation it beat to the Moon to fly its astronauts to orbit, using spaceships from exactly the moment when U.S. technological prowess swept past the Russians.

Whether or not that qualifies as failure, it certainly qualifies as disappointment. In terms of getting people into space, on the 50th anniversary of the Moon landing the dominant space power had less capacity than it had in 1965 and 1966, when during one stretch the U.S. launched 10 Gemini missions, one every eight weeks for two years.

So if the measure of Apollo's success was to do precisely what

Kennedy had challenged, to land a man on the Moon and return him safely to Earth, that mission was accomplished, with skill, with drama, even with a little panache, and with a thousand discoveries about the Moon, the Earth, and the art of spaceflight itself.

But in the 1960s the men and women racing for the Moon never considered the Moon as the final goal. Apollo wasn't just a game of Space Age capture the flag. As the legendary NASA historian Roger Launius put it, "The point of going to the Moon wasn't just to land on the Moon. It was to open the solar system to human exploration and settlement."[17] By that measure, we've been drifting backward, decade by decade.

But that's the wrong way of thinking about the race to the Moon, in terms of its impact both here on Earth and on the course of exploration that followed.

The significance of going to the Moon has gotten lost between the monumental reach of the event itself and what many perceive to be the trivia that we got in return. Ask people what going to the Moon got us, and the most common answers are Tang and Velcro, offered with exactly the wryness that those two innovations deserve. In fact Tang was created in 1957 by the man who also invented Cool Whip for General Mills, and Velcro was invented in 1948 in Switzerland and available in the U.S. in 1959. Velcro was indispensable to astronauts flying in zero gravity and flew on the first U.S. orbital mission with John Glenn, as well as on the Moon flights. Tang was tested by Glenn, and General Mills advertised the astronaut connection throughout the 1960s. (Tang sponsored ABC News coverage of the Apollo 8 mission, with the Tang logo on the anchor desk right in front of ABC's highly regarded space reporter Jules Bergman.) In the case of Tang, the NASA connection turned an indifferent product into a best-seller, but some of the astronauts didn't care for it. The crew of Apollo 11 specifically rejected Tang as part of their food supplies. Decades after the fact, NASA remains concerned enough about being given undue credit for inventing Tang and Velcro that it maintains a web page specifically to debunk that myth.[18]

The pace at which Americans moved on from landing on the Moon was extraordinary, even as the missions were still happening. The Apollo 11 Moon landing was the most-watched TV event in history at the time it happened—94 percent of U.S. households watched, and the

Moon walk didn't start until almost 11 p.m. Three years later, during Apollo 17, the last Moon mission, TV coverage of the Moon landing was watched by fewer Americans than watched that week's episode of *All in the Family*.[19]

Even in the space community, Apollo is often quietly accounted a sour disappointment. The greatest space achievement of humanity, the greatest engineering achievement, perhaps one of the greatest achievements of any kind, just a Cold War cul-de-sac.

Yet Apollo was anything but a failure. In fact we misunderstand our own achievement, regardless of its vividness in our national memory. And in misunderstanding it, we miss much of the significance of it. We are let down by the end of Apollo because we're looking in the wrong place for our success.

The success wasn't that we went on to Mars, that we created self-supporting settlements on the Moon and Mars, that we've extended human habitation across the solar system. We haven't done any of that, and we're decades from doing it now.

The success is the very age we live in now. The race to the Moon didn't usher in the Space Age; it ushered in the Digital Age. And that is as valuable a legacy as the imagined Space Age might have been. Probably more valuable.

That doesn't mean we didn't screw up the space part. We did. The failure of imagination wasn't in going to the Moon; it was in what we did next—in space. As much as getting to the Moon in eight years was a triumph of leadership, failing to figure out what to do next, and mustering the support for it, was an equally dramatic failure of leadership. That was the fault of the people running the space program in the 1970s and beyond, who couldn't pick the right course. One thing it's not is the fault of the people who made the Moon landing a reality.

On Earth, the race to the Moon did everything we could have imagined—and have all but failed to notice.

Space enthusiasts mope that it didn't open the solar system to human settlement. Space skeptics look at the billions of dollars spent and can't figure out how it was worth it, given the needs right here on Earth.

Those have always been the most pointed critiques of the U.S. space program, and the race to the Moon in particular. The spending of money in space instead of "at home" was denounced as soon as Apollo's spending started to ramp up. But understanding that Apollo helped create the digital revolution, in the U.S. and the rest of the world, deflates both critiques: it turns out that what we did in space didn't come at the expense of what we could have done on Earth. Just the opposite: it quietly revolutionized the way everyone on Earth lives. And far from being a dead end, little more than a grandiose Cold War gesture, Apollo opened a whole world of both exploration and innovation.

We've always looked at Apollo through the wrong lens, in that sense. It is both startling and unconventional to credit Apollo and NASA with helping create the digital revolution. NASA itself makes no such argument. The historians of NASA and its impact write constantly about "spinoffs" from the space program without ever taking the larger culture and economy into account. Historians of Silicon Valley and its origins skip briskly past Apollo and NASA, which seem to have operated in a parallel world without much connection to or impact on the wizards of Intel and Microsoft.

The space program in the 1960s did two things that helped lay the foundation of the digital revolution. First, NASA used integrated circuits—the first computer chips—in the computers that flew the Apollo command module and the Apollo lunar module. Except for the U.S. Air Force, NASA was the first significant customer for integrated circuits, and for years in the 1960s NASA was the largest customer for them, buying most of the chips made in the country. Microchips power the world now, of course, but in 1962 they were only three years old, and they were a brilliant if shaky bet. Even IBM decided against using them in the company's computers in the early 1960s. NASA's demand for integrated circuits, and its insistence on their near-flawless manufacture, helped create the world market for the chips and helped cut the price by 90 percent in five years.[20]

What NASA did for semiconductor companies was teach them to make chips of near-perfect quality, to make them fast, in huge volumes, and to make them cheaper, faster, and better with each year. That's the world we've all been benefiting from for the 50 years since.

For chip customers, NASA did something just as important: it established the unquestioned reliability and value of integrated circuits. NASA wasn't using the chips in a missile guidance system in a hundred missiles that might or might not be launched sometime in the future.

NASA was using the chips in spaceships that were the premier project of the entire nation, where the reliability of those computer chips was the key to success or failure. NASA was the first organization of any kind—company or government agency—anywhere in the world to give computer chips responsibility for human life. If the chips could be depended on to fly astronauts safely to the Moon, they were probably good enough for computers that would run chemical plants or analyze advertising data.[21]

NASA also brought the rest of the world into the era of "real-time computing," a phrase that seems redundant to anyone who's been using a computer since the late 1970s. But in 1961, when the Moon race started, there was almost no computing in which an ordinary person—an engineer, a scientist, a mathematician—sat at a machine, asked it to do calculations, and got the answers while sitting there. Instead you submitted your programs on stacks of punch cards, and you got back piles of printouts based on the computer's run of your cards—and you got those printouts hours or days later, depending on where you worked and how many other people were also using the computer.

But the Apollo spacecraft—command module and lunar module—were flying to the Moon at almost 24,000 miles per hour. That's six miles every second. The astronauts couldn't wait a minute for their calculations; in fact, if they wanted to arrive at the right spot on the Moon, they couldn't wait a second. In an era when even the batch-processing machines took up vast rooms of floor space, the Apollo spacecraft had real-time computers that fit into a single cubic foot, a stunning feat of both engineering and programming. On one floor of Mission Control was a computer complex that gathered all the data flowing in from the Apollo spacecraft and provided computing and displays for the flight control consoles in Mission Control. During Apollo, Mission Control relied on five IBM mainframes (working memory for each: 1 megabyte). The name of Mission Control's computer facility underscored its novel functioning: NASA called it the Real-Time Computer Complex. It was the first real-time computing facility IBM had ever installed.[22]

NASA revolutionized weather forecasting. NASA revolutionized global communications. NASA revolutionized rechargeable nickel-cadmium batteries. Would we have had advanced weather forecasting without NASA and the race to the Moon? Of course. Would we have had microchips and laptops without Apollo? Of course. But we would have had microchips and laptops without Intel and Microsoft and Apple, as well. Just because something would have happened anyway doesn't mean you take credit from those who actually did it. The race to the Moon took developments and technologies and trends that most of the rest of the world, most of the rest of the economy, didn't know about and magnified them, accelerated them, and helped make their significance and value clear well beyond space travel.

Just as important as NASA's impact on the actual technology of the digital revolution was NASA's impact on the culture. In 1961, when the race to the Moon kicked off, there was no sense in popular culture of "technology" as a force in the everyday lives of consumers as we think of it now. The 1950s and 1960s were the dawn of the consumer era in the U.S., and much of the consuming focused on home appliances, on creating convenience and comfort.

In 1953 the U.S. started counting the percentage of U.S. families that had various appliances. Between 1953 and 1960 the number of homes with air-conditioning jumped tenfold. By 1969 it had almost tripled again. From 1953 to 1960 the number of homes with dishwashers doubled, and the number with clothes dryers went up by a factor of five. (Both refrigerators and clothes washers were already common in 1953.) The only item that was being tracked in the 1950s and 1960s that could have been thought of as an electronic device was the TV, and although in 1953, 47 percent of homes already had a TV, by 1960 that number had doubled, to 90 percent.[23]

At the end of 1957 *Time* magazine did a cover story not on the booming consumer culture but on the frustrating scramble for scarce repairmen that the appliance boom had created. "The typical U.S. housewife who once considered herself lucky if she had a washing machine is now surrounded by 25 or more labor-saving electric yeomen worth $3,000," said the story. Not just refrigerators and washers, but vacuum cleaners, blenders, dishwashers, and the ultimate symbol of kitchen convenience: the electric can opener.[24] "No one can do without any of the marvelous new gadgets—therefore no one can do without the repairman to keep them going." Any housewife who had a reliable "Mr. Fixit," declared *Time*, had "a possession as chic today as the little dressmaker who could copy the latest Paris fashions." TV repairs in 1956, according to *Time*, cost $2 billion, exceeding the total spent on new TVs. (That $2 billion spent on TV repairs in 1956 is the equivalent of $18 billion in 2018.)[25]

In December 1959, the last month of the 1950s, *Time* did a cover story on the wonders of new waves of convenience and prepared foods, frozen and processed, mixes and cans, from brownie mix to au gratin potatoes to baby food. "Such jiffy cooking would have made Grandma shudder, but today it brings smiles of delight to millions of U.S. housewives" and "a bit of magic into the U.S. kitchen." Sales of frozen foods grew 2,700 percent in the 1950s.[26]

But Americans didn't think of this wave of conveniences as technology. The idea of "technology" was still linked closely to science. More than that, in the wake of World War II and during a decade of the Cold War marked by more than 150 open-air atomic bomb tests by the U.S., "technology" was thought of as largely military, in the form of radar and jet fighters, missiles and the nuclear bombs themselves.

The first truly personal electronic device was the handheld transistor radio, which debuted in 1954, but by 1961 was still relatively expensive—between $20 and $30 for an AM radio about the size of a modern smartphone, equal to $150 to $250 in 2018. Still, by the end of the decade millions of teenagers had gotten used to holding (or pocketing) a device that liberated them—and their music or their baseball game—from the living room or the bedroom. You could listen outside, on your bike, with your friends; you could listen privately with a single earphone; you could listen to whatever kind of music you could find on the dial, without arguing with (or irritating) your parents. The AM transistor radio was technology—it was quite literally named after its solid-state circuitry—and it was technology that you could buy with the money you earned by mowing lawns or babysitting; it was technology that was fun, that provided independence. Technology as freedom.[27]

Alongside all the other social transformation and upheaval, the 1960s served as a slow-moving prelude to the digital revolution that really got under way in the mid-1970s.

The IBM Selectric typewriter was introduced in the summer of 1961 and in the first six months sold four times as many as IBM had predicted. The first digital clock radio was introduced by Sony in 1968.[28] Touch-tone telephone service was introduced by AT&T at the

1962 World's Fair in Seattle and first offered in Pennsylvania in the fall of 1963. AT&T marketed push-button phones as much more efficient than waiting for the dial to circle around for each number, claiming that push buttons cut dialing time in half. But the company charged for touch-tone service (about $1 a month) and rolled the service out region by region slowly because upgraded technology had to be installed in every switching office. Washington, D.C., customers didn't get the touch-tone option until 1965. In 1969, as Collins and Armstrong and Aldrin were flying to the Moon, 96 percent of U.S. phones still had dials.[29]

The things we think of as heralding the age of consumer electronics were a decade or more from being in widespread use: the handheld calculator (1972), the Pulsar digital watch with the large red numbers (1972), the Sony Walkman (1979), the VCR (1984). The cordless phone didn't become a hit until the early 1980s. Even the ubiquitous microwave oven wasn't in half of U.S. homes until 1986.

But space and the technology it promised were present in 1960s America in a different way: as a promise for the future, especially in TV shows. In the 1950s there wasn't a single popular TV series that had to do with space. The 1960s had five shows with space as their setting or theme: *The Jetsons*, *Lost in Space*, *I Dream of Jeannie*, *My Favorite Martian*, and *Star Trek*. Three of those—*The Jetsons, Lost in Space,* and *Star Trek*—created whole worlds of technology, built especially around computers and robotic assistance of all kinds. Fifty years later we're still catching up to the home life of the Jetson family, and the USS *Enterprise* seems as fantastical an engineering achievement as it did during the Apollo age.

But those TV shows helped shape perceptions and attitudes. In all three, technology was in the service of people. It made food, navigated deep space, answered questions, provided instant video calling. In *The Jetsons*, technology cleaned house; made lunches for the children; and walked the dog, Astro. Computerized machines were occasionally frustrating; the robot on *Lost in Space*, voiced by Dick Tufeld, made famous the phrase "It does not compute!" But computers were easy to use and helpful and fit in seamlessly with everyday life.

Even the silly shows made the space program, and its technologies, a little more accessible. *I Dream of Jeannie* was just a sixties sitcom, in which Larry Hagman plays NASA astronaut Tony Nelson. When his returning space capsule splashes down in the Pacific Ocean far outside its planned landing zone, he discovers on the beach of a deserted island the bottle in which Jeannie is trapped. *Jeannie* ran for five seasons, set in and around Cocoa Beach, on the doorstep of Cape Kennedy. The space program was really just the backdrop for a goofy, now seriously dated romantic comedy. But *Jeannie* was so popular and identified so closely with the space program that its star, Barbara Eden, was invited to "Barbara Eden Day" in Cocoa Beach just three weeks before the launch of Apollo 11. During her visit she pressed the button that launched a rocket carrying a satellite into space and christened a street in Cocoa Beach named for her. By chance (at least according to the press reports), Eden met Buzz Aldrin while she was in Cocoa Beach. The two were photographed exchanging kisses—and that photo of Aldrin kissing the TV girlfriend of a TV astronaut just days before his own actual Moon mission made dozens of newspapers nationwide.[30]

When the U.S. manned space program began in the early sixties, Americans associated technology most readily with war. The invention of radar and the success of the Manhattan Project helped secure victory in World War II. Nuclear technology—in the form of intercontinental ballistic missiles (ICBMs), nuclear-powered submarines, the then brand-new and fearsome B-52 bomber—was technology.

But then we spent a decade watching ranks of men in white shirts and ties sitting for hours at computer consoles in Mission Control, flying space missions. As the astronauts created an aura of "the right stuff," the rest of NASA's staff, particularly the Mission Control crowd, introduced America to the geek, and the geek as someone cool, with superpowers as distinctive as those of the astronauts. The TV shows just reinforced that aura: George Jetson was a bit of a bumbler but was also an aerospace staffer at Spacely Space Sprockets; his son, Elmo, was a certified cartoon-boy-genius-tinkerer. On the bridge of the USS *Enterprise*, Sulu and Spock were nothing if not supercompetent geeks playing against the slightly hipper Uhura and Kirk.

We didn't take to handheld calculators and desktop computers 15 years later because we'd watched *Star Trek*. But space culture—out of Mission Control and from the bridge of the *Enterprise*—changed our perception of technology's appeal and usefulness. The space program and the aura of imaginative enthusiasm it brought changed the tone of technology, the attitude it presented to us, and the attitude we brought to it. That's the sense in which the culture of manned space travel helped lay the groundwork for the Digital Age. Space didn't get us ready for space; it got us ready for the world that was coming on Earth.

---

How far did we travel from May 1961 to July 1969, from the day President Kennedy first insisted we could fly people to the Moon to the day Neil Armstrong and Buzz Aldrin jumped the last 30 inches from the bottom rung of their spaceship ladder onto the gritty lunar surface?[31]

When Kennedy gave the speech that launched the space race, the United States was barely a spacefaring nation. On that Thursday afternoon the United States of America had exactly 15 minutes of manned spaceflight experience—of which just 5 minutes was in the weightlessness of space.[32] We had never sent an American into orbit. We had no idea how to fly to the Moon. Kennedy had vowed to do something that, at that moment, couldn't be done. Eight years later—eight years and two months—one astronaut was orbiting the Moon, and two were bouncing around on the surface. In eight years the spaceships were imagined, designed, constructed, tested, and then test-flown. The astronauts were chosen and learned to fly those spaceships, practicing so relentlessly that the routine procedures became instinctive. The spacesuits were designed and sewn; the problem of flying back through the atmosphere at 25,000 miles an hour without burning up was solved; a small group of determined engineers managed to get an electric car designed, built, and added to the flight manifest.

And those eight years weren't filled just with the intensity of the effort. There were events that might have ended the project. When the man who charged America to go to the Moon was murdered, NASA didn't flinch. Going to the Moon—safely, on schedule—became a double mission: fulfilling Kennedy's original vision and paying tribute to him as a leader. When NASA and the nation were rocked by the fire inside the Apollo 1 capsule in 1967—a fire so fast and so intense that the command module blew apart, knocking launchpad personnel off their feet—NASA mourned its three dead astronauts, and then reexamined every piece of equipment, every inch of wiring, and every assumption about safety. No less a figure than Democratic senator Walter Mondale led the blistering criticism of NASA's performance.[33] But the Apollo 1 fire didn't cripple Apollo or NASA; it transformed the agency and the project. As the assassination of President Kennedy in some measure safeguarded the politics of going to the Moon, the deep flaws the Apollo fire revealed in NASA's own performance guaranteed that subsequent spaceships would make it to the Moon safely.

But as the sixties gathered momentum, the space program was often overshadowed. The rest of America traveled just as far as the astronauts and space scientists between 1961 and 1969. In 1961 America was just on the verge of becoming the nation we think of it as being.

So much was about to happen, so much was about to change, that looking back it feels like the 1960s must have been composed of years accordioned in on themselves to accommodate all the revolutions. Every part of American society was transformed: race and race relations; sex and gender relations; politics and war and protest; the news media and television. Fashion. Music. Technology.

As Kennedy spoke on May 25, 1961, much of America remained deeply segregated: trains and buses, restaurants and hotels and water fountains. Most American schools had yet to be desegregated, although the Supreme Court had unanimously declared segregation unconstitutional seven years earlier. The U.S. Senate wouldn't see its first popularly elected African American member for another five years; the U.S. Supreme Court wouldn't get its first African American justice for six years. But by 1969, with the Freedom Rides and Freedom Summer, the March on Washington, the "I Have a Dream" speech, and the passage of the Civil Rights Act and the Voting Rights Act, the kind of discrimination that was thoughtless and almost universal in 1961 was illegal, and also increasingly unacceptable.

The birth control pill was approved by the FDA a year before Kennedy spoke, but in 1961 it was still illegal for unmarried women to buy it, or any kind of birth control, in 26 states. It was illegal even for married women to buy birth control in 25 states—and would remain so until a landmark Supreme Court ruling in 1965.[34] Betty Friedan's *The Feminine Mystique* was published in 1963; the National Organization for Women was founded in 1966. By the end of the decade, the Pill and the women's liberation movement would not only help usher in the sexual revolution—the era of "free love"—but would also transform the workplace, allowing women to postpone having children, to go to college and pursue careers. From 1960 to 1968 the U.S. birth rate dropped nine years in a row, falling 25 percent, and the absolute number of babies born dropped significantly (by 500,000 a year), even as the number of women age 18 and over grew. The change was dramatic. In the sixties, the number of women going to universities more than doubled, and in 1969, Yale and Princeton admitted women for the first time. A little more than 6 million men joined the workforce during the decade, but almost 8 million women did as well. The number of women in white-collar jobs grew 48 percent; the number of women in professional jobs jumped 56 percent. (But NASA would not fly a woman astronaut, Sally Ride, until 1983.)[35]

Sixties TV started with *Gunsmoke* in its fourth year in a row as the #1 TV show and ended with the satirical show *Laugh In* as #1. Sixties music started with Percy Faith and his orchestra setting a *Billboard* record with the sweet, waltz-like instrumental "Theme from *A Summer Place*" spending nine weeks in a row in 1960 at #1. The decade ended with "Aquarius / Let the Sun Shine In" from The 5th Dimension spending six weeks at #1 in 1969, along with the Rolling Stones' "Honky Tonk Women."[36]

The transformation of America during that nine years is clear just in the TV shows and the music: *Laugh In*, "Aquarius," and "Honky Tonk Women" wouldn't have made any sense to Americans in 1960, or would have been considered outrageous and offensive. But *Gunsmoke* and "Theme from *A Summer Place*" were still popular in 1969. (*Gunsmoke* still plays every day in 2018 on U.S. cable TV systems, and "Theme from *A Summer Place*" is still familiar background music.)

The sixties began with the legal pesticide DDT in such widespread use that its residue was showing up in people, and it was blamed for devastating the population of bald eagles in the U.S., which had fallen to just 500 breeding pairs. Air pollution was so serious that outbreaks of smog in New York twice killed hundreds of people, and smog was so bad in Los Angeles that at one point schools were closed for almost a month.[37]

The sixties ended with Rachel Carson's pioneering book, *Silent Spring*, published in 1962, having triggered the modern environmental movement. The Clean Air Act was passed by Congress in 1963, the Endangered Species Act in 1966; *The Whole Earth Catalog* was first published in 1968; in 1970 the Environmental Protection Agency was established; and in 1972 Congress passed the Clean Water Act and banned DDT.[38]

The sixties started with President John Kennedy and his ringing call to go to the Moon and ended with President Richard Nixon and his quiet cancellation of the last three Apollo missions.

The eight years from Kennedy's speech to Armstrong's first step were as transformative as any eight-year period in post–World War II American history: three presidents; a devastating and divisive war, a draft, and a nationwide protest movement; the revolution of civil rights across the country; the Beatles and the Rolling Stones; *The Flintstones*, *Batman*, and *2001: A Space Odyssey.*

During one terrible eight-week period, both Martin Luther King and Robert Kennedy were assassinated, shot in public, the leader of a civil rights movement that was finally bringing some measure of racial equality to the nation, and the leading Democratic candidate for president, who was vowing to end the war in Vietnam.

It would be hard to find a part of American society that was not revolutionized during the 1960s. The space program, though, often seems to exist outside those revolutions. The sixties were tumult and anger; the space program was quiet and orderly. The sixties were Woodstock and tie-dye; Mission Control was clipped radio communications and white shirts and ties. The sixties were student sit-ins, urban riots, civil rights protests, antiwar protests, and flag-burnings; the Apollo astronauts read Genesis from lunar orbit, planted an American flag on the Moon and saluted it, and then took a call from a near-giddy President Nixon from the Oval Office right into their spacesuit headsets.

We somehow don't associate going to the Moon with "the sixties," and we don't think of the race to the Moon when we think of the sixties. But the race to the Moon was as revolutionary as any other element of that decade. It was the largest single civilian project ever undertaken, dwarfing not just the Manhattan Project but the building of the Panama Canal and the building of the transcontinental railroad. (The construction of the interstate highway system ended up being much more expensive, but wasn't really a single project—the work and funding spread across 50 states, with 70 mainline interstates, and 35 years of construction.)[39]

In terms of staff and budget, Apollo was many times the size of those projects. It was a peacetime, civilian project with the scale, urgency, and impact of a wartime effort. In the three peak years of Apollo's employment, more Americans were working on the Moon mission than were fighting in Vietnam. In 1964, 380,000 people were already working on Apollo, and just 23,300 were deployed in Vietnam. In 1965 Apollo had 411,000 employees, and there were 184,300 U.S. soldiers in Vietnam. Even in 1966, when U.S. forces in Vietnam doubled to 385,300, back home there were 396,000 Americans working on Apollo.[40]

NASA's effort in the sixties was immense even compared to the corporate behemoths of that era. In the three peak years of NASA employment—1964, 1965, 1966—NASA and Apollo were bigger in terms of staff and contractors than every company on the Fortune 500 except #1, General Motors, with more than 600,000 workers. NASA was bigger than Ford and GE and U.S. Steel. Even in the shoulder years, when Apollo was staffing up and then starting to staff down—1963, 1967, and 1968—Apollo would have been the #4 organization in the country in terms of employees, ahead of every company except GM, Ford, and GE.[41]

The staffing gives a sense, in fact, of how revolutionary the effort was. More than 400,000 people were laboring to produce a relatively small fleet: 15 Saturn V rockets, 14 lunar modules, 13 command and service module combinations. To create and fly fewer than 15 fully equipped Moon ships, NASA needed a quarter-million people, six years in a row, and 60 percent more than that in the peak years. Apollo was an engineering and technology effort; it didn't require scientific breakthroughs, akin to the Manhattan Project, but everything it required was new.[42]

MIT was responsible for designing the flight computers for Apollo, writing their software, and then supervising the construction of the computers, the wiring of the software, and the training of astronauts. At MIT alone there were 700 staff working on Apollo, writing software for two computers for 11 missions—and that staff included none of the men and women who actually built the computers, gyroscopes, and navigation instruments at the companies that supplied them. In all, by

1966 there were 20,000 companies across the country making and assembling the pieces of Apollo.[43]

Part of the genius of Apollo, part of the accomplishment, was NASA's management of the project. In some ways, NASA had to invent large-project management for the modern era, while supervising the invention and perfection of the technology to do something that had never been done before, all inside an agency that was itself not even three years old when Kennedy charged it to go to the Moon, and wasn't having much success to that point. NASA's own staff rose from 10,000 in 1960 to 24,000 in 1962, and then to 33,000 in 1964. That year, the 33,000 NASA staff were riding herd on 350,000 contractor employees. The scale of the ramp-up for Apollo and the speed with which it happened were astonishing, even for senior managers who had experience with World War II efforts.

In political terms, NASA was managed to be the inverse of Vietnam: every state benefited. Literally. Ten percent of companies working on Apollo—2,000—were considered prime contractors, and every one of the 50 states got some NASA prime contractor dollars (although the three smallest amounts, to Nebraska, North Dakota, and Wyoming, together totaled only $1.7 million). Prime contracts, in dollar terms, were heavily concentrated in six states: California, New York, Louisiana, Alabama, Florida, and Texas together accounted for 78 percent of prime contractor spending. But again, NASA was careful. The top 100 largest companies receiving contracts were spread across 22 states where two-thirds of Americans lived—and that was just 100 of 20,000 companies. Apollo's economic benefit reached into communities from one side of the country to the other. Americans weren't just watching the space program on TV; they saw it bring jobs right into their towns.[44]

Apollo was the opposite of Vietnam in another way. NASA was a government agency that did what it set out to do and did what it said it would do. NASA in the 1960s had a clear goal and a comprehensible plan for reaching the goal; both the goal and the steps were public; and the effort to execute them was played out before the public as well.

In contrast, the Vietnam War created a nationwide antiwar movement that fractured the whole country because the war looked both pointless and hopeless. It wasn't clear what American troops' ultimate goal was; it wasn't clear how they were supposed to achieve that goal;

and the promises and pronouncements of the people running the country and the war turned out to be hollow or wrong or purposefully misleading. The stakes were brutal—180 dead a week in 1967, 280 a week in 1968, 181 a week in 1969—and the financial cost staggering. The war in Vietnam formally lasted 11 years, a year less than the race to the Moon. It cost $138 billion, six times what Apollo cost.[45] But Vietnam was a mess. It destroyed the country we set out to save, and it shredded the political culture of the country that set out to do the saving.

If the race to the Moon is captured in the single image of an American astronaut in a gleaming white spacesuit, standing on the gray dust of the Moon, saluting the American flag, the Vietnam War is captured in the single image of a U.S. helicopter perched atop a building in Saigon, as dozens of people climb a ladder to the roof for a chance to board and be evacuated as the city falls to the North Vietnamese.

The stakes in both Vietnam and space were global—nothing less than the standing, the credibility, the power of the United States. Indeed, Vietnam and Apollo were both Cold War contests. In Vietnam we were defeated. In space we triumphed. Vietnam wasn't just a defeat, of course; it was a failure. It was the result of incompetence: the strategy was incompetent, the war-fighting tactics were incompetent, the politics was incompetent.

That's the contrast. Right alongside Vietnam's slow-moving global display of incompetence was Apollo. The very same government that couldn't figure out how to fight the Vietnam War, or even end it with dispatch and dignity, that very same government, at that very same moment, flew 27 men all the way to the Moon and back. Even when disaster struck on Apollo 13, the determined rescue effort and the courage of the mission's astronauts, all playing out hour after hour on live TV, only underscored the cool, fearless, implacable competence of NASA's staff. In a near-hopeless situation, in which they didn't know what the right thing to do was, NASA's engineers and scientists, its technicians and astronauts and managers, dissected and solved one problem after another, right up to the moment the Apollo 13 capsule and its astronauts were dropping toward the Pacific Ocean under three orange-and-white parachutes.

That reputation, that halo of confidence has lasted a long time. In an era when it can take eight years to build a bridge, when it can take two years to bring a murder suspect to trial, when any highway trip more than 100 miles encounters the ever-present orange barrels of interstate lane closures and reconstruction, Apollo stands as a testament to the power of clear focus and of enlisting smart, talented, determined people behind even the most audacious goal. (Yes, Apollo had an ample budget, but so did the Vietnam War.) Fifty years after Apollo, NASA remains the second most popular federal agency, after only the Centers for Disease Control, ahead of the Environmental Protection Agency and the Defense Department. The public's confidence in NASA has never faded since the Moon landings.[46]

Indeed, the phrase "If we can put a man on the Moon . . ." still has such power and such currency—it appears in print as often now as in the sixties—in part because the leap to the Moon represents the opposite of the bureaucratic tangles we've come to expect.[47] On the eve of the Apollo 11 Moon mission, *Fortune* magazine suggested that Apollo's greatest breakthrough wasn't the hardware but the technology of managing the sprawling project itself, and that if America could learn to apply NASA's techniques to other big projects, "then the $20-odd billion price of Project Apollo could turn out to be a splendid bargain."[48]

President Lyndon B. Johnson, who in the end made sure NASA was able to keep the promise to get to the Moon, was six months out of office when Apollo 11 was launched from Cape Kennedy, but he was in the VIP viewing stand on the morning of the launch. In his memoir of his presidency, *The Vantage Point*, Johnson writes that while watching Apollo 11 "rise on a pillar of flame . . . I could not help remembering that earlier vigil, twelve years before, when we strained to see the Soviet Sputnik orbiting overhead. In the short span of time between those two events, we wrote a story that will be told for centuries to come. We developed the ability to operate in space with both men and machines." Space, wrote Johnson, "was the platform from which the social revolution of the 1960s was launched. We broke out of far more than the atmosphere with our space program. We escaped from the bonds of inattention and inaction that had gripped the 1950s. New ideas took shape."[49]

It is perhaps too imaginative a leap to suggest that NASA and the space program inspired the social revolutions of the 1960s, and perhaps too convenient to overlook the Vietnam War—Johnson's war—which was such a galvanic force in those revolutions. And it can certainly be hard to see men in white shirts, narrow ties, and sport coats, talking the language of math and physics and orbital rendezvous, as the vanguard of a revolution. But Johnson was right in reminding us that the race to the Moon was as revolutionary as anything else in the 1960s. The mobilizing of a nationwide effort behind a single expedition—a single adventure—was revolutionary. The execution of Apollo was revolutionary—a sprawling government program that was done on time, on budget, without scandal or corruption or simple incompetence. The management of the project was revolutionary—a blend of private and academic innovation with government oversight, weaving thousands of companies, hundreds of thousands of employees, and millions of individual parts into a system that required absolute quality and reliability. And that worked.

Most of all, the ambition itself was revolutionary. Kennedy picked the leap to the Moon for all kinds of reasons having to do with politics and the Cold War. The race to the Moon was the result of two converging arcs of history. The first was the global rivalry with communism and the Soviet Union, which had an urgency and immediacy in the 1960s that is hard to recapture today. Political leaders, business leaders, and ordinary people all had the sense that the struggle with the Soviet Union was a struggle for the very survival of democracy and liberty, around the world and in the U.S. itself. In fact "rivalry" is too sporting a way of describing it, as if the stakes were simply on a global scoreboard. Every move the U.S.S.R. and the U.S.A. made with relation to each other, and with relation to every other nation, seemed to have significance in how the battle would turn out. The space programs of the Russians and the Americans were a vitally important field where that competition played out, and at the start of the Space Age the Russians used their own space achievements much more effectively than the Americans did, and in the process transformed the world's opinion of Russian technical competence and skill.

The second arc was the development of the technology to go to space, the missiles—the rockets—and their guidance and support systems. The rockets were weapons of war, invented to deliver nuclear weapons across the globe from Russia to the U.S. and vice versa. The first man in space, Russia's Yuri Gagarin, rode to orbit on the Soviet

Union's first ICBM, the R-7 rocket. The second man in space, America's Alan Shepard, was launched atop a modified Redstone rocket, one of the U.S. Army's well-tested missiles that wasn't as powerful as an ICBM. Both men rode on missiles designed to deliver nuclear warheads that were adapted to carry tiny capsules instead. The Cold War created the technology that made the civilian space program in the U.S. possible, and then the Cold War energized the civilian competition into space.

But it was Kennedy's genius and boldness that created the race to the Moon. Out of frustration and political necessity, he concluded that the only way to reassert American leadership in space wasn't with individual launches or steadily matching Soviet achievements or patient explanations of the sophistication of American satellite technology. Kennedy wanted a single leap that was distinctly American in ambition. Putting people on the Moon was it. (Robert McNamara, Kennedy's secretary of defense, was so worried about the Russian head start that he pressed NASA officials to recommend to Kennedy a mission straight to Mars.)[50]

The Moon was so vivid a presence and so dramatic a destination that simply announcing the goal did exactly what Kennedy wanted to do at that moment: reset the terms of the competition, reset the meaning of success in space. With the Moon as the destination, any particular Russian accomplishment short of that could be shrugged off. *We're going to the Moon.*

In September 1962, at Rice University in Houston, which had provided the land that became the Manned Spacecraft Center, Kennedy gave a speech devoted to explaining the power and purpose of going to the Moon. "No nation which expects to be the leader of other nations," he said, "can expect to stay behind in the race for space." Going to the Moon was an almost insurmountable challenge that would "serve to organize and measure the best of our energies and skills." It would prove to be "the most hazardous and dangerous and greatest adventure on which man has ever embarked."[51]

It was a Cold War mission, but Kennedy transformed it—the reach for the Moon itself transformed it—into something larger. It was an adventure, an expedition, like Lewis and Clark to the West and Robert Peary to the North Pole. The sheer audacity of a country that hadn't even been to orbit declaring it was landing on the Moon: that was a new version of Manifest Destiny. Going to the Moon was a test and a demonstration of America's resolve, its ability, its strength, its brilliance.

---

Photos and artifacts from 1969 convey their age; they often seem quaint or dated, old-fashioned or precious. Style and technology have moved on. But what hasn't aged at all are the photos of Apollo and the tools of Apollo. The photos taken in space are vivid and memorable: of the lunar module floating over the Moon, of the astronauts motoring along in their electric car on the Moon's surface, of Aldrin standing by the American flag in his spacesuit with the lunar module behind him, of the command module floating down to the Pacific Ocean beneath its orange-and-white parachutes. Those photos have an immediacy and a modernity that hasn't faded. The actual equipment, on display in museums and NASA facilities around the country, is vivid and serious and intriguing in exactly the same way. You want to reach out and touch the charred metal surface of the Apollo capsule and maybe see if it smells burned; you want to put your hand in an astronaut's glove, see how it feels on the inside, see if you can flex it; you want to hop onto the ladder on the leg of the lunar module, climb up and see what the cockpit and controls of an actual spaceship look like.

It's not just the objects that have a certain charisma; going to the Moon provided a shared sense of purpose, a national mission that still has a powerful appeal. Unlike the artifacts, that sense of shared purpose, even of patriotism, feels passé. At the 50-year mark from going to the Moon, the whole enterprise seems like something from a different era. Today Americans don't tackle such vast undertakings. American confidence in the ability of government to get things done is near an all-time low, going back to the presidency of Dwight Eisenhower.[52]

What did the America of 1961—or 1969—have that today's America does not?

That's the question that makes examining the race to the Moon more than just intriguing or compelling. Because while the pictures and the video snippets are familiar, the events have receded. Seventy percent of Americans today weren't born, or were five years old or younger, when Armstrong and Aldrin walked on the Moon.

We need to rescue the race to the Moon from American mythology.

The myth of Apollo has gradually infused and taken over the story, the history of Apollo. Real people did it. It was heroic, but heroic in the way of real life, not mythology. The America of 1969 has plenty to tell the America of today.

It wasn't just the spacesuits, for instance, that required assembly by hand. The parachutes—a total of a half-acre of nylon fabric—were cut and sewn by seamstresses sitting at black Singer sewing machines, sliding the fabric through by hand. That blend of craftsmanship and high technology was a part of every element of getting Apollo to the Moon, in some ways as emblematic of the 1960s as anything, the collision of the 1950s with the 1970s and 1980s to come. The high-strength nylon fabric for each main parachute weighed about 55 pounds; together the three parachutes slowed a command module weighing 11,000 pounds from its plunge at 160 miles per hour to 20 miles per hour just before it splashed into the ocean. The parachutes, once stitched, were also folded and packed by hand, with the help of a hydraulic ram to compress them so they took up as little space as possible.[53]

For Apollo's two onboard spacecraft computers, the programs weren't software; they were hardware—wires and tiny metal rings woven together with absolute precision to create the 1s and 0s of the digital code of a particular program, hard-wired for each computer, for each mission. The knitting was done by women using long needles, on the factory floor at Raytheon in Waltham, Massachusetts, each program taking weeks of work, each 12-inch-long memory module requiring half a mile of wire, and all that work producing, in each module, 65,000 bits of information, just 8.125 kilobytes.[54]

The batteries for the lunar module were assembled by hand in the Eagle-Picher battery factory in Joplin, Missouri, a place that started out so rumpled and disreputable that the senior engineer responsible for the design and construction of the lunar module saw technicians letting their cigarette ashes fall directly into the interior of the spacecraft batteries they were assembling. Needless to say, the factory's operations didn't stay that rumpled.[55]

The heat shield on the Apollo capsules was made of fiberglass sheets of honeycomb cells, filled with a protective putty-like resin developed by Avco Corporation, able to buffer the capsule from the 5,000-degree heat of reentry. The heat shield for an Apollo command module had

370,000 cells, each filled one at a time by a technician squirting in the resin using a custom-designed "gun," a sophisticated type of caulk gun, at the factory in Lowell, Massachusetts. The process was so exacting that each "gunner" trained for two weeks to properly fill the cells.[56]

In the end, there were 11 Apollo flights with crews—Apollo 7 through 17. Those spacecraft and their 33 crew members spent a total of 2,502 hours in flight—104 days—from the moment of launch to the moment of splashdown.[57] Every hour of spaceflight required more than 1 million hours of work on the ground—an astonishing level of preparation. A person who lives to the age of 80 lives 700,000 hours. A person who works until the age of 70—a 50-year career—spends 120,000 hours at work. Every hour of spaceflight required the equivalent of the work done in the entire work lives of eight people.[58]

It's possible no other project in history has demanded the sheer density of preparation required by Apollo. Understanding what work went into getting to the Moon only magnifies what those hundreds of thousands of people accomplished.

From the moment of his speech in May 1961, President Kennedy's ambition to send men to the Moon, and to do it with speed and determination, gripped the imaginations of Americans.

It still does.

# 2

# The Moon to the Rescue

Ah, you may leave here, for four days in space,
But when you return, it's the same old place.

> **Barry McGuire**
> *"Eve of Destruction," the 1965 protest song that reached No. 1 on the U.S. pop charts*

The world's first spaceman returned to Earth separately from his spaceship. He floated down through the last four miles of sky, a strange descending figure in a white helmet and a bright orange spacesuit, swinging gently under a pair of white parachutes. He wasn't just the first person to go to space; he was the only space traveler to land safely in just his spacesuit. So he got a view no other human ever has, drifting down from space back to Earth, over a landscape he recognized from practice jumps: a railroad bridge, the Volga River.

He was about 200 miles short of his planned landing site, in part because the rocket motor that slowed his spaceship shut down one second early. It was Wednesday morning, April 12, 1961. He was far enough off course that the only people to greet him as he settled into the field beneath his parachutes were some Russian potato farmers.

Yuri Gagarin had just looped the Earth a single time, 180 miles up, soaring through space at 17,000 miles an hour. The first human being to go to space was a Russian, and when he returned, he landed in the field of a collective farm 16 miles southwest of the town of Engels, named for Friedrich Engels, coauthor with Karl Marx of *The Communist Manifesto*.

The Russians didn't yet know how to slow down their spacecraft enough to land them safely with a person inside, so as the cannonball-shaped capsule blazed back through the atmosphere, the hatch blew off at an altitude of 23,000 feet, and Gagarin was ejected, still in his seat. He and the seat and the capsule all landed separately. Coming down without his spaceship and its radio gear meant that once Gagarin had ground under his feet, his first thought was to get to a telephone. "I had to do something to send a message that I had landed normally," he said.

The cosmonaut set off, loping through the field in search of a telephone, a man wearing a bright orange spacesuit and a white helmet, leaving behind a cascade of parachute. He spotted a woman and a girl in the distance, coming toward him. The woman and her five-year-old granddaughter had been planting potatoes when they saw Gagarin coming down. "I walked to her to ask where I could find a telephone," he said. But as he drew closer, the little girl got frightened and turned and ran the other way.

"When I saw that, I began to wave my hands and shout, 'I'm a friend. I'm Soviet!' " The woman helped Gagarin wrestle his helmet off, gave him some milk she had brought to the field for lunch, and told him where he could use a phone. A group of men nearby, tractor drivers and mechanics, told him they were hearing news about his flight announced on state radio, although at that moment Soviet authorities had no way of knowing whether or not Gagarin had landed safely.[1]

Gagarin knew how momentous his mission had been. He asked the woman to guard his parachute, not to let anyone touch it, while he used the phone. When he was able to meet up with members of a nearby military unit who spotted him parachuting down, he gave them his orange outer suit, his watch, handkerchief, and a pistol he carried to space for safekeeping.[2]

Gagarin's flight lasted 1 hour and 48 minutes, from 9:07 a.m. to 10:55 a.m. Moscow time. In Washington, D.C., that was 1:07 a.m. to 2:55 a.m. John Kennedy, the U.S. president who would have at least as much impact on the history of spaceflight as Gagarin himself, slept through that first space mission.

U.S. officials knew enough about the progress of the Soviet program that Jerome Wiesner, President Kennedy's science advisor, had stepped into the Oval Office on Tuesday evening and warned Kennedy that the flight might take place that night, within hours. Kennedy's military aide, Major General Chester Clifton, asked the president if he wanted to be woken up if the Soviets launched a man into space while he slept. "No," Kennedy replied. "Give me the news in the morning."

At 1:35 a.m., 28 minutes after launch, the Pentagon called Wiesner to tell him the Russians had sent up a big rocket and that the U.S. military was tracking it.

The news broke through publicly in the middle of the night because the Russians announced the flight on Moscow radio while Gagarin was still in orbit, 30 minutes after the Pentagon's private alert and just halfway through his flight. It was a remarkable leap of confidence given that the very rocket configuration Gagarin rode to orbit had been launched a total of 16 times and had failed on eight of those.

The *New York Times* bureau in Moscow picked up the radio report of Gagarin's flight, and someone from the *Times* called Kennedy's press secretary, Pierre Salinger, at home at 2 a.m. for confirmation.[3]

The result was that—almost in real time, on the very day that Gagarin's monumental flight took place—the *New York Times* was able to announce the flight, just four hours after it was finished, in the late city editions of the paper, with an eight-column, three-deck headline, in all-capital italic letters:

*SOVIET ORBITS MAN AND RECOVERS HIM;*
*SPACE PIONEER REPORTS: "I FEEL WELL";*
*SENT MESSAGES WHILE CIRCLING EARTH*

The headline was the same size as the headlines the *Times* used to announce the attack on Pearl Harbor and the dropping of the first atomic bomb.

That's the front page Kennedy woke up to Wednesday morning. The first sentence of the *New York Times* story cast the stakes: "The Soviet Union announced today it had won the race to put a man into space."[4]

Yuri Gagarin had just changed the world. With a single 108-minute lap around the globe, spaceflight went from science fiction story to news story. People didn't need to dream up what going to space was like, they could actually go.

But that wasn't the point of the first sentence in the *New York Times*. The person who first made spaceflight a reality was a Russian. The country that made human spaceflight a reality was the Soviet Union.

That gave Gagarin's mission a double-boom back on Earth—his flight carried as much political as scientific significance.

---

It isn't possible to understand Gagarin's flight outside the political and military rivalry that had been building for more than a decade between the U.S. and the U.S.S.R. and that colored everyday life in the 1950s and 1960s in those countries and many other corners of the world. Gagarin's flight took place on April 12, 1961, a Wednesday. The following Monday, CIA-backed rebels launched an invasion of Cuba at the Bay of Pigs, and within two days that invasion had collapsed into a hugely public debacle—not just with Fidel Castro and his regime untouched, but with 1,400 invaders surrendering to Cuban forces and Castro crowing that he and his military had crushed forces supported by the U.S.[5] In the space of a week, the Kennedy administration suffered two global humiliations at the hands of the communist world. (Four months later East German soldiers would erect the Berlin Wall.)

The Russian leap into space was a spectacular achievement of engineering and science, but it was also a resounding global statement. The U.S. had been playing catch-up in space since the launch of the first spacecraft of any kind, the Soviet satellite Sputnik in October 1957. The problem was, after four years the U.S. didn't appear to be catching up, and the Russian space achievements were steadily shifting global opinion on which country was leading the world in science and technology.

More than a year before Gagarin's breakthrough flight, in early 1960, the Gallup Organization had done a poll in 10 countries around the world. The question: "Looking ahead 10 years, which country do you think will have the leading position in the field of science?" The choices were the United States, Russia, "other," or no opinion.

In every country but two, those polled by Gallup thought the Russians would be leading the world in science by 1970. The British voted for the Russians over the Americans, 48 percent to 17 percent; the French, by 59 to 18; the West Germans, by 36 to 29; the Indians, by 46 to 8. In Greece, 29 percent thought the Americans would be ahead, and

27 percent thought the Russians would be. In the United States, confidence was undiminished: 70 percent of Americans thought the U.S. would be ahead, and 16 percent thought the Russians would be.

A few months later the U.S. Information Agency did confidential polling for the U.S. government in Britain, France, Germany, and Norway, where people "overwhelmingly considered the Soviet Union to be ahead of the United States in both science and space technology."[6]

These weren't measures of actual scientific and technological mastery, of course; they were just measures of how people around the world thought the U.S. and the U.S.S.R. were performing. But in the Cold War that kind of perception was what much of the battle was about. It was a stunning reversal of positions for both countries, and a worrying one for U.S. officials.

As it happened, on the Wednesday afternoon of Gagarin's flight, Kennedy had a news conference already scheduled. He had been president only 83 days; the April 12 press conference would be his ninth.[7]

Kennedy's press conferences were full-dress affairs. He prepared the night before with briefing books laying out 20 to 30 likely questions and their answers, and did a practice run-through with senior staff the following morning. The press conferences were either in late afternoon or early evening, and Kennedy typically took a nap beforehand. So many reporters wanted to cover them that they were held at the auditorium at the U.S. State Department. The smallest gathering for the first eight was 297 reporters.[8]

That Wednesday afternoon there was no doubt Gagarin's soaring victory was going to come up, in part because Kennedy himself had turned space into a symbol of the presidential campaign that got him elected. During that campaign against Richard Nixon—and against what Kennedy portrayed as the indolent presidency of Dwight Eisenhower—Kennedy was sharply critical of Eisenhower's slow-motion approach to space. Eisenhower hadn't even been goaded by the indignity of Russian dogs going to orbit and returning safely.

"The first vehicle in space was called Sputnik," Kennedy said in one campaign speech. "The first country to place its national emblem on the Moon was Russia, not America. The first passengers to return safely from a trip through space were named Strelka and Belka, not Rover or Fido."

For most of the 20th century, Kennedy said, the people of the world "have admired the wonders of American science and education and economic growth, but now they are not at all certain as to which way the future lies."[9]

Kennedy's campaign promised to shake Americans out of the sleepy 1950s, to revive American energy and imagination. He used an old New England word often: "vigor." He conveyed it, and he aimed to inspire it. "I believe in an America that is on the march," said Kennedy. "If we do not soon begin to move forward again, we will inevitably be left behind. And I know that Americans today are tired of standing still—and that we do not intend to be left behind."[10]

In accepting the Democratic Party nomination for president in Los Angeles in July 1960, Kennedy delivered what became the signature speech of his campaign, portraying a nation on the verge of a "New Frontier." A Kennedy presidency would seize that New Frontier, as Americans had for centuries. "Some would say . . . that all the horizons have been explored," Kennedy said. "That there is no longer an American frontier. But . . . we stand today on the edge of a New Frontier: The frontier of the 1960s."

Kennedy wasn't so much predicting that the sixties would be a revolutionary decade as he was prescribing that they *must* be. "The New Frontier of which I speak is not a set of promises, it is a set of challenges," he said, ticking them off: "uncharted areas of science and space, unsolved problems of peace and war, unconquered pockets of ignorance and prejudice."

The only thing Kennedy seemed to miss on that summer night in Los Angeles was rock 'n' roll. And the cultural and political tumult all those challenges would unleash.

In choosing between Kennedy and Eisenhower's vice president, Kennedy said, Americans were choosing "between national greatness and national decline, between the fresh air of progress and the stale, dank atmosphere of normalcy."

Nixon, at 47 years old, was just four years older than Kennedy. "The Republican nominee, of course, is a young man," said Kennedy. "But his approach is as old as McKinley" (who was elected president in 1897). Nixon's speeches "are generalities from *Poor Richard's Almanack*," and the Republican Party "is controlled by men who believe the past is bright."[11]

Just the year before, Nixon had famously squared off against the Soviet premier Nikita Khrushchev at an exhibition of American life in Moscow, in a wide-ranging public exchange that became known as the "Kitchen Debate" because it took place in a model American kitchen of the late 1950s, featuring an array of the latest American appliances. During that impromptu debate with Khrushchev, which was so dramatic and so unscripted that it made the front pages of newspapers across the country, Nixon at one point said, "There are some instances where you may be ahead of us—for example in the development of the thrust of your rockets for the investigation of outer space. There may be some instances, for example, color television, where we're ahead of you."

In the campaign, Kennedy turned the exchange back on Nixon, pointing out during one of their debates, "You yourself said to Khrushchev, 'You may be ahead of us in rocket thrust but we're ahead of you in color television'—in your famous discussion in the kitchen. I think that color television is not as important as rocket thrust."[12]

The space gap "symbolized the nation's lack of initiative, ingenuity and vitality under Republican rule," as Ted Sorensen, Kennedy's speechwriter and White House counselor, put it.

Achievements in space were powerful, visible, dramatic, and easily understood examples of technological excellence, and the Russians were using space as a Cold War battleground. "With East and West competing to convince the new and undecided nations which way to turn, which wave was the future, the dramatic Soviet achievements, [Kennedy] feared, were helping to build a dangerous impression of unchallenged world leadership generally and scientific pre-eminence particularly," recalled Sorensen.[13]

The United States had stepped into World War II with unprecedented industrial and engineering ability—building 85,000 warplanes in 1943, and 95,000 in 1944—and finished off the war in a fearsome blaze of technological brilliance. Now it was losing a step to the Soviet Union, which had finished World War II victorious but in tatters.[14]

Once Kennedy became president, though, the urgency of the space race appeared to be more symbol than passion. On the day after his inauguration, he presided over the simultaneous swearing-in of 10 of his cabinet members, including the postmaster general. But there was not a head of NASA. It was the highest unfilled job as Kennedy took office.

At that moment there wasn't even a candidate for NASA administrator. The outgoing administrator, T. Keith Glennan, who had assembled the space agency during the previous two years, was allowed to resign, and then to drive home to Cleveland in the family station wagon on Inauguration Day. The man who created the nation's space agency had been allowed to leave Washington without anyone from the new administration bothering to even have a conversation with him.[15]

The President Kennedy who took the podium for the 4 p.m. press conference the day of the first human flight into space was not the John Kennedy of the New Frontier, at least not for that 30 minutes, in front of the assembled 426 members of the press. Gagarin's flight was just 12 hours old, and Kennedy wasn't interested in racing the Russians anywhere that afternoon. In fact he was doing just the opposite.

Three of the 20 questions asked were about the first man in space. Question 2 was anodyne: What did Kennedy think about the day's achievement and "what it would mean to our space program, as such"?[16]

Kennedy called Gagarin's flight "a most impressive scientific accomplishment" and "an extraordinary feat." As to the state of the U.S. space program, he reminded the press that his transition team had acknowledged that the Soviets were ahead in space and had predicted they might be first to launch a human into space. "We are carrying out our program, and we expect to, hope to, make progress in this area this year ourselves." It was an answer designed to be ignored.

Ten minutes later a reporter asked a much more thoughtful question that cut to the heart of the day's achievement—not in scientific or engineering terms but in political terms: "Mr. President, this question might better be asked at a history class than a news conference, but here it is anyway. The Communists seem to be putting us on the defensive on a number of fronts, now again in space. Wars aside, do you think that there is a danger that their system is going to prove more durable than ours?"

This was precisely the point Kennedy had made so often on the campaign trail. But his answer that afternoon was almost diffident. "We're in a period of long, drawn-out tests to see which system is . . . more durable," Kennedy began. "A dictatorship enjoys advantages in this kind of competition, over a short period, by its ability to mobilize its resources for a specific purpose." The U.S. had made its own important contributions to science in the past decade, he continued, "not as spectacular as the man in space, or as the first Sputnik, but they are important."

Kennedy then launched into an aside, not about the virtues of Americans pushing their own space frontier but about finding a way to desalinate water. Space missions are showy, but desalination would be a scientific breakthrough with real meaning, real impact, in helping the world, Kennedy told the room. To find a cheap way to get fresh water from salt water "would really dwarf any other scientific accomplishment. And I'm hopeful that we will intensify our efforts in that area."[17]

The Russian space achievements were a warning, he said. "I do not regard the first man in space as a sign of the weakening of the free world. But I do regard the total mobilization of men and things for the service of the Communist bloc over the last years as a source of great danger to us."

The Soviet Union's spaceflight itself wasn't either surprising or disturbing. But the fact that the Soviet Union could bring its resources and its energy into such focus—that was a reminder about the quality of the opponent.

One question later a reporter invoked the withering criticism that Kennedy would read in the papers the next morning, not aimed so much at Kennedy himself but at the American space enterprise that was now his responsibility: "Mr. President, a member of Congress said today that he was tired of seeing the United States second to Russia in the space field. . . . What is the prospect that we will catch up with Russia and perhaps surpass Russia in this field?"

Kennedy remained determined, on that day at least, to leave the heavens to the Russians: "However tired anybody may be—and no one is more tired than I am—it is a fact that it is going to take some time, and I think we have to recognize it."

"We are behind," Kennedy continued. "I am sure they are making a concentrated effort to stay ahead. . . . The news will be worse before it is better."

Kennedy didn't invoke the brilliance of American scientists and engineers; he offered no rallying cry on behalf of the challenges and opportunities of space exploration; he did not call Americans to the New Frontier—a New Frontier that had just been defined by America's lone rival. Just five months earlier Kennedy had insisted, "Americans today are tired of standing still [and] . . . do not intend to be left behind." But on a day when America appeared to be standing still and was unequivocally being left behind, he offered neither energy nor even reassurance.

He sounded all too much like his 70-year-old predecessor. Indeed the just-retired Eisenhower, asked about Gagarin's history-making spaceflight, replied, "It is not necessary to be first in everything."[18]

On April 12 Kennedy certainly didn't sound like a man about to lead his nation, and the world, on a 100-month race to the Moon.

---

The Soviet Union's exuberance in its own triumph was mixed with the irresistible reflex to taunt the United States. One of the first people cosmonaut Gagarin spoke with after landing back in the Soviet Union was his country's leader, Premier Khrushchev. Part of the telephone call was broadcast on Russian radio.

> ***Khrushchev:*** You have made yourself immortal because you are the first man to penetrate into space.
> ***Gagarin:*** Now let the other countries try to catch us.
> ***Khrushchev:*** That's right. Let the capitalist countries try to catch up with our country, which has blazed the trail into space and launched the world's first cosmonaut.[19]

Speaking to the most powerful man in his own country, the world's first spaceman didn't accept congratulations; he turned his triumph into a direct challenge to America. And Khrushchev broadened it into a victory for communism over capitalism.

The first headline in the *Washington Post* about Gagarin's flight underscored Khrushchev's point, across the full width of the front page: "Soviet Lands Man after Orbit of World; K Challenges West to Duplicate Feat." The competitive pride was built into the mission from the very start: Gagarin's ship, and his mission, were called Vostok 1. *Vostok* means "east" in Russian, a name chosen in part, the *Washington Post* explained, "to counter the political and cultural prestige associated with the West."[20]

Everywhere the reaction to Gagarin's flight was a blend of awe and congratulations, Cold War politics and American humiliation.

Jules Bergman, who would become one of the most recognizable American TV space journalists, said on the ABC News evening broadcast, "Tonight, all Russia has gone wild with joy. Delirious crowds in the streets of Moscow, Leningrad, and other cities, hailing the triumph of Soviet science over the West."

Professor Bernard Lovell, a legendary British astronomer and director of Britain's first radio telescope observatory, declared, "This is the greatest scientific achievement in the history of man." Gagarin had reduced the chance of the U.S. beating the Soviets to the Moon to "negligible." The French astronomer Paul Couderc called Gagarin's flight "an exploit comparable to Lindbergh [crossing the Atlantic], carried to the sixth power."

Some of the praise spilled into ridiculousness. East German leader Walter Ulbricht said, "What Columbus 500 years ago achieved when he discovered the new continent pales before this gigantic deed."

The post–World War II world had sorted itself into pro-Soviet and pro-American poles, and for pro-Soviet leaders, April 12 was an occasion to reinforce that socialism was winning. Communist Chinese premier Zhou Enlai said Gagarin's mission showed "the incomparable superiority of the Socialist system."[21]

Two days after the spaceflight, Khrushchev hosted a national celebration for Gagarin and for his nation's achievement; it was the Russian equivalent of a New York City ticker-tape parade. The red carpet that greeted Gagarin's arrival in Moscow was 150 feet long. Two million jubilant Russians lined the streets to cheer Gagarin, and hundreds of thousands poured into Red Square. Khrushchev brushed away tears after giving Gagarin a bear hug, then declared that the spaceflight had given the Russians "colossal superiority" over the United States and the West. He presided at the largest banquet in Kremlin history in Gagarin's honor. The celebrations were broadcast live across Europe, including in Paris and London, the first time the Soviets had allowed live TV coverage of any event to air in the West.[22]

The Soviet Union's official statement cast back half a century to underscore exactly what the Communist Revolution had done for Russia: "In the past, backward Tsarist Russia could not even have dreamt of achieving such exploits in the struggle for progress of competing with technically and economically more advanced countries." Gagarin's flight "embodied the genius of the Soviet people and the powerful force of socialism."[23] The Soviet Union hadn't just bested the United States of America. It had bested its own history, its own inferior self.

The Soviets didn't *seem* to be seizing the future. They were.

---

Whatever the shock of Gagarin's flight—which, as Kennedy pointed out, had been predicted and discussed—it was nothing compared to the reaction in the U.S. to the Russians' first space launch, which was also the world's first-ever space mission. The launch of Sputnik did more than rattle the U.S.; it changed how the world saw the Soviets, and it changed, briefly, how Americans thought about themselves.

Although that launch had happened just three and a half years earlier, it took place in a world that seemed very different. On the evening of October 4, 1957, when the Russians put up the world's first space satellite, Dwight D. Eisenhower was president, Richard Nixon was his vice president, and America was firmly in the grip of the 1950s, a decade whose impact and character has always been underestimated. The U.S. was using its vast economic power after World War II to transform itself into a vision of capitalist consumerism.

The 1950s were the decade TV grabbed hold of the American family and American culture. In 1950, just 9 percent of American homes had a TV. Ten years later the number had jumped to 90 percent, and *I Love Lucy* and *Gunsmoke* were the top-rated shows for most of the decade.

The 1950s were also when the car grabbed hold not just of the American family but of the American landscape. The number of cars on the road grew 50 percent, three times faster than the number of families. The Interstate Highway System, which wove the U.S. together and also opened the continent in ways never before seen, was authorized by Eisenhower in 1956, and by the time he left office it was a quarter finished—high-speed, multilane highway being laid at a pace of 44 miles a week.

Businesspeople quickly caught on to the appeal and the power of Americans in their cars. McDonald's was born in 1955, and three years later the sign on the Golden Arches reported 100 million burgers sold.

Harlan Sanders franchised the first Kentucky Fried Chicken restaurant in 1952, and by 1960 there were 200. The first Denny's opened in 1953 (as Danny's), the first International House of Pancakes in 1958. Holiday Inn was started in 1952 specifically to offer a consistent, family-friendly alternative to the uncertain quality of "road" motels; by 1959, 100 Holiday Inns had opened across the U.S., more than one a month.

And Americans journeyed to all kinds of new places. During the 1950s the annual number of visits to state parks doubled. Visits to national parks tripled.

The 1950s saw the dawn of the great American shopping mall: the first enclosed mall opened in Edina, a suburb of Minneapolis, in October 1956. In just the last four months of 1956, 17 big regional malls opened in the U.S. that had more shopping space than all previous malls combined. It was the decade that shopping went from chore to pastime. Retail spending in the 1950s grew twice as fast as the population.

That was driven by an almost decade-long economic boom. The U.S. economy grew by 40 percent, in real terms, during the 1950s. The average family's income, adjusted for inflation, also grew 40 percent. Across the board Americans were making more money than ever before, and they were looking for ways to spend it.[24]

The boom reshaped where Americans lived and how they spent their time. The 1950s saw the invention of what Americans think of as the classic suburb, at Levittown, in Long Island, where at one point 30 homes a day were being constructed, assembly-line style. So many homes were built in American suburbs in the 1950s that by the end of the decade, one in four single-family houses in the country had been built in just the previous 10 years. Between 1951 and 1955 sales of barbecue grills increased eightfold, causing the *Washington Post* to observe, "Outdoor cooking has become as popular as golf."[25]

TV culture and car culture, mall shopping and the suburban home with its patch of green lawn—the 1950s saw the birth of what we consider key elements that have defined modern America.

But the 1950s were hardly the era of placidity and suburban contentment that has somehow lodged in popular memory and imagination. We think of the civil rights movement as part of the 1960s, but Rosa Parks refused to give up her seat on the city bus in Birmingham in 1955. The unanimous Supreme Court decision that would lead to the integration of schools across the country, *Brown v. the Board of Education*, came down in 1954. The first national school integration crisis took place in Little Rock, Arkansas, when nine black students enrolled in the city's Central High School for the school year in 1957. After weeks of white resistance that was fanned by Governor Orval Faubus of Arkansas, President Eisenhower sent 1,200 troops from the U.S. Army's 101st Airborne Division to safely escort the nine students into school, causing an angry racial confrontation that lasted more than a year and made the news for weeks.[26]

The Cold War was in full force and gave both international relations and ordinary life a shadow of anxiety. Senator Joseph McCarthy launched his corrosive campaign in 1950, insinuating communist infiltration of the movie business, the federal government, and the military, and the accusations didn't end until 1954, with his censure by the Senate. The suspicion and paranoia his campaign stirred lasted far longer.

Fear of the Soviets had plenty of real-life events to stoke it. Julius and Ethel Rosenberg were convicted of selling U.S. atomic weapons technology to the Soviets in 1951. Hungarians mounted a popular revolution against Russian domination of their country, which the Russians crushed after just 18 days with a middle-of-the-night invasion of Budapest by Soviet tanks and troops in November 1956.

But the biggest shadow was cast by the rivalry over nuclear weapons. In the early 1950s the atomic bomb was still new, a mark of technological prowess and a symbol that the world had stepped into a new, terrifying arena of warfare.

The "A-bomb," as it was referred to, was a kind of celebrity weapon. For the whole decade of the 1950s, the U.S. and the U.S.S.R. tested atomic weapons in the open air, dropping them from planes or mounting them on towers hundreds of feet tall—tests both nations thought were necessary for a rapidly developing technology that wasn't well understood. But the open-air tests, especially in the U.S., also became public spectacles, with an odd blend of military secrecy and subtly brilliant military promotion. They were managed in such a way as to stoke both fear and pride.

For the earliest U.S. tests at the Nevada Test Site, about 65 miles north of Las Vegas, reporters were kept at a distance, and the predawn explosions were described by how bright they were and how they felt.

A test in January 1951 was "visible halfway across Arizona," "a brilliant flash of light [that] brightened the sky in four states." One six days later "lit up skies for hundreds of miles with an eerie sun-gold glow" and shattered a plate-glass window in Las Vegas, where the explosion produced "a blast of air like a windstorm." The fourth bomb was strong enough to shake Las Vegas "like a quake."[27]

That open-air test, on January 27, 1951, was only the second time an atomic bomb had been detonated in the U.S., the first since the Trinity test in New Mexico just 21 days before the bomb was dropped on Hiroshima in 1945. When the U.S. started a wave of testing in 1951, the Soviet Union had detonated only a single atomic bomb, in 1949.

Today, in the 21st century, nuclear weapons have retreated to an almost symbolic power, their hard-to-imagine destructiveness something to learn about in school or the media. But that first series of tests in Nevada in 1951 was a bold display of both military force and technological prowess. The U.S. detonated atomic bombs, dropped by a B-50 bomber, over open Nevada desert on January 27, January 28, February 1, 2, and 6. Four of them were smaller than the bombs dropped on Japan, the fifth one bigger, and three of the five detonations made the front page of the *Washington Post*. At least one of the bombs was bright enough to be visible across the Southwest, simultaneously in Las Vegas and Phoenix, Reno and San Diego, and at Los Angeles International Airport. Imagine heading out to catch an early-morning flight and seeing the flash from an atom bomb exploding in the east.[28]

The Atomic Energy Commission was responsible for conducting the U.S. tests and quickly learned to stage-manage the publicity around them. Before 1951 was over, the AEC was announcing tests in advance, with the caution that technical or weather issues could delay them. Just the announcement of upcoming tests was for years considered big enough news that it made the front pages of major newspapers, as did the tests themselves.

When the Soviet Union started testing its weapons more routinely, in 1951, 1953, 1956, and 1957, news of those tests, too, made the front pages, most often announced not by the Soviet government but by the U.S. government, once even by President Eisenhower himself.[29]

The AEC grew gradually more confident and more creative in its media relations. In 1952 Los Angeles TV station KTLA broadcast an April 22 test live from Nevada, and broadcast live again on May 1. "There it is, that brilliant flash!" the KTLA news commentator announced. "Almost blinding, that white light. . . . A beautiful sight, the typical mushroom shape, now blossoming out like a big ball of cotton, an amazing sight even from 40 miles away." There was far more awe and admiration than anxiety, or even caution, in the commentator's A-bomb play-by-play.

The following year, the AEC permitted the big three news networks to broadcast a test nationwide, and the *Washington Post* TV critic reviewed both the coverage and the blast itself in a story headlined "Incredibly, 'Bomb' Was Tame on TV." "The blast was a bust on television," wrote Sonia Stein. "There was nothing frightening, dramatic, illuminating, clarifying or even very interesting about the March 17 atomic blast at Yucca Flat, Nev., as seen through television's eyes."[30]

The routine explosion of atomic bombs, out in the open, over the American landscape, had become boring. It was a foreshadowing of the space program itself, where Americans would start off captivated, but without fresh spectaculars our attention would quickly fade.

---

The A-bomb tests were designed to see not just how the bombs worked but what their impact was.

The military put troops in trenches within a few miles of explosions and flew remote-controlled aircraft loaded with monkeys and rats through the mushroom cloud of a test bomb. Members of Congress were invited to Nevada for an April 1953 test. The explosion broke the lightbulbs and windowpanes in buildings near their viewing stand and blew the hats off all 14 men. The congressmen issued a joint statement saying, in part, "We were not prepared for what we saw, heard and felt today."

At least the congressmen had the humility not to treat the atom bomb explosion like a sporting event. In Las Vegas, however, the casinos took advantage of the advance notice of the tests to host A-bomb-watching parties, at dawn on their rooftops, for which some guests simply stayed up all night.[31]

In March 1955 the military loaded five print reporters on a B-25 bomber and took them airborne for the A-bomb detonation, then spent three hours, with the reporters still aboard, chasing the drifting mushroom cloud, sampling the radioactivity at its edges.

By this point A-bomb explosions were being seen as far away as Pocatello, Idaho, and the Black Hills of South Dakota, 800 miles from the Nevada test site.

In May 1955 federal civil defense officials constructed a town to test the impact of an A-bomb on various kinds of buildings, including five different types of homes; each had a pantry and refrigerator stocked with food to determine whether it could be eaten after a nuclear explosion. The homes were populated with mannequin families, the mannequins provided by J.C. Penney. The test site was named "Survival Town," but it was anything but. The May 5 bomb—twice the power of the Hiroshima bomb—flattened much of Survival Town, reducing the one-story wooden house to splinters and the two-story brick house to rubble in seconds as cameras rolled.

At the same bomb blast, the military arrayed 57 Patton battle tanks, each weighing 50 tons, each with crews in their seats, less than two miles from ground zero. The tanks and crews "emerged in fighting trim" and advanced in attack formation four minutes after the explosion, as if maneuvering in response to a tactical nuclear attack.[32]

It wasn't just the tests themselves that got publicity. After a few early explosions, radioactive snow was detected far east, in Rochester, New York, and Cincinnati, Ohio. City officials in International Falls, Minnesota, including the mayor, warned at a public meeting that the Defense Department appeared to be testing "tea-cup sized" atom bombs on a bombing range over the skies of northern Minnesota. The AEC denied they were testing atomic weapons of any size in that state, but the *Washington Post* put a story about the teacup-size A-bombs on its front page. (The U.S. Defense Department eventually developed nuclear weapons the size of a backpack or an artillery shell, but never the size of a teacup.)[33]

The prospect of nuclear war was part of everyday life. The movie that taught American kids to save themselves from atomic attack by learning to "duck and cover" under their school desks came out in 1952, financed and supervised by the Federal Civil Defense Administration. Over the next three years it was shown to millions of schoolchildren.[34]

In 1953 and 1954 there were all-time record outbreaks of tornadoes across the U.S., accompanied by fierce thunderstorms. The belief that the tornadoes were caused by the atomic testing in Nevada was so common among the public that Gallup conducted a poll on the topic, which showed that 29 percent of Americans thought the A-bombs were causing the tornadoes many states away; 20 percent couldn't say if they were or not. Members of Congress required the military to provide data on any connection. The myth was so persistent that at the opening session of its 1954 meeting, the American Meteorological Association devoted a session to debunking the link between the atomic testing and the tornadoes, while acknowledging that, at that point, meteorologists didn't actually know what did cause tornadoes. In fact it turned out there weren't really more tornadoes in those years; it was just that tornado reporting and tracking had gotten much more effective and thorough.[35]

The U.S. conducted 188 nuclear weapons tests from 1951 through 1958, detonating, on average, 24 nuclear weapons a year for eight years straight. As scary and sobering as the vivid demonstration of the power of atomic bombs was, the very public testing was clearly designed, in the U.S., to instill not fear but confidence—that the U.S. nuclear arsenal was unsurpassed.[36]

---

And then the nuclear race took an unsettling turn that didn't have to do with the bombs themselves but with the ability to deliver them.

On August 26, 1957, the Soviets announced that they had successfully launched an intercontinental ballistic missile—a rocket, designed to carry a large nuclear warhead, which could leap halfway around the world in 15 minutes. The ICBM did this by traveling at 15,000 mph and by arcing up almost into orbit, tracing the curve of a high pop-fly in baseball, before rocketing back to Earth and its intended target. The Russian statement had the oddly elusive language typical of the Cold War: "a super-long-distance" missile was launched "a few days ago"; "the missile flew at a very high, unprecedented altitude," "covering a huge distance in a brief time." Despite the lack of detail, the statement pointed out, "The results obtained show that it is possible to direct a missile into any part of the world." The ICBM had become known around Washington, D.C., policy circles as "the ultimate weapon" for its ability to deliver nuclear destruction anywhere, at almost a moment's notice.[37]

That same day, the U.S. government granted permission to the Russians to fly two passenger jets filled with Russian diplomats to the September meeting of the United Nations General Assembly in New York. The Russian planes belonged to the Soviet airline Aeroflot, which had the only passenger jets in regular service anywhere in the world.

Newspapers twinned the stories: Americans woke to the news on that summer Tuesday that the Russians had tested a weapon that could drop nuclear bombs on any point on Earth and that their passenger jets—"the envy of Western commercial aviation," as the *New York Times* described them—would be landing in a few weeks in New York City.[38]

Just four days later U.S. intelligence would reveal that the Soviets hadn't tested a single ICBM, as their own announcement said. They had tested six, over a period of weeks, with the missiles flying 4,000 miles or more within the territory of the Soviet Union. U.S. intelligence claimed, now, to have known about the tests before the Russian announcement. Neither the Russians nor the Americans revealed any details—the range of the missiles, their carrying capacity, or, most important, their accuracy.[39]

Americans had tested their own first ICBM, the Atlas, just a few weeks earlier that summer, from the new launchpads of Cape Canaveral, Florida. The Atlas launch, with thousands of people watching, lasted less than a minute. The missile's first stage lost first one engine, then a second, pitched horizontal, and was blown up by remote control in gales of orange flame and black smoke over the Atlantic Ocean. It was the sixth failed missile launch for the U.S. in the first six months of 1957.[40]

And it was just the beginning.

Just 39 days after their ICBM announcement, on October 4, 1957, the Soviets launched Sputnik, the first satellite, into orbit. Sputnik weighed 184 pounds; it was a metal sphere about the size of a big beach ball, with four swept-back antennas, a very 1950s design. NBC and CBS interrupted regular programming on both their television and radio networks to broadcast the sound of Sputnik's watery beep-beep-beep that Friday night, the signals captured from space. Sputnik swooped around the Earth every 96 minutes, at an altitude of up to 560 miles, moving at 18,000 miles an hour.

The first U.S. satellites were designed and set for launch in the spring of 1958, and they weighed 21.5 pounds. Sputnik, already in orbit, was eight times heavier than the U.S. satellites that hadn't yet been launched. By the second day of its orbit, U.S. newspapers were publishing timetables with major cities listed and the time Sputnik would race overhead, like train timetables: Detroit, 9:30 a.m.; Washington, D.C., 1 minute later.[41]

The Soviet Union congratulated itself by saying that the launch of the first artificial Earth satellite showed that "the new socialist society turns even the most daring of man's dreams into reality."[42]

Remarkably, with just a single satellite in orbit, a metal ball carrying nothing but batteries and a pair of beeping radio transmitters, the Moon was immediately on everyone's mind. The day after the launch, Russian scientists said their rockets would soon be headed for the Moon. And a grumpy U.S. rocket scientist said, "Maybe the Russian-American competition hasn't been so bad on the satellite if it encourages us into beating them to the Moon."

The politics in the U.S. wasn't neat. Democrats had control of both the House and the Senate, but Republicans had the White House.

Senator Henry Jackson, a prominent Democrat from Washington State, described Sputnik as "a devastating blow to the United States' scientific, industrial and technical prestige in the world." Senator Stuart Symington, a Democrat from Missouri, warned that if America let the Russians continue to surge ahead in space technology "the future of the United States [could] well be at stake." Senator Styles Bridges of New Hampshire, a Republican, connected Russian supremacy to American consumer culture: "The time has clearly come to be less concerned with the depth of the pile on the new broadloom rug or the height of the tailfin on the new car and to be more prepared to shed blood, sweat and tears if this country and the free world are to survive."

Was America's education system inadequate? Was America's missile funding inadequate? Was America's respect for and celebration of science inadequate? Was America's sense of urgency inadequate? Americans suddenly feared the answer to all these questions was yes—that the inadequacy was proven by the beep-beep-beep.

President Eisenhower was visiting his farm in Gettysburg, Pennsylvania, the evening Sputnik was launched, and the White House reported that he played golf the next day. He said nothing. His press secretary said Sputnik would not cause the U.S. to alter the carefully planned schedule of U.S. satellite development.[43]

Eisenhower had a perspective that inclined him not to panic. He was 67 at the launch of Sputnik, not particularly old, but he had been born in 1890, in the horse-and-buggy era. Two of the most influential members of his cabinet spanned the same period: Secretary of State John Foster Dulles was born in 1888, and his brother, the CIA director Allen Dulles, was born in 1893. This meant that three officials in critical positions to shape American space policy were born even before electricity was an everyday convenience. Eisenhower had seen one transformative technological revolution after another: electricity, the car, radio, motion pictures, television, the telephone, the airplane, the jet plane, the vacuum tube, the rocket. Not to mention radar, sonar, and the atomic bomb. A single metal ball whirling overhead was not going to fluster the man who led the invasion of Normandy, the liberation of France, and victory in Western Europe in World War II.

Eisenhower's calm notwithstanding, the launch of Sputnik was galvanic. That first evening, the NBC News anchor introduced Sputnik's beep-beep-beep to the public by saying, "Listen now for the sound which forever more separates the old from the new." The dawn of the Space Age had shifted the geopolitical axis back on Earth. On the very first day of the Space Age, the *New York Times* said in an analysis piece, "The Soviet Union is thought to be making a conscious effort to persuade people, especially in Asia and Africa, that Moscow has taken over world leadership in science."[44]

In a classified assessment for Eisenhower, U.S. government officials told the president the same thing—not that the Soviet Union *had* surpassed the U.S. in technological achievement, but that it looked like they had. "Soviet claims of scientific and technological superiority over the West and especially the U.S. have won greatly widened acceptance," the memo reported. Countries friendly to the U.S. were suddenly worried "over the possibility that the balance of military power has shifted or may soon shift in favor of the U.S.S.R." Countries pondering whether to align with the U.S. or the Russians would take the satellite very seriously "as a demonstration that the Soviet system has gained scientific and technical superiority," particularly in places "that view their problems as requiring the rapid achievement of a higher technological level."

And, the memo concluded, reaction in the U.S. wasn't helping. "American anxiety, recrimination, and intense emotional interest have been widely noted abroad, and assiduously reported by the Soviet media. . . . The American reaction . . . has itself increased the disquiet of friendly countries and increased the impact of the satellite."[45]

Sputnik became an instant cultural touch point. New York City department stores reported a run on the sale of binoculars after the Russians said in their official statement that Sputnik could be seen in orbit "in the rays of the rising and setting sun with the aid of the simplest optical instruments such as binoculars." Newspapers referred to it as "the red Moon." Wry humor came quickly into play. A bartender in San Diego devised what he called the "Sputnik cocktail": one-third vodka, two-thirds sour-grape juice. In Poland the joke was that the Soviet Union finally had a satellite smaller than Albania. *Life* magazine, in a story headlined "Soviet Satellite Sends U.S. into a Tizzy," said Sputnik's beep "sounded like a cricket with a cold." The Vatican had perhaps the least generous assessment of Sputnik. Referring to it as "this baby moon," the Vatican called it "a frightening toy in the hands of childlike men who are without religion or morals."[46]

In some ways, Sputnik was both the birth of the Space Age and the earliest riff on the Digital Age. The beep-beep-beep from Sputnik is almost identical to the beeping of 60 years of electronic devices that came after it. It was the first major electronic beep most people came to know.[47]

Day after day, for weeks, the Russian satellite whirled and beeped overhead. The sense of foreboding about what Sputnik meant didn't dissipate as the days passed, it gathered force. NASA's history of the early rocket era says of the weeks following Sputnik, "Gone forever in this country was the myth of American superiority in all things technical and scientific."[48]

Then it happened again.

Thirty days after Sputnik, early on the morning of November 3, 1957, the Soviets launched Sputnik 2. This second satellite managed to make Sputnik 1 look like a practice flight. Sputnik 1 weighed 184 pounds; Sputnik 2 weighed 1,120 pounds—half a ton. Sputnik 1 carried some batteries and a pair of radio transmitters. Sputnik 2 had a pressurized cabin and a passenger, a 12-pound terrier named Laika.[49] Sputnik 1 was a beach-ball-size satellite. Sputnik 2 was a real spaceship—a rudimentary one, but a spaceship nonetheless.

Sputnik 1's batteries had run out of power a week earlier, but it remained in orbit. The world now had two artificial moons, both Russian.

Sputnik 2 was launched on a Sunday. Three days later, on Wednesday evening, the Soviet Union began an epic celebration of the 40th anniversary of the Bolshevik October Revolution that had brought communism to power. On stage celebrating with Khrushchev, among others, were Mao Zedong, the leader of China, and Ho Chi Minh, the leader of Vietnam.

After Sputnik 2, scientists weren't talking idly about missions to the Moon; the speculation was that there was a third Sputnik already headed for the Moon, perhaps carrying a hydrogen bomb that would explode there to mark the Bolshevik Revolution in truly spectacular fashion, with the first lunar "firework." As it happened, that Thursday was also the date of a lunar eclipse. On its front page the *New York Times* explained the timing of the eclipse relative to the arrival of a possible Soviet Moon rocket, and that scientists concluded that the explosion of an atomic weapon against an eclipse-darkened Moon "would create an illumination . . . brighter than the light of a full moon."[50]

Sputnik 2 was shrouded in some mystery. The Russians reported day after day that the dog's vital signs were "satisfactory": heartbeat, respiration, and blood pressure were all measured by sensors and radioed down to Earth. But the Russians didn't grasp either the public relations upside or downside of sending the first living creature—an adorable ginger-colored terrier—into space.

For days the international press didn't learn what the dog's name was, or whether it was a male or a female. At one point, three days after Sputnik's launch, four names had all been offered: Damka ("Little Lady"), Kudryavka ("Curly"), Limonchik ("Little Lemon"), and Laika ("Barker").

Meanwhile dog lovers worldwide protested what was clearly a one-way flight for the Pupnik. A march at the United Nations included dogs, one of which was a Russian wolfhound, wearing protest signs. Londoners marched on the Soviet embassy and presented a letter of protest to the ambassador. The Soviet government replied that the British protesters were "barking at the Moon."

On day 4 of the mission, the Russians had clarified that the dog's name was Laika and that she was female.[51]

Come Thursday and the 40th anniversary of the October Revolution, there turned out to be no rocket delivering atomic fireworks to the Moon. Khrushchev gave a speech that lasted three hours and seven minutes to 17,000 people. "Now our first Sputnik is not lonely in its space travels," he said. He poked fun at the word "vanguard," the optimistic name of the U.S. satellite program. "Life has shown it was the Soviet Sputniks which were ahead, in the 'van.' Our Sputniks are circling the world and are waiting for the American . . . sputniks."[52]

The most striking messages of Sputnik 2 were unspoken but unequivocal. First, scientists, engineers, and defense experts agreed that the rocket necessary to get a 1,100-pound satellite 1,000 miles into orbit had more than enough power to deliver a nuclear weapon to any spot on Earth. It might not have had the accuracy that ICBMs require, but it had the oomph. And Sputnik was no fluke. The launch of a single small satellite was an achievement; the launch of a second one, six times larger and carrying a live passenger, within 30 days, was the sign of an ambitious space program. The Soviet Union was making itself a spacefaring nation.

And given the global politics, it was doing so at the expense of the United States. Worldwide the U.S. was a laughingstock. At the Moscow Circus, Karandash, Russia's most famous and beloved clown, quickly incorporated Sputnik into his act. Karandash would enter the circus arena carrying a balloon, which would pop as soon as he entered. A fellow clown would ask, "What was that?" Karandash's reply: "The American Sputnik!" And the circus audience would explode in gales of laughter.[53]

Eisenhower had been scheduled to give a series of speeches on science, security, and public policy, an effort to boost national morale in response to Sputnik 1, starting on November 13. The talks had earned the informal nickname "chins up" speeches. But with Laika overhead, November 13 was too leisurely even for Eisenhower. He moved the first speech up six days. Even so, it had more the air of an FDR fireside chat (with the occasional touch of a Jimmy Carter lecture) than of a rallying cry. "Let me tell you plainly what I am going to do in this talk," Eisenhower starts. "I'm going to lay the facts before you—the rough with the smooth. Some of these facts are reassuring; others are not—they are sternly demanding."

It was a good speech—a tour of the entire world of nuclear weapons (the U.S. Navy, it turned out, had nuclear depth charges) and of missiles (the U.S. had 38 different kinds of missiles in 1957, both deployed and under development). "The United States is strong," Eisenhower said, with enough military power "to bring near annihilation to the war-making capabilities of any other country."

And Eisenhower was blunt about how the world had changed. The president hadn't just seen World War II; he'd been in charge of a lot of it. "One B-52," he said, "can carry as much destructive capacity as was delivered by all the bombers in all the years of World War II combined."

As to the Sputniks—Eisenhower referred to them as "Earth satellites"—they were "an achievement of the first importance." But with "no direct present effect upon the nation's security." He didn't agree that the U.S. suffered any kind of "gap" when it came to its ability to defend itself, its allies, or freedom anywhere in the world, the fretfulness from Capitol Hill notwithstanding.

Eisenhower made one bold move that Thursday night, but it was bureaucratic: he announced the creation of a new position in the White House, a special assistant to the president for science and technology, so he would have the best guidance when it came to making decisions about science and policy. The science advisor would report directly to the president. What was bold wasn't so much the job itself as the man he had recruited: James Killian, then president of MIT.[54]

It wasn't inspiring. It wasn't an answer to Laika circling overhead at 18,000 miles an hour. It was reasonable and reassuring. Eisenhower wasn't interested in chasing the Russians into orbit, and he wasn't worried. Just a few weeks earlier—before Laika's mission—Eisenhower had said at a press conference that the U.S. satellite program had "never been conducted as a race with other nations." Asked specifically if Sputnik worried him, he replied, "So far as the satellite itself is concerned, that does not raise my apprehensions one iota."[55]

In private Eisenhower simply didn't get what the hubbub was about. If you have a fleet of 180 brand-new B-52 Stratofortress bombers, who cares about a dog zooming around in circles? "I can't understand why the American people have got so worked up over this thing," he said at a meeting of White House advisors. "It's certainly not going to drop on their heads."[56]

In his nationwide speech, Eisenhower didn't advocate speeding up the U.S. space program; he didn't discuss rising to the challenge of the Russian satellites with U.S. satellites. In a speech conceived as a response to Sputnik anxiety, and then delivered six days sooner in response to a second wave of Sputnik anxiety, Eisenhower didn't mention the U.S. space program or future U.S. space plans or even U.S. space ambitions *at all.*

*Life* magazine, for one, was unimpressed. The week after Eisenhower's speech, it published a caustic essay by a former Manhattan Project scientist headlined, "Arguing the Case for Being Panicky." George R. Price wrote that a country "is apt to get the things it values most. And so we will probably continue to have the world's best TV comedians and baseball players, and in a few years Russia will have the world's best teachers and scientists. . . . We will not stay free simply by appointing a science coordinator."[57]

The first real U.S. satellite was set for launch sometime in early 1958. That was the 21.5-pound Vanguard 1, which, despite being tiny compared even to Sputnik 1, was expected to carry an array of scientific instruments.

But missile launches in the early days of the space program were notoriously subject to failure, and so the U.S. Navy, which was in charge of Vanguard, had a series of three test launches set up to make sure the three-stage Vanguard launch rocket would work when the real satellite was on top.

The first full-up Vanguard test launch was scheduled for Wednesday, December 4, 1957, a month after Laika's flight and Eisenhower's "chins up" speech. It was unquestionably a test launch: the first time all three stages of the 72-foot-tall Vanguard rocket would be fired together. Atop the stack: a miniature satellite weighing three pounds.

As the first U.S. effort to launch a satellite in the Space Age, the launch of that Vanguard became a media sensation. Newspapers around the country published front-page stories for days in advance of the launch. On Monday, December 2, the *New York Times* set up the launch with a front-page story that put the word "test" in quotes. "This week's Vanguard shoot cannot be divested of its historic quality," the *Times* wrote, "whether it is officially described as a preliminary or as the main event."[58]

Technical problems delayed the launch from Wednesday to Thursday, and then from Thursday to Friday morning. By Friday, December 6, 127 reporters were on hand to cover the launch of a satellite the size of a soccer ball, which weighed not even half what Laika the dog had weighed. All three broadcast networks too were on hand. A young Harry Reasoner had rented a beach house for CBS News to give the network a particularly good angle on the launchpad.[59]

The countdown reached zero at 11:45 a.m., and a young technician threw a switch in the Cape Canaveral control room to send the Vanguard rocket on its way. The engines ignited, for a moment the rocket was still, then it started to rise. After just two seconds the rocket hesitated, then stalled at an altitude of three feet. Then the rocket started sinking backward into its own skirt of flame, slowly toppling over—and was suddenly engulfed in an orange fireball twice as tall as the rocket itself. The last view of the rocket as it toppled to the right and disappeared in the flames and black smoke was of the cone-shaped third stage, containing the satellite, toppling free.

In Washington the director of the Vanguard program, John P. Hagen, was on an open phone line to the Cape, listening to a second-by-second narration of the launch from his deputy, J. Paul Walsh: "Zero. Fire. First ignition." A pause. "Explosion!" At Hagen's end of the line, a single word: "Nuts!"[60]

The satellite itself had been blown clear, landing on the ground about 75 feet from the flaming launchpad. Sensing that it had been released from the rocket, the satellite commenced broadcasting its radio signal, which was instantly picked up in Vanguard control. The satellite was unharmed and functioned so well that it continued broadcasting from the beach for hours after the disaster. As Deputy Director Walsh explained at a postlaunch press conference, "It's still operating. . . . You have to take it apart to turn it off."

The legendary columnist Dorothy Kilgallen, from the *New York Journal-American*, said, "It seems almost inhuman to let the poor thing go on. Someone should go out there and kill it."[61]

The embarrassment, the indignity, the vaudevillian slapstick failure of the world's greatest nation having its first satellite launch ballyhooed for a week and then ending in a spectacular, televised explosion was almost too much to take. And also completely irresistible.

Senator Lyndon B. Johnson, the Senate majority leader, said, "What happened this morning is one of the best publicized and most humiliating failures in our history." The *Los Angeles Herald & Express* ran the headline "9-8-7-6-5-4-3-2-1-PFFT." The *New York Times* ran a roundup of slashing editorials from 22 different papers around the U.S. The foreign press was particularly wicked.

> *London Daily Herald:* "Oh, What a Flopnik!"
> *London Daily Express:* "U.S. Calls It Kaputnik"
> *Tribune de Lausanne* (Switzerland): "If ridicule could kill, America would be dead today."

The official newspaper of the Polish Army wrote a thank you note to the United States for providing "a moment of merriment in the dull grayness of our everyday life." And at the United Nations on the afternoon of the Vanguard failure, the Russian delegation reminded members of the U.S. delegation that the Soviet Union had a program providing technical assistance to developing nations. Would the U.S., perhaps, be interested in applying?[62]

The global mockery was all in good fun. Except the Soviet Union wasn't a rival or a competitor, and the Cold War wasn't a sporting event. The Soviet Union was an adversary, often an antagonist. Despite all the good humor, the Russians were almost, but not quite, the enemy. Yet. All the war planning assumed they might become the enemy one day.

---

At just about the time of Sputnik 1, Sputnik 2, and the Vanguard flopnik, a dinner meeting came together at a legendary Boston restaurant. Locke-Ober had been opened in 1875 and was an institution, a place, as the *New York Times* once described it, of "old silverplate, old woods, old manners and old waiters." It was the kind of place where people had lamb chops or roast beef for lunch. It was a favorite haunt of Senator John F. Kennedy and his brother Robert; it had also been a favorite of their grandfather, John Francis "Honey Fitz" Fitzgerald, who had been the mayor of Boston in the early 1900s. Another regular at Locke-Ober was Charles Stark Draper, an MIT professor and engineer, who was as legendary and influential in the world of aerospace engineering as the

Kennedy brothers were in Massachusetts politics. It was Draper, in fact, who with his MIT colleagues had refined and perfected the technology that would ultimately allow humans to navigate their way in space, even to the Moon. Doc Draper liked to adjourn to Locke-Ober after a morning in the classroom and an afternoon in his research labs at MIT, often bringing a crowd of fellow scientists with him for dinner.[63]

One evening, before John Kennedy had formally announced his run for the presidency, a mutual acquaintance arranged a dinner for him, Robert, and Draper at Locke-Ober. Draper's low-key mission was to try to intrigue the Kennedy brothers with the possibilities of spaceflight. Although Draper was something of a raconteur himself—and a great salesman of cutting-edge technology—the Kennedy brothers just didn't seem interested in space. As Draper would later recall, they treated his ideas "with good-natured scorn." The Kennedys "could not be convinced that all the rockets were not a waste of money, and space navigation even worse."[64]

It was a remarkable dinner nonetheless, because three or four years into the future, Kennedy would make possible the most significant achievement to come from Draper's work—a Moon landing—and Draper's work would make possible the most dramatic legacy of Kennedy's presidency: that same Moon landing.

But at that first dinner Draper came away with the distinct impression that John Kennedy didn't know that much about space and didn't care that much about it.

Later, during the political campaign when he defeated Nixon, Kennedy was eloquent about the space race. The embarrassing and indolent U.S. performance in space was the perfect emblem of the administration and the era Kennedy was running against. And dramatic achievements in space were the perfect promise of his New Frontier. But when it came to doing more than talking about space—when it came to reimagining and energizing the U.S. space program—the John Kennedy of Locke-Ober was a better indicator of his true intent than the John Kennedy of campaign speeches.

Despite being a theme of the campaign, in fact, space didn't seem like it was going to warrant much attention from a Kennedy presidency.

Kennedy had asked his vice president–elect, Lyndon Johnson, to be in charge of space policy, his first assignment being to find a new NASA administrator, the replacement for NASA's founding chief, Keith Glennan. By Inauguration Day Johnson had tried to persuade no fewer than 19 people to take the job, without success.[65]

The world of space is a relatively small community, much smaller in 1961 than it is today, and the world of experienced officials capable of managing a major government agency is also a relatively small community. When a man as persuasive as Vice President–Elect Lyndon Johnson can't convince any of his first 19 choices to take the nation's senior space job, the word begins to get around, and the lack of interest can become self-reinforcing.

"When it became known that people were turning down the top job at NASA," said Robert Seamans, the agency's associate administrator when Kennedy became president, "it was not good for agency morale. Why would anybody not want to run NASA? It must mean that the agency doesn't have a high priority in the administration's planning."[66]

It's not clear if he was, in fact, the 20th person to be interviewed for the job of NASA administrator, but when he was approached, James Webb fully expected to be the 20th person to turn it down. Webb was a businessman and lawyer and a Washington veteran; he had been head of the office of budget for President Harry Truman, and then second-in-command at the State Department. On Friday, January 27, 1961, Kennedy's science advisor Jerome Wiesner called Webb in Oklahoma and asked if he would come talk to Vice President Johnson on Monday about the NASA job, and then perhaps the president. Webb apparently tried to demur in that first telephone call, but Wiesner was insistent.

Webb was a smart enough Washington hand to go to Washington that very Friday to spend the weekend talking to people about what the NASA job might involve. Nothing he learned made him want the job; as he headed to meet the vice president on Monday morning, he had concluded in his own mind, "I would not take the job if I could honorably and properly not take it."[67]

What followed next was a deft and classic Johnson squeeze. Outside Johnson's office was Hugh L. Dryden, a highly respected NASA scientist and administrator who was then the acting head of NASA. Webb told him he was there to talk about the NASA administrator's job but that he didn't think he was the right person to do it. He asked Dryden to convey that message to the vice president, but Dryden replied, "I don't believe he wants to listen to me on that."[68]

When Webb spoke to Johnson, he said he would accept the job only if he were asked directly by the president. And so Johnson arranged for Kennedy to see Webb that afternoon. Webb had been explaining to each person in turn that he didn't have the technical background or proficiency to be the head of a space agency, and he tried that in the Oval Office as a way of declining Kennedy's offer. According to Webb's account, the president immediately parried: he didn't want a technical person; NASA was loaded with smart scientists and engineers. "There are great issues of national and international policy involved in this space program," Kennedy said. "I want you because you have been involved in policy at the White House level, State Department level."

Webb had an old-fashioned sense of both honor and obligation: if the president of the United States wants you to take a job, and thinks you can do it well, and you don't have a good reason to say no, you have a duty to say yes. And so the man who had spent the past three days trying to figure out how to decline the NASA job left the Oval Office having accepted it.[69]

That was Monday, January 30. Webb's Senate confirmation hearings were held three days later, on Thursday, February 2. He gave an opening statement, was confirmed unanimously, and was sworn in on February 14. Webb, the 20th person down the list to be the head of NASA, would sit at the center of some of the most momentous and memorable events of the 1960s. His stewardship of NASA would ultimately be something people studied in graduate school. But at his confirmation hearing that Thursday, his appointment didn't seem particularly significant. After delivering his opening statement, there was not a single question for him from the assembled senators.[70]

# 3

# "The Full Speed of Freedom"

> To those of us who had watched our rockets keel over, spin out of control, or blow up, the idea of putting a man on the Moon seemed almost too breathtakingly ambitious.
>
> **Eugene Kranz**
> *Apollo flight director*[1]

It was Representative James G. Fulton (R-PA), then an eight-term congressman, who said on the day of Gagarin's flight, "I'm darned well tired of coming in second-best all the time." That was the comment a reporter waved at President Kennedy at his press conference and that got Kennedy to say, with world-weariness, "No one is more tired than I am."[2]

But that's the point: Gagarin's flight on April 12, 1961, wasn't a new triumph for Russia or a new indignity for the U.S. Every six months, for three and a half years, the story had been the same: a Russian space program that, in so many important and visible ways, was the world's leading space program.

Kennedy wasn't that interested in space, but he understood how important space was becoming—strategically and symbolically. Nuclear missiles and long-range bombers were indispensable, but they weren't persuasive and inspiring and visible the way astronaut heroes were.

The Russians launched the first satellite and the first satellite with living creatures aboard (Sputniks 1 and 2). The Russians flew the first spacecraft to the Moon (Luna 1). The Russians launched the first spacecraft that hit the Moon, including delivering tiny metal Soviet flags to the surface (Luna 2). The Russians launched the first spacecraft to visit the dark side of the Moon, which had never before been seen by humans. It took photos using conventional film; the film was developed onboard the space probe while it was streaking around the Moon, and those negatives were scanned, digitized, and radioed back to Earth—all in 1959 (Luna 3).[3] In 1960, the Russians launched the first animals into space who orbited and then landed back on Earth alive and safe—the dogs Strelka and Belka (Sputnik 5)—before Kennedy was elected president.

And then the Russians launched the first astronaut into orbit and brought him home safely and to worldwide acclaim.

The years of frustration were, at last, boiling over. Americans didn't like being second—which in this case was the same as being last—and were particularly irritated by being told they would catch up eventually, but it would take a while. It had been a while since that first eerie beep-beep-beep, and since Laika, and there was no catching up. In fact the Russians had sent Gagarin into orbit on their first manned flight; the U.S. too was scheduled to put its first astronaut into space, Alan Shepard, but he would go just to the edge of space in a 100-mile-high arc, landing just 300 miles from Cape Canaveral in the Atlantic Ocean, without going into orbit. The U.S. was systematically planning to underperform the Russians, even after the Russians had shown what they could do.

The day after Gagarin's flight, NASA administrator James Webb and his second-in-command, Hugh Dryden, were called before the Science and Astronautics Committee of the House of Representatives for what was dubbed an "autopsy." It was conducted in the largest committee hearing room available to the House. But the congressmen weren't that interested in what Webb and Dryden had to say.

"I want to see this country mobilized to a wartime basis because we are at war," said Representative Victor Anfuso (D-NY). "I want to see schedules cut in half. I want to see what NASA says it's going to do in 10 years done in five.

"I want to see some 'first' coming out of NASA such as a landing on the Moon."

The chairman of the House Committee, Overton Brooks (D-LA),

insisted, "It's time we stopped making excuses on why we are behind and have been for three years. . . . The nation that controls space may well control Earth."

Representative Fulton made Webb an offer: "Tell me how much money you need and this committee will authorize all you need."[4]

That Friday night, April 14, 1961, as the regular workday wound down, President Kennedy gathered a group of advisors in the Cabinet Room, just off the Oval Office, to talk through the space problem. The question they confronted wasn't how to build a rational, step-wise, carefully conceived, and science-driven program. The problem was clear on the front pages of the nation's newspapers and the TV news broadcasts each evening: the Soviet Union kept demonstrating over and over its initiative, its ambition, its technological preeminence, at least in space, at least in space spectaculars. If you were the leader of a nation trying to decide which of the great powers to follow, which of the two systems produced great results, which way was the future, the Russian space program was one way of seeing the future.

Present in the Cabinet Room that Friday evening were seven people in addition to the president: James Webb, who had been head of NASA for 59 days, and his deputy, Hugh Dryden, a scientist deeply experienced in both NASA and rocket technology; Jerome Wiesner, Kennedy's science advisor and a strong opponent of expensive manned space programs; David Bell, the head of the federal budget office; and Ed Welsh, a longtime space policy advisor. Ted Sorensen, one of Kennedy's closest aides, was there, after running an hours-long preliminary meeting that afternoon with the same group. The seventh person was Hugh Sidey, the White House correspondent for *Time* and *Life* magazines, who had been invited to watch Kennedy hash out the space question, to get a sense of what the response of the New Frontier would be to Gagarin's flight, for use in a future magazine story.

The group assembled around the cabinet table. Wiesner had his pipe, unlit. Kennedy entered, and the group rose as he took one of the chairs. He pushed it away from the table, then rocked back on the two rear legs and propped his right shoe on the edge of the table.

"As I understand it," Kennedy said to the group, "the problem goes back to 1948, when we learned how to make smaller [nuclear] warheads that could be carried with smaller boosters." The U.S. didn't have the big rockets that were powering the Russian space program because the U.S. didn't need them to deliver nuclear bombs. "What can we do now?"

The president went around the table looking for recommendations on the possibility of catching up to the Russians. As he listened, Sidey says, he rocked on the two legs of his chair and absentmindedly tapped his front teeth with his fingernails.

The reports were not encouraging; when it came to human spaceflight, the Russians probably still had a two- or three-year lead. Sidey says the president found the conversation frustrating.

"Now let's look at this," the president said. "Is there any place where we can catch them? What can we do? Can we go around the Moon before them? Can we put a man on the Moon before them? . . . Can we leapfrog?"

The Cabinet Room discussion was a mirror of the one Sorensen had conducted a few hours earlier, which had concluded there was only one path to beating the Russians.

Now Dryden spoke up. He told Kennedy that putting a man on the Moon was the one way to best the Russians, but that it would require a Manhattan Project–style effort in terms of both cost and intensity. Dryden believed it might cost up to $40 billion, an extraordinary sum at a moment when the entire federal budget was less than $100 billion a year. Even with that kind of effort, Dryden said, the chance of beating the Soviets, the chance of success, was probably only 50 percent.

As Sorensen had imagined at the afternoon meeting, the idea of landing on the Moon captured Kennedy. "The cost, that's what gets me," he said. "When we know more, I can decide if it's worth it or not." He was clearly talking about the only serious option on the table: an all-out push to the Moon. "If somebody can just tell me how to catch up. Let's find somebody—anybody. I don't care if it's the janitor over there, if he knows how." Kennedy paused. "There's nothing more important."

Kennedy let his chair rock back to the floor, rose, thanked the men, then turned and headed for the Oval Office, trailed by Sorensen and Sidey. Sorensen and Kennedy stepped into the Oval Office to confer, while Sidey waited outside.[5]

Sorensen says Kennedy was nervous about the idea of going to the Moon, but animated. "He immediately sensed that the possibility of putting a man on the Moon could galvanize public support for the exploration of space as one of the great human adventures of the twentieth century." The president told him to figure out if it was really possible.

Sorensen stepped out of the Oval Office to find Sidey waiting for him. The U.S. response to Gagarin, he told Sidey, "would be strong and dramatic."

"We're going to the Moon," Sorensen said. He was exultant at the idea. He was talking to a reporter, though, and immediately qualified his excitement, telling Sidey that Kennedy wanted the question thoroughly researched in terms of logistics and cost.[6]

That Friday evening, though, was the moment Kennedy concluded that if the U.S. was going to race the Russians, the only finish line was the Moon itself.

---

The six weeks from that Friday evening meeting to May 25, and what the White House called Kennedy's second State of the Union address for 1961, were packed with a series of interlocking and momentous events.

The next morning, Saturday, April 15, the long-planned Cuban exile effort to overthrow Fidel Castro by invading Cuba began with a somewhat hapless attack on Castro's air force by American B-26 bombers, piloted by Cuban exiles, painted to disguise their origin and flown from Nicaragua.

Early Monday morning, amid worldwide attention to the air attack and accusations that the U.S. was behind it, 1,400 Cuban exiles motored ashore in landing craft at Bahía de Cochinos—the Bay of Pigs. They had been trained in Guatemala for a year by the CIA, and the attack itself had been planned, coordinated, supplied, and executed under the direction of the CIA, an operation approved by President Eisenhower, and then inherited, reviewed, and approved by Kennedy.

Within hours the invasion turned into a debacle, both slapstick and deadly. Two of four freighters loaded with ammunition and supplies for the invaders were sunk by Cuban air force jet fighters, and the other two freighters fled back south to international waters. CIA planners didn't know enough about the Bay of Pigs to know about its coral reefs that crippled and capsized some small boats taking troops ashore. Radios and weapons were immersed or lost overboard, leaving whole platoons with no weapons and no communications. Although Kennedy staunchly refused to provide support from the U.S. military to the invading exiles—to avoid the U.S. military actually invading Cuba, which might have provoked a direct military response from Russia—any hope that the invasion would be seen as an organic Cuban effort, a rebellion unconnected to the U.S., unraveled as quickly as the invasion itself.

By Wednesday the invading force was so low on ammunition and so desperately outmanned by 20,000 Cuban soldiers who surrounded them—soldiers led by Castro himself—that the invaders surrendered.

The Bay of Pigs invasion was over about 60 hours after it started.[7]

On the previous Wednesday, the U.S. had been humiliated on the world stage by the triumphant spaceflight of the Russian Yuri Gagarin. One week later the U.S. was humiliated by the hapless collapse of a military effort to overthrow Fidel Castro. In both cases the communist world triumphed, and wasn't shy about the triumph. The reputation of Khrushchev and Russia was immeasurably enhanced by the first human spaceflight. Castro and his revolution were immeasurably strengthened, both literally and in terms of worldwide reputation, by swiftly defeating invaders backed by the might of the United States.

The failed Bay of Pigs invasion was such a series of compounded errors, and such an international embarrassment, that amid all the post-invasion statements, press conferences, and consultations, Kennedy decided to quietly meet with his predecessor at Camp David on the Saturday after the invasion failed. At Camp David, Eisenhower, out of office just three months, urged unity behind Kennedy and then confided, "It is nice to be in a position where you are not expected or really even allowed to say anything."[8]

The hurried space program review that Kennedy had ordered just the previous Friday night had managed to gather some momentum during the week of the Bay of Pigs, despite the disaster.

Kennedy met with NASA chief James Webb and Vice President Johnson on Wednesday afternoon, April 19, just as the Cuban exiles were getting ready to surrender. Johnson had played a key role in the Senate in the wake of Sputnik, convening hearings on the space program, and now he was head of a group Kennedy had revived called the National Aeronautics and Space Council. Kennedy asked Johnson to take the lead on figuring out what the U.S. should do in space, and to figure it out quickly—a role Johnson clearly relished.

The next day Kennedy sent Johnson a memo that has become famous as a kind of foundation stone of the U.S. race to the Moon. It was drafted by Sorensen but has the conversationally inquisitorial tone of Kennedy's voice. The whole note—it's more a note than a memo, although the main points are numbered—is 12 sentences long, and nine of them are questions, direct, challenging:

> 1. Do we have a chance of beating the Soviets by putting a laboratory in space, or by a trip around the moon, or by a rocket to land on the moon, or by a rocket to go to the moon and back with a man. Is there any other space program which promises dramatic results in which we could win?
> 2. How much additional would it cost?
> 3. Are we working 24 hours a day on existing programs. If not, why not? If not, will you make recommendations to me as to how work can be speeded up.
> 4. In building large boosters should we put [our] emphasis on nuclear, chemical or liquid fuel, or a combination of these three?
> 5. Are we making maximum effort? Are we achieving necessary results?

The note ends, "I would appreciate a report on this at the earliest possible moment."

These were the vice president's marching orders:

Can we beat the Russians?

Is going to the Moon the way to do it?

Is there any other way to win?

Are we running the race with maximum intensity—a question Kennedy asks twice in five bullet points.

And by the way, could you please hurry up, Mr. Vice President, and get these questions answered—including what will this all cost?

That note became the foundation of the Apollo program, and it's important to pause and notice that it's really about just one thing: how to beat the Russians in space. And perhaps it's about the reverse as well: how to use space to beat the Russians on the global stage.

It's an internal memo, briskly summarizing a conversation that lasted an hour. It's a way for Kennedy to give Johnson clarity and focus on what he wanted from Johnson's review in the next several weeks. But it's not a memo about space science or technology development, about the challenge and adventure of exploration. It's a memo about the role of space in the Cold War.[9]

The next day, Friday, April 21, Kennedy held another press conference, although his most recent one had been just the previous Wednesday. It was the end of the worst week of Kennedy's young presidency. The failed invasion had only aggravated Cold War tensions around the world, Khrushchev vowing to provide "all necessary assistance in beating back the armed attack on Cuba," with well-organized anti-American protests in Moscow and Warsaw, Cairo and Mexico City. On Thursday, Kennedy had given a speech about the Cuba invasion to the American Society of Newspaper Editors, declaring that "Cuba must not be abandoned to the Communists" and that "the forces of communism are not to be underestimated in Cuba or anywhere else in the world."[10]

Kennedy opened the Friday morning press conference by saying he wouldn't take questions on Cuba, preferring to let his speech to the newspaper editors "suffice for the present." But that morning he got the sharpest questions he'd ever faced about space, five questions out of twenty-six.

A reporter asked, "Mr. President, you don't seem to be pushing the space program nearly as energetically now as you suggested during the campaign that you thought it should be pushed. In view of the feeling of many people in this country that we must do everything we can to catch up with the Russians as soon as possible, do you anticipate applying any sort of crash program?"

Kennedy immediately plunged into a list of his increased funding of various rocket booster projects and stated that everything was being studied: "I don't want to start spending the kind of money that I am talking about without making a determination based on careful scientific judgments as to whether a real success can be achieved or whether, because we are so far behind now in this particular race, we are going to be second in this decade."

At another point a reporter asked, "Isn't it your responsibility to apply the vigorous leadership to spark up this program?"

That was just what Kennedy had been trying to do in the past week, but he again said only that ways of beating the Russians were being studied.

The key moment was a question from William McGaffin of the *Chicago Daily News*:

> ***McGaffin:*** Mr. President, don't you agree that we should try to get to the Moon before the Russians, if we can?
>
> ***President Kennedy:*** If we can get to the Moon before the Russians, we should.

Kennedy had never said anything like that before. At his press conference on the previous Wednesday, the day of Gagarin's flight, he had reminded the press what he had said before: "The news will be worse before it is better." The meetings, conversations, and events of the previous 10 days were changing his thinking, his tone, and also his willingness to start talking publicly about "beating the Russians" in space.

That line—"If we can get to the Moon before the Russians, we should"—was the news out of the press conference. In a week of incredible events piled up one after another, beating the Russians to the Moon made headlines across the country.[11]

Vice President Johnson, meanwhile, plunged into the task of answering Kennedy's list of questions with LBJ-style gusto. He started conducting meetings on Saturday; that first one included Webb, Dryden, and the legendary rocket scientist Wernher von Braun, up from the rocket development center in Huntsville, Alabama. On Saturday afternoon Johnson met with Defense Secretary Robert McNamara. On Monday he convened what he called a "hearing" in his office, to try to sort through everything he'd heard and walk through Kennedy's five questions once more. He assembled a dozen or so people, including Webb and Dryden and von Braun; Wiesner, Kennedy's very skeptical science advisor; senior officials from Defense and the budget office; and Ed Welsh, executive director of the National Space Council. Johnson also invited three outsiders to provide perspective: George Brown, of the huge Texas construction firm Brown and Root; Donald Cook, vice president of the electric utility American Electric Power; and Frank Stanton, CEO of the TV network CBS.

There were some presentations. There was discussion of where the U.S. could—possibly—leap ahead of the Russians. Dryden had told Johnson that it was possible the U.S. could circumnavigate the Moon before the Russians, robotically return a sample of Moon soil to the Earth, or land astronauts and return them. All were so far beyond the capacities of either nation that the Russians' head start didn't amount to much. Von Braun said the same.

Wiesner recalls how the meeting wrapped up: "Johnson went around the room saying, 'We've got a terribly important decision to make: Shall we put a man on the moon?' And everybody said yes. And he said, 'thank you' and reported to the President that the panel said we should put a man on the Moon."

Welsh, the staff person for the National Space Council, which Johnson chaired, says Johnson "listened a great deal in the first few of the meetings, finding out what Dryden believed could be done and what Von Braun . . . and others . . . thought could be done." But as the meetings continued, Johnson became more confident in the course to take, says Welsh, and when anyone in the meeting "seemed to be a little hesitant [Johnson] would go around the room, and he would point to that individual and say, 'Now, would you rather have us be a second-rate nation or should we spend a little money?' "[12]

Webb was, quietly, a little more deliberate than the vice president and a little surprised at his approach. "He just picked up the phone and called everybody that he thought was tops, independently," said Webb. Johnson looped in von Braun, who worked for Webb, without asking Webb first. He looped in several senior Defense officials without consulting McNamara first.

Webb was a deeply experienced manager, both inside and outside government; he'd help run Sperry, a key aerospace technology company, during World War II when it grew from 800 to 33,000 employees; he'd been President Truman's director of the federal budget, and then assistant secretary of state, helping Dean Acheson reorganize the State Department. Webb grew up in Oxford, North Carolina, and during the early years of his childhood, his family relied on a horse and buggy to take them over the dirt roads of Granville County. He'd been a Marine Corps pilot during the Depression. He had seen exactly what the U.S. could do—organizationally, industrially, militarily—during World War II.

"I'm a relatively cautious person. I think when you decide you're going to do something and put the prestige of the United States government behind it, you'd better doggone well be able to do it." Webb wasn't going to challenge the vice president head-on, but he did want to be able to deliver what Johnson, and then Kennedy, promised. On the day of the "hearing" Johnson conducted, when he polled the men in the room, Webb had been head of NASA for only 69 days.[13]

On Friday, April 28, 1961, just a week after getting Kennedy's 12-sentence memo, Johnson delivered a five-page reply, summarizing his week of research and meetings, and then answering the president's questions directly. Manned exploration of the Moon, he wrote, "is not only an achievement of great propaganda value, but it is essential as an objective whether or not we are first." During Johnson's week of meetings, landing on the Moon had become "essential" to national policy.

Johnson went on to warn that urgency was also essential, because soon the lead would swing so far to the Russians, both technologically and in the minds of people and leaders around the world, "that we will not be able to catch up, let alone assume leadership."

The last sentence of Johnson's memo is this: "We are neither making maximum effort nor achieving results necessary if this country is to reach a position of leadership."[14]

---

Johnson's reply—he considered it preliminary, with many details still to be filled in—went to the president on a Friday. That was two Fridays after Kennedy's Cabinet Room meeting where he wondered if the White House janitor knew what to do in response to Gagarin's flight. It was 10 days after the unraveling of the Bay of Pigs invasion, and a week after the press conference where Kennedy said, "If we can get to the Moon before the Russians, we should."

On that same Friday, April 28, 1961, on Launchpad 5 at Cape Canaveral, the rocket that would take America's first astronaut into space was receiving final inspections and preparations. The launch of the first Mercury mission—the pop-fly-style flight, up to the edge of space and back down in a long arc—was scheduled for four days hence, Tuesday, May 2.

The previous three weeks had upped the stakes. A successful flight, even though it would just touch the edge of space and not orbit the Earth as Gagarin had, could shake loose the sense of stagnation and inadequacy in the U.S. It would give Americans something to cheer about. Given the conversations of the previous two weeks, the stakes for NASA and for James Webb couldn't have been higher. Success would be a demonstration for the new space agency of competence, predictability, reliability.

A failure would be bad in so many ways, it was discouraging to even think them through. It would be disastrous for the U.S. to seriously injure, or even kill, the first person it tried to launch into space. It would be a worldwide mortification that would make the satellite launch failures look simply pathetic in retrospect.

Any kind of failure, even one in which the astronaut was rescued unhurt, would make it hard for President Kennedy to call for a crash national effort to put astronauts on the Moon, if NASA couldn't even get one 60 miles up to the edge of space and back safely.

The launch of Alan Shepard's Freedom 7 Mercury capsule was a perfect mirror of the December 6, 1957, launch of the tiny Vanguard satellite that plopped out onto the beach at Cape Canaveral, beeping away. And Shepard's flight, although a little later than Gagarin's and a little less of an achievement, would also be done in public for the world to see.

In a moment that certainly has no equivalent, President Kennedy personally made the decision to launch Freedom 7. At a meeting in the Oval Office on Saturday, April 29, 1961, a group of Kennedy aides—including Sorensen, National Security Advisor McGeorge Bundy, Wiesner, and Welsh—talked through the pros and cons of the pop-fly flight. According to John Logsdon's account in *John F. Kennedy and the Race to the Moon*, the sentiment in the room was against the launch as scheduled.

"There was a hesitancy there," Welsh remembered. "One of his staff people raised the question about whether it should be postponed [because of fears] of another disaster."

Kennedy had been assured by Webb and, perhaps more important, by Webb's deputy Dryden, who knew NASA's capabilities, that NASA would launch only if everything was ready. Pressure from the schedule, from the assembled hundreds of reporters, from global politics, would not move the countdown clock one tick toward zero if the Redstone rocket and capsule weren't ready to go. In the end, a single question seemed to give Kennedy his resolve. Welsh asked, "Why should we postpone a success?"[15]

Weather did postpone the launch, from Tuesday to Thursday, then to Friday, May 5, at 7 a.m.

Navy Commander Alan Shepard, 37 years old, one of the original Mercury 7 astronauts, was woken at 1:10 a.m. for a breakfast of scrambled eggs and filet mignon wrapped in bacon. With him was his fellow astronaut and backup flyer, John Glenn. Shepard got yet one more physical exam—which quite likely lasted longer than the flight itself would—and then shrugged into his spacesuit. The Mercury suits had a silvery exterior. At 5:15 a.m., Shepard, carrying his portable air-conditioning unit, stepped into the small gantry elevator, and at 5:20 a.m. he was helped into the Mercury capsule, its interior as tight as the cockpit of a jet fighter.

Among the indignities Shepard was subjected to, and which was duly reported, his full array of body sensors included a rectal thermometer that made the ride to space with him, in place.[16]

The first manned spaceflight offered watchers what became a hallmark of U.S. spaceflights for the next 50 years: delays. Countdown holds. There was a hold because clouds made photographic conditions poor. An electrical component in the booster needed to be replaced. An IBM computer wasn't working right, and it and its fellow computers had to be completely recycled.

Finally, at 10:17 a.m., the countdown resumed and headed for ignition at 10:34 a.m.[17]

The launch and the mission were broadcast live on radio and TV, and the whole nation watched and held its breath. At the White House, President Kennedy was conducting a National Security Council meeting about policy toward Castro's Cuba, which was interrupted so the members of the NSC could pile out and watch the launch on a black-and-white TV perched on a table behind the desk of Kennedy's longtime secretary, Evelyn Lincoln. Lined up in a semicircle around her desk were the chief of staff of the navy; Vice President Johnson; Attorney General Bobby Kennedy; Abraham Ribicoff, secretary of Health, Education, and Welfare; Deputy Secretary of Defense Paul Nitze; and the speechwriters Richard Goodwin and Arthur Schlesinger. A series of White House photos shows the group watching, all with intensity and seriousness. The president stands at one end of the semicircle, hands in his pockets, eyes on the screen. Jacqueline Kennedy is to her husband's left, wearing a suit and a pillbox hat. There's only one picture in which Kennedy has broken into a smile.[18]

The flight was narrated nonstop on CBS by Walter Cronkite, but more important, by NASA's public affairs chief John Powers, who repeated, almost verbatim, everything Shepard said, about one second after he said it.

For the first time in history, anyone in the country, and in much of the world, who wanted to could follow the flight of a human being into space from the moment of launch to the moment of splashdown. Just five people watched Frank and Orville Wright make the first airplane flight. The U.S. and the world learned of Lindbergh's safe crossing of the Atlantic to Paris only by telegraph, and then by radio and newsreel. For Freedom 7, NASA had credentialed 350 reporters.[19]

The first manned U.S. launch was a particularly pointed contrast to the Russians', whose space achievements were announced hours after the fact, with no detail, no film, no voices, just statements of official excitement. Soviet failures never happened, as far as the world was concerned, because no one outside the Russian program itself ever knew about them.

The BBC broadcast NBC's audio coverage of Shepard's flight live to Great Britain. Japanese radio and TV covered it live. The U.S. government's Voice of America broadcast live updates around the world—in 35 languages—including to the Eastern European Iron Curtain countries. In New York City the mayor's office broadcast the radio feed over loudspeakers to a crowd of hundreds gathered in City Hall Park.[20]

It was a suspenseful and flawless 15 minutes.

The rocket launched. Shepard provided his own play-by-play.

"Roger, lift off and the clock is started," he said, starting his onboard elapsed-time clock. Then, with a test-pilot's instincts, he gave his call sign and started providing data, reviewing the status of his onboard systems. "This is Freedom 7. The fuel is go, 1.2 G, cabin at 14 p.s.i., oxygen is go."

Shepard didn't stop talking for the whole 15 minutes, but he offered only a couple sentences of observation, four minutes into the flight, looking out a submarine-style periscope: "What a beautiful view. Cloud cover over Florida. . . . Can see Okeechobee, identified Andrus Island, identified the reefs."

The flight was so short that except for that one line—"What a beautiful view"—and a brief mention that the flight was "a lot smoother now" as his rocket rose, Shepard didn't convey any sense of his experience at all. He simply radioed the status of indicator lights, equipment, and altitude.

Freedom 7 traced an arc through the sky off Florida. At the very top of the curve, Shepard was 116.5 miles high.

Before you knew it, he was floating back to Earth. Exclaimed Walter Cronkite, "The parachute is open and the spaceflight is a success!"[21]

To ensure Shepard's safe retrieval NASA had stationed six U.S. Navy destroyers in a kind of necklace in the Atlantic from Cape Canaveral to the splashdown point, 300 miles east. In the end, Shepard landed close enough to the aircraft carrier *Lake Champlain* that the 1,200 crew members cheering on deck could see the splash as he hit the water.

A Marine Corps helicopter hovering just a few hundred feet away hooked the capsule. Shepard popped his hatch and was winched aboard the chopper, which wheeled off to the deck of the *Lake Champlain.* On the brief flight Shepard told the marine pilots, "Boy, what a ride!"

An hour later, Shepard was summoned from his debriefing in the admiral's cabin to the bridge for a call from President Kennedy.[22]

The nation reveled in the event. The flight, reported the *New York Times* on its front page, "roused the country . . . to one of its highest peaks of exultation since the end of World War II." On an ABC News special report that evening, anchor Bill Shadell said, "The country's faltering prestige received a strong booster shot."

The evening of the launch, TV viewers could hear Shepard's own voice from space. The openness, which would become a hallmark of NASA launches through the sixties and beyond, was an obvious risk. But the rewards from unqualified success were, in some ways, as great as those Russia reaped for going first, 23 days earlier. Not much happened during that 15 minutes—which was the best possible outcome—but it felt like NASA was determined to release every detail it had.

The newspapers, starting with the afternoon papers printed five or six hours after splashdown, published transcripts of the radio exchanges between Shepard and ground control. There were diagrams of the capsule and the flight path, photographs of Shepard eating, having his cardiac sensors attached, riding the elevator to the capsule, bounding from the helicopter to the deck of the *Lake Champlain.* During NBC's evening special report, correspondent Frank McGee reported that Shepard's pulse had been 105 during reentry. "It was a great privilege to be allowed to participate in Shepard's flight," said Leonard J. Carter of the British Interplanetary Society. "I was pretty well up there in the capsule with him."[23]

America's first manned spaceflight gave people a taste of triumph and a vicarious sense of what real space travel would be like. It also provided a burst of anticipation and ambition about the future. Louise Shepard, the commander's wife, speaking from the front lawn of their home in Virginia Beach, told reporters, "This is just a baby step, I guess, compared to what we will see."[24]

---

Jerome Wiesner, the MIT professor who became Kennedy's science advisor at age 45, and went on to be president of MIT, had run a group that looked at U.S. space policy for Kennedy during the transition. The resulting Wiesner Report was bluntly skeptical of America's manned space effort and of any "race" in space, a reminder that even inside the White House there was doubt about the Moon race. Wiesner thought manned spaceflight in general, and a leap to the Moon in particular, were poor science, even bad science, and a waste of money that could be more smartly used on robotic probes.

"By having placed highest national priority on the Mercury program," the Wiesner Report said, "we have strengthened the popular belief that man in space is the most important aim of our non-military space effort." But the publicity and focus on manned spaceflight "exaggerates the value of that aspect of space activity." The U.S. "should stop advertising Mercury as our major objective in space" and should "diminish the significance of this program to its proper proportion before the public."

Forget going to the Moon: the Wiesner Report said that even "a crash program aimed at placing a man into an orbit at the earliest possible time cannot be justified solely on scientific or technical grounds" and might even hinder a smartly thought-through manned space program "by diverting manpower, vehicles and funds."[25]

Wiesner was in many of the meetings to discuss how to respond to Gagarin, and his skepticism did not waver. It's worth saying that within the terms that Wiesner and his committee framed questions about space, they were probably right: in purely scientific and technical terms, if you were designing a space program without regard to public support or public understanding of space, without regard to the politics of funding by Congress or international politics, you would design a different space program than the one either the U.S. or the U.S.S.R. ended up pursuing.

But that's also a silly, perhaps even an irresponsible argument for a scientist working at the highest level of public policy. If not for politics and public support, of course, antipoverty programs would be designed differently, and so would funding for mass-transit systems, and research into disease and medicine, and the priorities for weapons purchases by the Defense Department. Columbus's voyage to America involved politics and national aspirations; Lewis and Clark's journey from St. Louis to the Pacific Ocean involved politics and national aspirations; and so too did the earliest space efforts.

Wiesner, to his credit, had a firm grasp on the political reality, even if he didn't like it. "Kennedy found himself confronted with three choices," he said. "Quit, stay second, or do something dramatic." Quitting wasn't practical. Continuing to come in second, as Wiesner put it, "was even worse."

Wiesner told of a moment that captures well how Kennedy was wrestling with the decision, and with its impact well beyond space. On the Wednesday before Shepard's flight, the president and first lady hosted the first state dinner of the Kennedy administration, for Habib Bourguiba, the president of Tunisia.

Wiesner was off to one side, talking to Bourguiba, when President Kennedy wandered up. As Wiesner recounts the story, Kennedy said to

Bourguiba, "You know, we're having a terrible argument in the White House about whether we should put a man on the Moon. Jerry here is against it. If I told you you'd get an extra billion dollars a year in foreign aid if I didn't do it, what would be your advice?" Bourguiba thought for a long moment. Then he told the president, "I wish I could tell you to put it in foreign aid, but I cannot."[26]

Kennedy had yet another press conference, on May 5, 1961, a few hours after talking directly to Shepard aboard the aircraft carrier. That call, similar to the one Gagarin had had with Khrushchev after his flight except that there was no gloating and no politics, had apparently not been thought of in advance. It was spontaneous, and with the technology of 1961, it turned out to be challenging to patch a telephone in the Oval Office through to the bridge of a U.S. Navy ship at sea. It ended up being a minute-long exchange of congratulations and thank-yous.[27]

At the press conference—his third in 23 days, each one making space news—it was clear Kennedy's push to go to the Moon was on his mind, and he seemed almost at pains not to brim with enthusiasm for Shepard's accomplishment.

Would the president expand on his personal reaction to Shepard's historic first U.S. flight into space? "As an American," Kennedy replied, "I am of course proud of the effort that a great many scientists and engineers and technicians have made, of all of the astronauts, and of course particularly of Commander Shepard and his family."

Then, without missing a beat, Kennedy added, "We have a long way to go in the field of space. We are behind. But we are working hard, and we are going to increase our effort."[28]

---

The speech that launched the United States to the Moon almost didn't happen. Kennedy's address to a joint session of Congress on May 25, 1961, was unusual. Presidents didn't typically travel to Capitol Hill and address both houses of Congress except on three occasions: their inaugurations, to deliver the annual State of the Union address, and in the case of war. In 1961 Kennedy had already given an inaugural address and, 10 days later, a State of the Union speech.

But April had gone so badly for the Kennedy administration that the president clearly sensed he was losing both the substance of the Cold

War with Khrushchev and also the symbolism, and he decided to try to reset the administration's sense of priorities, and also its momentum, with a fresh message to Congress. He would request a new wave of programs to show his determination to counter the Russians—money to modernize the military, for foreign aid, for an all-new effort at a civil defense program to protect Americans from nuclear attack. And money for a dramatically expanded and dramatically accelerated space program.

But while there was going to be a *message*, and a fresh set of requests for hundreds of millions of dollars, the plan was to send the written message up to Capitol Hill via courier, where it would be read by clerks to the Senate and House chambers.[29]

The section of Kennedy's speech that lays down the challenge for Americans to go to the Moon has become one of the iconic moments of NASA's history, of the Moon mission itself, and of Kennedy's time as president. It's hard to believe his words would have gotten the attention, not to mention providing the momentum and the lasting historical resonance, if he hadn't actually said them. In fact it's possible the decision to actually deliver the speech is what gave the Moon mission its first, critical burst of momentum.

Vice President Johnson had delivered a much more thorough analysis of the space issues to Kennedy, written and edited in a furious burst of activity the weekend after Shepard's flight, and delivered to Kennedy the day Shepard visited the White House. It was 32 pages of history and reasoning, well argued but without much eloquence.

The report was blunt on U.S. space performance to date: "Our results have, despite many excellent achievements, been disappointing in many ways. Nearly half our attempted launchings failed to achieve orbit. Certain programs achieved success, real success, on fewer than a third of all attempts." On the question of racing the Russians to the Moon, the report argued, "Even if the Soviets get there first, as they may, and as some think they will, it is better for us to get there second than not at all." And there was a line aimed directly at Kennedy's inclination for action: "If we fail to accept this challenge it may be interpreted as a lack of national vigor and capacity to respond." "Vig-ah" was not something Kennedy wanted to be lacking.

But the key moment in the document was this: "We recommend that our National Space Plan include the objective of manned lunar exploration before the end of this decade. . . . The orbiting of machines is not the same as the orbiting or landing of man. It is man, not merely machines, in space that captures the imagination of the world."[30]

Kennedy would be considerably more eloquent. He formally agreed to the recommendation at a meeting two days later, on Wednesday, May 10.[31]

Racing the Russians to the Moon was big news. It was a complete reversal of Kennedy's, and his administration's, lack of enthusiasm for space. It would cost what in the 1960s was a huge amount of money. The spending estimates in the Webb-McNamara report were surprisingly accurate: between 1961 and 1963 NASA's budget would quadruple, and then it would hover at $4 billion a year for five years in a row (the equivalent of $32 billion a year in 2018 dollars).[32]

The *Washington Post* broke the news five days before Kennedy announced it, on Saturday, May 20, with the bold front-page headline "U.S. to Race Russians to Moon." The story, by John G. Norris, opened, "President Kennedy has definitely decided to try to put men on the Moon ahead of Russia under a greatly accelerated space program controlled by civilians." The story had two interesting qualifications. "Officials decline to describe this as a crash program, because it brings to mind the secret, cost-is-no-object Manhattan atomic bomb project." And, Norris wrote, "in deciding to try to beat the Russians, officials said there has been no firm determination that this can be done, but rather that it is worth trying." On Wednesday, the day before the speech, the *Post* refreshed the story after Kennedy's weekly meeting with congressional Democrats. "Mr. Kennedy told the group the United States either has to get all the way into the space race or get out, and that his decision is to give it all the Nation has."

That same day the *New York Times* also had the news in advance of the speech, but framed it with a Cold War emphasis. "President to Ask an Urgent Effort to Land on Moon" was the headline on the lead story on the front page, which opened, "President Kennedy is expected to tell Congress Thursday that there is an urgent need for the United States to land a man on the Moon—and to do it first, if possible." Congress, wrote W. H. Lawrence, would be asked for "a vast expansion and speed-up of the entire space program in the context of a race for survival with the Communist world."[33]

It wasn't just a race to the Moon. It was a race for survival.

In Kennedy's hands, the pitch was not quite as instrumental and was more persuasive. But it was clearly a Cold War challenge.

Why the White House decided not to send the message to Capitol Hill but have Kennedy deliver it in person is a little vague. That Thursday morning—just hours before the speech at 12:30 p.m. from the well of the House of Representatives—the reversal made news. Johnson was credited, in part, with urging Kennedy to make the speech in person. Kennedy, said the *Washington Post*, needed "a revival of spirit in Washington. Ever since the Cuban invasion fiasco the bloom has been off the bright rose of the early days of the new Administration."[34]

Kennedy's presence and delivery gave the speech an eloquence and an impact it could never have had if read aloud by a clerk. (It was a long speech: 5,800 words as delivered, 46 minutes. Kennedy's working text, with handwritten edits, was 81 pages.)

It was not, in fact, a speech about space. It was a speech about the Cold War. It was considerably less romantic than his Inaugural.

In just the first five minutes, Kennedy described a worldwide conflict: "The great battleground for the defense and expansion of freedom today is the whole southern half of the globe—Asia, Latin America, Africa and the Middle East, the lands of the rising people. Their revolution is the greatest in human history."

Kennedy didn't name the Soviet Union, but he enumerated Russian strategy and techniques: "Their aggression is more often concealed than open. They have fired no missiles. And their troops are seldom seen. They send arms, agitators, aid, technicians and propaganda to every troubled area. But where fighting is required, it is usually done by others. By guerrillas striking at night, by assassins striking alone, assassins who have taken the lives of 4,000 civil officers in the last 12 months in Vietnam alone." It is, said Kennedy, "a battle for minds and souls as well as lives and territory. And in that contest, we cannot stand aside."

"There is no single, simple policy which meets this challenge," said Kennedy, introducing the heart of the speech, a 30-minute list of proposals for increasing U.S. strength around the world. The need to "turn recession into recovery" in the U.S. to provide economic strength. More economic aid for emerging nations. More military aid for emerging nations. A tripling of Voice of America broadcast hours across Latin

America, where daily broadcasts from Russia and Red China dramatically exceeded U.S. broadcasts.

Before he got to the subject of space, the very last part of the speech, 30 minutes in, Kennedy had listed 21 specific proposals for countering the Soviets.

"All that I have said makes it clear that we are engaged in a worldwide struggle in which we bear a heavy burden." Seven minutes later: "This battle is far from over. It is reaching a crucial stage."

Indeed, the section on the need for a bold expansion of U.S. space ambitions began exactly the same way: "Finally, if we are to win the battle that is now going on around the world between freedom and tyranny, the dramatic achievements in space which occurred in recent weeks should have made clear to us all, as did the Sputnik in 1957, the impact of this adventure on the minds of men everywhere who are attempting to make a determination of which road they should take."

Kennedy's advocacy for space was alternately soaring and specific, cast in terms of rivalry and then explicitly rejecting the rivalry.

"Now it is time to take longer strides, time for a great new American enterprise, time for this nation to take a clearly leading role in space achievement, which in many ways may hold the key to our future on earth." Acknowledging the head start of the Soviets, Kennedy said, "We . . . are required to make new efforts on our own. For while we cannot guarantee that we shall one day be first, we can guarantee that any failure to make this effort will make us last."

Space "is not merely a race. Space is open to us now, and our eagerness to share its meaning is not governed by the efforts of others. We go to space because whatever mankind must undertake, free men must fully share."

That line got full-throated applause from the assembled members of Congress. Of course, in the very passage where Kennedy insisted that space was not merely a race, he promptly reasserted the opposite: he would not leave space to the communists to conquer.

Then he launched into specifics: "First, I believe this nation should commit itself to achieving the goal, before this decade is out, of landing a man on the Moon and returning him safely to the earth. No single space project in this period will be more impressive to mankind or more important for the long-range exploration of space. And none will be so difficult or expensive to accomplish."

The House chamber seemed oddly unmoved by this moment, the one that went down in history. There was no applause.

"In a very real sense," Kennedy continued, "it will not be one man going to the Moon. We make this judgment affirmatively. It will be an entire nation. For all of us must work to put him there."

Again, the audience simply waited for Kennedy's next point.

Kennedy issued a warning and a challenge: "Let it be clear that I am asking the Congress and the country to accept a firm commitment to a new course of action. A course which will last for many years, and carry very heavy costs. . . .

"If we are to go only half way, or reduce our sights in the face of difficulty, in my judgment, it would be better not to go at all."

With acute political instincts and an acute sense of the impact of the Moon race on federal spending, Kennedy was imagining the day when Congress wanted to know why the project cost so much, and yet no spacecraft were on the Moon. If you start spending billions to go to the Moon, you have to go, because otherwise you simply waste the billions. There is no "halfway to the Moon," as there might be 500 miles of interstate highway instead of 1,000. You cannot run out of patience on the way to the Moon.

What's more, Kennedy was saying, to start and then give up would do much more damage to the sense of resolve and technical skill of the United States than never setting out in the first place.

Kennedy paused to remind everyone in Congress of their own sharp frustration at each Soviet achievement in space. "All of you have lived through the last four years. And have seen the significance of space and the adventures in space. And no one can predict with certainty what the ultimate meaning will be of mastery of space."

"I believe we should go to the Moon," he declared. But it will require "a heavy burden, and there is no sense in agreeing or desiring that the United States take an affirmative position in outer space unless we are prepared to do the work, and bear the burdens to make it successful."

This was the point at which the audience applauded; it was the second longest interruption by applause in the whole speech, lasting

13 seconds. The members of Congress who would have to vote for the Moon were ready.

Of the 81 pages of Kennedy's text, the space section—the last before his conclusion—consumed 10 pages. He ended his rallying cry with a single sentence that is a remarkably prescient description of the culture that the Moon race would require.

America will not succeed, Kennedy said, "unless every scientist, every engineer, every serviceman, every technician, contractor and civil servant gives his personal pledge that this nation will move forward, with the full speed of freedom, in the exciting adventure of space."

The Moon will require "the full speed of freedom." That's a splendid, original phrase that captures the innovation of capitalism and the determination of democracy, unleashed. It was an echo of what the U.S. had done economically, technologically, and militarily in World War II, which was at that moment only 16 years and two presidents in the past. Kennedy was, perhaps, remembering his own wartime service, for World War II was without question won in part with "the full speed of freedom." It's a phrase that fully anticipated exactly the "failure is not an option" culture NASA created: getting to the Moon before the decade was out would indeed require the full speed of freedom.[35]

---

Kennedy's speech didn't mention what we would learn by going to the Moon. He didn't mention the science we would have to master to get there, or the technology we would develop to do it, or how that technology would find its way into daily life. He didn't mention the way the almost unbelievable years-long enterprise would inspire a generation of kids to become engineers and scientists and astronauts.

If you read the whole speech, it's clear the Moon proposal isn't about the reach of humankind or the power of curiosity or the irrepressible adventurousness of the human spirit.

Going to the Moon was about beating the Russians and about the impact that beating the Russians would have on "the minds of men [and women] everywhere," trying to pick between freedom and communism. Even more pointedly, going to the Moon was about the impact that America losing space to the Russians was having on the minds of men and women everywhere. Space was another hemisphere in the geopolitics of the Cold War. Kennedy was not going to let the communist banner fly over Vietnam, and he wasn't going to let it fly over the Moon, either.

And Kennedy was going to beat the Russians on a deadline. Perhaps the most memorable detail from the space passage is the promise that America would land on the Moon "before this decade is out." Those five words ended up having incredible power as the sixties progressed. They stuck in the minds of NASA managers and engineers, but also in the minds of senior officials who had no intention of letting the first Moon landing slide into the 1970s. The original text of Kennedy's speech—written by Sorensen, the space section sent to NASA in advance for review—announced that the Moon landing would take place by 1967. "We were aghast," said Robert Seamans, who was NASA's associate administrator and part of the three-person team of senior leaders, with Webb and Dryden. "Jim [Webb] called Ted Sorensen and convinced him and later, the President, that the stated goal should be *by the end of the decade*. In the final version, President Kennedy changed the deadline to 'before this decade is out.'" The year 1967 had been mentioned at congressional hearings, in Johnson's weeks of space policy meetings, even at the press conference after the speech, but Webb was well aware how little NASA knew about getting to the Moon, and the whole nation had seen how prone to delay and unexpected problems early spaceflight was. Webb didn't want a dramatic late-decade success to seem like a failure because it happened in 1969 instead of 1967. Indeed, the general reading inside NASA was that "by the end of the decade," the phrase NASA suggested, meant 1969 or sooner, but that "before this decade is out" could be interpreted to include 1970, if the extra time were necessary (or even until 1971, a decade from Kennedy's decision). Given the Apollo 1 fire in January 1967 that grounded NASA's spaceflights for 18 months, Webb's caution was astute.[36]

President Kennedy would give a richer, more inspiring, more textured speech about going to the Moon—"Those who came before us made certain that this country rode the first waves of the industrial revolutions, the first waves of modern invention . . . and this generation does not intend to founder in the backwash of the coming age of space"—but that speech was 16 months in the future.[37]

At this moment Kennedy's call to go to the Moon more than did the trick. Despite hundreds of millions of dollars in requests for military spending and foreign aid, the leap to the Moon dominated coverage of the so-called second State of the Union in dramatic eight-column banner headlines: "Kennedy Calls for 'Crash' Space Effort" (*Baltimore Sun*); "U.S. Is Going All-Out to Win Space Race, Land on Moon in '67" (*Washington Post*); "Kennedy Asks Billions for Man-to-Moon Shot" (Battle Creek, Michigan, *Enquirer and News*); "Moon by 1970 Is Goal of JFK" (*San Mateo* [*California*] *Times*); "Shoot for Moon, Kennedy Urges U.S." (*Palm Beach* [*Florida*] *Post*).

Vice President Johnson, asked to summarize the theme of the speech, replied, "Peace through space."

The speech was accompanied by detailed budget proposals for increased spending on military helicopters, civil defense shelters, and also spaceships, and the many billions of dollars required to go to the Moon got as much attention as the goal. One anonymous Republican lawmaker said, "Kennedy's deficit is going to reach the Moon before we do."[38] But Kennedy's push for actual legislation to make the Moon race a reality was critical. Many U.S. presidents since Kennedy have given dramatic speeches about space policy, with soaring visions, specific ideas, and deadlines.

In 1984 Ronald Reagan charged NASA with creating a spectacular space station, and doing it within 10 years. By the time the International Space Station had its first permanent crew, Reagan's second term had been over for 11 years. On the 20th anniversary of the Apollo 11 Moon landing, July 20, 1989, George H. W. Bush gave a speech on the steps of the Smithsonian Air and Space Museum declaring that the U.S. was going back to the Moon, would establish a permanent Moon base, and then move on to Mars. A year later Bush set a firm goal of a Mars landing by 2020. In 2004 George W. Bush announced a return of U.S. astronauts to the Moon by 2020 for long-term stays and as a staging base for trips to Mars. In 2010, with the ambitious blueprints of his two predecessors unused, Barack Obama told an audience at the Kennedy Space Center that NASA and the U.S. needed a new generation of advanced spacecraft to allow astronauts to land on an asteroid, then to allow people to orbit Mars by the mid-2030s, and land on Mars shortly after. "I expect to be around to see it!" Obama said.[39]

Presidents have been exhorting NASA and the nation to do the next great thing for 40 years, but not one of them mustered the political will, the congressional support, or the public enthusiasm to make it happen. Kennedy lived in a different world, but he also put enough presidential muscle behind the race to the Moon to keep it going even after he wasn't there any more to push it.

The brilliant success of Apollo has washed out two important elements of the story almost to invisibility. First, Americans don't associate the Moon landings with the Cold War or see them as a dramatic victory over the Soviet Union. The Moon landings have enduring, iconic resonance, but in a way different from the Cuban missile crisis or the fall of Saigon or President Reagan's 1987 speech in Berlin calling on Mikhail Gorbachev to "tear down this wall."

In a way that Kennedy could not anticipate, the mission itself took over. The mission, and the deadline—before the decade was out—motivated and inspired. It's also true that, at least in the popular imagination, Russian space successes started to fade in drama after Gagarin's orbital flight, even as U.S. space missions accelerated in drama and frequency and accomplishment, and the ability of Americans to follow them and have a sense of involvement increased. NASA leaders continued to fear that the Russians' abilities were close to those of the U.S. right through 1968. That, in fact, is why we sent a single space capsule around the Moon at Christmas 1968. With the U.S. so close, but the lunar module not quite ready, NASA wasn't going to let Russia "lasso the Moon" before the U.S. did.

But for the rest of the world, by Apollo 8 and beyond, the spark of beating the Russians, which lit the fuse on the Moon missions, was replaced with the all-consuming effort that getting to the Moon required. The effort transcended the original purpose.

By the time they happened, the Moon landings had become a singular achievement that didn't require racing the Russians for their motive force. And so it can be easy for that original spark to fade in significance. But the race to the Moon was born in the Cold War and wouldn't have happened when it did, with the urgency it did, without it.

The second thing it's easy to lose track of is how completely unready to fly to the Moon NASA and the nation were on May 25, 1961.

As Kennedy spoke, the United States had 15 minutes and 22 seconds of experience flying an astronaut in a spacecraft. Of that, the actual time in space of Shepard's spacecraft was 5 minutes and 4 seconds. Shepard spent his entire flight doing tasks, changing the position of switches, checking the status of equipment, operating the capsule's control jets, and communicating nonstop with Mission Control. He was so busy during his 15 minutes, in fact, that he didn't notice he was weightless—he didn't know he was *in space*—until he saw a washer floating up alongside him in the capsule. The point of all that busyness was to see if the human brain could function normally during weightlessness. With half a century of spaceflight experience, with half a dozen astronauts living and working in space full time now, that question seems almost silly. But it was a genuine medical question and spaceflight concern: How would the brain respond to weightlessness? Would astronauts be able to think?[40]

Every question about how to fly to the Moon was unanswered, and many of the questions themselves hadn't even come up yet. What was the surface of the Moon like? How would you land on it? Could two spaceships flying in orbit rendezvous? What kind of math and controls would be required to do that? How do you protect a capsule coming back into Earth's atmosphere from the Moon—at 25,000 miles per hour—from the incinerating temperatures of reentry?

The senior-most officials at NASA thought there was a 50-50 chance NASA could beat the Russians to the Moon by the end of the sixties. Much of the rest of NASA—which already had 17,600 personnel—was as surprised by Kennedy's challenge as the rest of the nation. "To those of us who had watched our rockets keel over, spin out of control, or blow up, the idea of putting a man on the Moon seemed almost too breathtakingly ambitious," said Eugene Kranz, who was developing flight rules for Mercury flights and went on to be a legendary Apollo flight director.

Said Chris Kraft, the flight director for Shepard's flight, "When [Kennedy] asked us to do that in 1961, it was impossible."[41]

In his speech Kennedy explicitly called on Americans (and on Congress) to decide with him to go to the Moon, not once but several times. "I am asking the Congress and the country to accept a firm commitment to a new course of action—a course which will last for many years and carry very heavy costs," he said. Then, a minute later: "I believe we should go to the moon. But I think every citizen of this country as well as the members of the Congress should consider the matter carefully in making their judgment . . . because it is a heavy burden."[42]

Kennedy was asking not just for a commitment, but for a leap of faith. Put aside spaceflight. In 1961 passenger jets had been in regular service in the U.S. for only a little more than two years. And in 1961 most Americans had never taken an airplane flight of any kind. They'd never been airborne, let alone headed for the Moon.[43]

# 4

# The Fourth Crew Member

How reliable does a computer have to be? . . . I said, "It has to be as reliable as a parachute." [Laughing] I thought that was a real brilliant off-the-cuff response. And I was never asked again.

> **Robert Chilton**
> *the NASA official who helped select the organization that designed the Apollo flight computer*[1]

The first landing on the Moon was a bumpier ride than everyone would have liked. Three sets of problems cropped up in the last seven minutes as the lunar module *Eagle* flew toward the Moon's surface, its descent engine cushioning it as it dropped to landing from orbit.

The problems were quite serious.

Neil Armstrong discovered passing through about 2,000 feet that the place he and Aldrin were aiming to land was strewn with large, dangerous boulders—between 5 and 10 feet in diameter—and he was going to have to fly *Eagle* to another spot, either pull up short and land sooner, or hover and fly beyond that spot.

Not a big deal—that's why the astronauts practiced in simulators, that's why they had the ability to take manual control, that's why, in fact, every crew ended up flying the last part of the landing themselves, taking control from the lunar module's computer. But as Armstrong searched for a good spot, *Eagle* was gulping down fuel. Both Mission

Control in Houston and Armstrong and Aldrin in the cockpit thought *Eagle*'s fuel was getting perilously low.[2]

How could the first spaceship to head for the Moon get low on fuel? Weight was a huge problem for the entire Apollo program, because every pound of supplies you launched to the Moon from Cape Kennedy required three pounds of fuel on the launchpad. The lunar module in particular suffered constant weight growth and weight reduction "scrubs." On the LM, to save weight, the seats for the two crew members were eliminated early, when John Glenn sat in a mock-up and suggested the seats were unnecessary for astronauts flying mostly in zero gravity.[3] The aluminum skin of the crew compartment was milled down to 0.012 inches, the thickness of three sheets of ordinary aluminum foil. The lunar module itself weighed about 9,200 pounds empty and carried almost 24,000 pounds of fuel. And fuel was just like everything else: adding a pound of fuel to extend the landing time required adding three pounds of fuel to carry the extra fuel. So enough was provided to fly from orbit to the landing site on the Moon, with a margin of safety—but it was not a luxurious cushion. (Fuel to return to orbit and dock with the waiting command module was in separate tanks in the upper stage, the crew compartment portion of the lunar module, and that fuel was just for the ascent engine.)[4]

In the back of his mind, Armstrong was thinking about how close he had to be to the surface so he could run out of fuel and still drop down safely without any power at all in those last few feet. There's no air on the Moon; the moment the engine's thrust stops, your spaceship drops like a rock. Twenty-five feet up, no problem. Forty feet up, a hard landing, probably okay. Running out of fuel at 70 feet was considered the edge of safety, the point at which you might damage the spacecraft so much that it couldn't take off again, and you ended up trapping yourself on the Moon.[5]

Armstrong and Aldrin had flown this landing hundreds of times in the simulator at the Kennedy Space Center. But as the LM powered down through 1,000 feet, Mission Control went quiet and the CapCom—the capsule communicator, the astronaut Charlie Duke, who was the only person talking directly to the astronauts—also went silent.

From 1,000 feet down, Armstrong and Aldrin were doing their own flying, relying on Aldrin's familiarity with the flight computer, which was telling him *Eagle*'s descent rate, altitude, and speed, and relying on Armstrong's familiarity with the LM's flight controls. During that last three minutes, Houston radioed *Eagle* only twice: once to say there was a minute of fuel left, then to say there was half a minute of fuel left. Even those communications were compressed. All Duke said was, "Sixty seconds." Then, "Thirty seconds." Armstrong and Aldrin knew what he was talking about.

Neither of those problems—having to pick a fresh landing site or flying with near-empty fuel tanks—ruffled Armstrong and Aldrin. They were going to put their spaceship down on the Moon.[6]

But in the background was a more troubling, more puzzling problem. *Eagle*'s flight computer kept sounding an alarm about its status, then pausing in its work and restarting itself. This wasn't an obvious problem, like boulders at the landing site, nor was it a trivial one. The computers on the Apollo command module and the lunar module were remarkable pieces of technology for their time.

The *Eagle*'s computer pulled in data from the lunar module's gyroscopes, its accelerometers and radar and the radio link to Mission Control, and the computer controlled the main engine, the small reaction control jets, and displayed all kinds of data for the astronauts about where the LM was in space and how fast it was moving, while also accepting commands and instructions from the astronauts via a keyboard. The computer had the ability to do the complicated math necessary for spaceflight almost instantly.

The Apollo guidance and navigation computer (known as the AGC) did all that with less computing power than a typical microwave oven has today, and it was indispensable in getting from lunar orbit to the Moon's surface and then back to the safety of the command module.

When the flight computer kept announcing that it was having a problem, people paid attention, in Mission Control and in the *Eagle* cockpit. Five times in four minutes, just as Armstrong was getting ready to pick a place to land and take manual control, the navigation and guidance computer somehow got overloaded with work, sounded an alarm, dumped the work it couldn't handle, and then restarted itself.

As the lunar module passed 1,400 feet, the control panel and display screen for the computer went completely blank for 10 seconds. It's not clear Mission Control knew this had happened. Aldrin and Armstrong don't mention it. But when you're flying the most sophisticated spaceship ever created, 1,000 feet over the surface of the Moon, dropping at 30 feet per second (20 mph), and searching for a new place to land that spaceship, having your flight computer go completely dark for 10 long seconds would test the cool of any pair of space jockeys.

"It seemed like a long time," Armstrong would say later with restraint. "I never expected it to come back."[7]

Meanwhile the whole world was watching. The landing was broadcast live, except it was 1969, and there were no external cameras on the *Eagle*, so it was really more like radio. The action was in the audio feed of the back-and-forth between Mission Control and the astronauts. If something had happened, it would have been right there with everyone watching and listening.

Except, in fact, things were happening.

The remarkable thing is, you can listen to the 16 minutes of radio communications during the landing flight—from about nine miles (47,000 feet) over the surface to touchdown—and as an ordinary person, you wouldn't know that there was even a single problem, let alone three. Except at the very end, there's an almost nonstop exchange of information, but it's crisp, clipped, and mostly indecipherable. It has exactly the tone of the cockpit crew of a jet talking to an air traffic controller. Occasional references to pitch and yaw. Lots of numbers called out, with critical meaning for the astronauts but no context for those eavesdropping.

The most common thing that gets said is "You're looking good" and "You're looking great"—Charlie Duke from Houston, telling Armstrong and Aldrin that their ship and they were doing well. Duke offers "looking good" six times and "looking great" six times, almost once a minute, outside the three minutes of radio silence at the end.[8]

Houston had a lot more data about *Eagle*'s condition than Armstrong and Aldrin did, plus a lot more people to absorb and understand that data. And Armstrong and Aldrin were very busy trying to fly their spaceship straight in. No go-arounds for the lunar module; with the limited fuel capacity, they had one shot to get from orbit to the lunar surface. They could have aborted at almost any moment; the abort would have used the fuel in the upper stage of the lunar module, which was there precisely to take them back to orbit and the command module, whether in an emergency or a routine takeoff after visiting the Moon.

Equally remarkably, you can listen to Walter Cronkite's narration of the landing for that night's CBS News television broadcast, co-anchored by just-retired Apollo astronaut Wally Schirra (who had flown on the test flight Apollo 7), and also have no inkling of the problems. The computer was setting off what were called "program alarms," and the second time Cronkite hears the phrase "program alarm," he asks Schirra, "What's this alarm, Wally?" Schirra, without sounding particularly convincing, replies, "It's just some function that's coming up on the computer."

But there is a hint of the problems, one that may sound familiar. The actual landing goes like this:

***CapCom:*** Sixty seconds.
***Aldrin:*** Forty feet, down two and a half. Picking up some dust.

For those listening in Houston and at MIT in Cambridge, where the computers were designed, that was an arresting moment. After hundreds of simulations, it was something unanticipated, something never heard before, something vividly real. Dust. The engine floating the lunar module 40 feet off the surface was blasting up dust. Moondust.

***CapCom:*** Thirty seconds.
***Aldrin, nine seconds later:*** Contact light.

The lunar modules had contact probes, five-and-a-half-foot-long aluminum tubes dangling from three of their landing pads, and the moment any of them touched the Moon's surface, a blue light in the cockpit went on.

Four seconds after Aldrin's "Contact," with *Eagle* settled on the Moon, Aldrin says, "Okay. Engine stop."

They are on the Moon, but there's no romance, no poetry, no rejoicing. Instead there is a series of technical call-outs as the astronauts secure the ship, its engine and controls.

***Aldrin:*** ACA out of detent.
***Armstrong:*** Out of detent. Auto.
***Aldrin:*** Mode control, both auto. Descent Engine Command override, Off. Engine arm, Off. 413 is in.

That last call-out—"413 is in"—was Aldrin confirming he had entered code 413 to tell the computer that the lunar module had landed on the Moon. That way, in an emergency, if they needed to take off, the computer would start out knowing where the LM was and be able to navigate accordingly. Then,

***CapCom:*** We copy you down, *Eagle*.
***Armstrong:*** Houston, Tranquility Base here. The *Eagle* has landed.

Then the second most famous lines from the Moon landing, which come 20 seconds after the actual touchdown.

***CapCom:*** Roger, Tranquility. We copy you on the ground. You got a bunch of guys about to turn blue. We're breathing again. Thanks a lot.
***Aldrin:*** Thank you.[9]

*You got a bunch of guys about to turn blue.*

The natural thing is to imagine they were turning blue over the tension and excitement of having landed the first people, ever, on the Moon. In fact, they were all holding their breath, having landed on the Moon while jockeying through problems the rest of the world didn't realize existed. The folks in Mission Control were holding their breath because Armstrong had been hovering above the Moon's surface on the thrust from his rocket engine, looking for a good place to set down, while getting to the very last vapors of gas in his fuel tanks.

But to a lot of people listening, particularly at the Massachusetts Institute of Technology, which had designed and programmed the flight computers, it was the computer alarms that had them holding their breath. They were worried that something was going seriously wrong with the computer.

One of those listening was Don Eyles, a 25-year-old computer programmer at MIT who had written, specifically, the lines of computer code that landed the lunar module and who knew exactly what he was hearing—at least to the point of understanding that the computer was being asked to do something mysterious it couldn't do and was pausing and restarting itself in the middle of its most critical role in the eight-day Moon mission. "I know too much," said Eyles of what he was thinking. He and his colleagues were listening to the Mission Control feed live on squawk boxes in a second-floor classroom in the MIT building devoted to the Apollo effort. What Eyles thought was, "If it were in my hands I would call an abort."

But Eyles had another colleague, a man named Jack Garman, also 25, who was one of the NASA-side technical staff working with MIT on the computer. Garman was in a support room in Mission Control precisely for this moment. The lunar module guidance officer, Steve Bales, in Mission Control proper, punched through to Garman on the internal voice loop to find out what the alarms meant.

Could the first Moon landing proceed, or should Armstrong and Aldrin press the abort button and head back to orbit? The success of the entire program that had begun with Kennedy's speech to Congress eight years and $20 billion ago suddenly came down to answering this question, and doing it with confidence in the space of 10 or 20 seconds.

Garman, who was three years out of the University of Michigan, knew the computer's functioning almost as well as Eyles did, and he had a handwritten list of computer alarm codes and notes on how serious they were tucked underneath the Plexiglas desktop in front of his computer monitors—29 codes in all—where he could quickly consult it. Garman told Bales that if *Eagle* were flying normally, and if the alarm didn't recur too often, it was safe to keep heading for the Moon. Twenty-seven seconds after Armstrong called the first alarm, CapCom said, "We got you. We're go on that alarm."

When it occurred again, Garman didn't wait to be asked; he simply yelled into his headset, "Same type!" He could hear Bales yell, in turn, "Same type!" to Flight Director Gene Kranz, and then hear CapCom call up to *Eagle*, "We're go, same type, we're go." The lunar module was at 2,000 feet, having dropped through 19,000 feet in less than three minutes. Touchdown was 200 seconds away, and a 25-year-old with a handwritten cheat sheet was giving a thumbs-up to the first Moon landing.[10]

It would turn out that the computer was doing just what it was designed to do—a whole series of things that were, in fact, quite extraordinary.

---

In 1969 there wasn't another computer like the Apollo navigation and flight computer in use anywhere in the world.

In an era when computers were just coming down from room-size to closet-size, the Apollo computers were small: one cubic foot, a single box that a person could carry. In an era when all routine computing was done by creating stacks of punch cards, which then had to be run in batches through a computer while the person who needed the work waited hours or days for the results, the Apollo computers worked instantly. They were what were known then as "real-time" computers.

The people at MIT who created the Apollo flight computer and wrote its software did not have access to "real-time" computers for that task. They too relied on piles of punch cards. Savvy MIT programmers worked late at night and on weekends—when fewer people were around—to get their card stacks run faster. NASA's Mission Control computers, of course, did operate in real time. They were IBM 360 mainframes that took up a whole floor in the Manned Spacecraft Center building 30. It was known as the RTCC, the Real-Time Computer Complex, the distinctive speed of those computers built right into the acronym.

In an era when the people using computers almost never interacted directly with them (that was the point of handing over your stack of punch cards and receiving back a stack of green-and-white fanfold paper), and when the people interacting directly with the computers were running them but not actually using them, the Apollo flight computer was something completely novel. The command module and the lunar module each had a keypad and a display right in the middle of the spaceship control panel.[11] The astronauts talked directly to their spaceship computer using what became known as the DSKY ("dis-key," display and keyboard): typing in commands, requests, navigational information, and on rare occasions even keying in fresh computer programming instructions, and reading responses and data from the display. User and computer interacted directly with each other in a way that would be routine a decade later but was then all new, using a computer designed specifically for the astronauts, down to the fact that the keys on the keyboard were slightly oversized so they could easily be typed on while wearing spacesuit gloves.

And the Apollo computer had two qualities which have become part of the foundation of computing, so routine we don't notice them, but which in 1969 were precedent setting, each a critical masterstroke of insight and genius that helped make the Moon missions possible, and also occasionally saved them from failure.

First, the computer had built-in decision-making ability. It was programmed in such a way that it could look at all the work that needed to be done to fly the lunar module at any given millisecond, and do the work in order of importance. It had "executive decision-making capability," as it was called. We take that for granted today; modern computers, right down to the smartphones in our pockets, make dozens, even hundreds of decisions every second about what task to do and in what order. But in 1969, a computer that made its own decisions was unique. The Apollo computer picked the most mission-critical tasks to do first, and did less important tasks later, or skipped them altogether, if necessary—making those judgments, and doing the work, all within tiny fractions of a second, just like modern computers.

And the Apollo computer could fail gracefully and fully recover, almost without missing a beat—as a result, say, of a brief spacecraft power interruption. The Apollo computer kept track of what it was working on at all times, and if something bad happened, it wiped its working memory clean and restarted, picking up almost exactly where it had been, including having preserved any data and calculations it happened to have been in the middle of at the moment of failure.

Finally, in an era when people were just figuring out what you could depend on computers to do besides math, the Apollo flight computer was the first anywhere to have responsibility for human lives. We take the indispensability of digital computing for granted; today, chips and software are embedded everywhere from birthday cards to lightbulbs, and they control everything from the electric grid and the brakes and airbags in our cars to the flight controls of passenger jets and the invisible pulsing of cardiac pacemakers. Our very lives routinely rely on computers without our even thinking about it. But there was a first time, and it was Apollo.

In the 1960s, in the era before microchips, computers were discrete machines that did the work they were asked to do but weren't connected to other things they were responsible for running. Except in Apollo. In the capsule and the lunar module, the Apollo Guidance Computer (AGC) was at the center of flying both ships. It monitored and controlled almost all the vital functions of the spacecraft having to do with the actual flight. Even when astronauts took control of flying using joysticks and throttle controls to maneuver the spaceships, they were "flying by wire," giving instructions to the computer, which was in turn giving those instructions to the engine and thrusters. Without a properly working AGC, you couldn't land on the Moon or safely reenter the Earth's atmosphere on your way home. The nation was betting its Moon ambition on the reliability of the computer, and the astronauts were betting their lives on it.

The Apollo computer—one in the command module and one in the lunar module, identical but programmed differently—is often described with a certain condescending awe at its primitive physical limitations.

A single AGC like the one on *Eagle* that Armstrong and Aldrin were using had 3,840 bytes of erasable memory, what we call random access memory (RAM) today. It had 69,120 bytes of fixed memory, or read-only memory (ROM). So the astronauts flew to the Moon with a computer that had 73 kilobytes of memory. A single email of the day's headlines from your local newspaper might require twice that space.[12]

The AGC could execute 85,000 instructions a second, which sounds busy. But an iPhone Xs, introduced in 2018, can handle 5,000,000,000,000 (5 trillion) instructions per second. The Apollo computer had 0.000002 percent of the computing capacity of the phone in your pocket: two-millionths of 1 percent.[13]

The miracle isn't that your dishwasher has more computer brainpower than the computer that flew to the Moon; it unquestionably does. Yet few of us would depend exclusively on our occasionally erratic iPhones to fly us to the Moon, let alone depend on one of our kitchen appliances. The miracle is just the opposite: it's what the engineers, scientists, and programmers at MIT were able to do with such austere computing resources; it's the amount of work they were able to wring out of the AGC and the amount of reliability they were able to build into it. And in the process the Apollo computer became an example and a foundation for the digital work and the digital world that followed.

---

The flight to the Moon was born aboard an aging but specially equipped B-29 bomber, on a flight from Bedford, Massachusetts, to Los Angeles International Airport, on a Sunday in early February. It was 1953, four years before Sputnik, eight before Kennedy's "go to the Moon" speech, 16 years before Apollo 11. The B-29 flew nonstop from east to west across the United States before there was any nonstop airline service flying that route.

But that wasn't what was so remarkable about the flight. The B-29 Superfortress, the same kind of plane that delivered both atomic bombs at the end of World War II, was part of a special fleet maintained by MIT at Hanscom Air Force Base for testing cutting-edge instruments and navigational equipment. That Sunday morning the B-29 had a pair of air force pilots in the cockpit, one of whom doubled as MIT's chief pilot, along with nine scientists, engineers, and air force officers as observers. As the bomber lifted off the runway at Hanscom, it was in the control of an all-new device the size of a washing machine; it weighed 2,700 pounds and had been mounted toward the rear of the fuselage.

The device was an inertial navigation unit. It had the ability to sense every movement of the airplane: its velocity, altitude, even the smallest changes in direction or attitude. The magic of inertial navigation is that all the instruments that do the sensing and the guiding and the calculations are packed inside the device. They are gyroscopes, accelerometers, and, in the case of this first experimental unit, a pendulum and a clock, connected to an early computer aboard the plane. The inertial navigation unit doesn't need any inputs from any other instruments; it doesn't need radio signals or beacons, or even a compass. It didn't need clear weather—it didn't care about the weather at all. If you tell the unit where the airplane is, at the foot of the runway on Hanscom Air Force Base in Bedford, Massachusetts, and where you want to go, Los Angeles

International, it can fly the plane there all by itself, without any input from other technology or from the people on the plane.

That, at least, was the theory. If the technology could be perfected, inertial navigation would be able to guide any kind of craft—airplane or missile, submarine or ship or even a truck—anywhere in the world, in any weather, with pinpoint accuracy. The inertial navigation system would, in fact, be doing the piloting. In some ways, inertial navigation is a super-sophisticated version of the human inner ear: it perceives every motion, every shift, every acceleration and deceleration. And if you know the starting point of your craft and can account for every motion over time, you always know exactly where you are, and you can get wherever you need to go. Importantly in the 1950s, as the Cold War was amplified by nuclear weapons and increasingly long-range missiles, you can guide and target your weapons with precision, and also without fear of interference with that guidance from the enemy.

And inertial navigation was a window to the future: It was really the only technique to use in order to fly away from the Earth into space. If you wanted to go to the Moon, or on to Mars and beyond, the only way to have the precision and the speed of measurement you would need would be inertial navigation.

The man in charge of the B-29 that February Sunday was Charles Stark Draper, known to his colleagues as Doc and to his friends, since childhood, as Stark. Doc Draper was then 51 years old, the head of MIT's aeronautical engineering department and also the head of an MIT division he had created, the Instrumentation Laboratory (IL).

Draper was a man of enormous intellect, capable of connecting the curves and equations of advanced math to physical problems in the real world, and often the person who would invent or refine the technology to take advantage of those equations. He loved to fly—he went to flight school on his own after washing out of the Army Air Corps, and then bought a plane with a colleague. And he connected his own experiences with piloting planes, with motion through the sky and those planes' primitive flight instruments, along with math to solve big puzzles of navigation and guidance.

He was also a man of enormous personality, sociability, appetite—and occasional impatience and irritability. Draper was a familiar presence at Boston's legendary downtown restaurant Locke-Ober, often with a party of colleagues from the Instrumentation Lab. During the 1960s he would be sure to take astronauts who were training at MIT along on those dinners. He occasionally wore a white suit, including to at least one Apollo launch, and favored bow ties. Even at the height of the crush to get the Apollo work finished, he would sometimes slip away for weekend ballroom dancing competitions. Draper was both a genius and a raconteur.[14]

He grew up in Missouri, entered the University of Missouri at Rolla as a 15-year-old, and after two years went to Stanford, where in 1922 he got a BA in psychology. With a group of friends, he drove across the country—quite an undertaking in 1922 America—and when the group arrived in Boston, the then-new campus of MIT caught Draper's attention. MIT had just a few years earlier moved from Boston to Cambridge. While Draper's friends went on to visit Harvard, the story goes, Draper poked around MIT. He was captivated, and enrolled for the fall, earning a second bachelor's degree, this one in electrochemical engineering, in 1926. Even as he pursued practical research involving engines and aircraft instruments, Draper earned a master's and then a PhD in physics at MIT, started teaching, and by 1939 had become a full professor. Legend has it that Draper holds the record for taking the most courses for credit as an MIT student.[15]

During World War II, Draper and his colleagues and students at MIT worked on the problem of providing accurate gun sights to protect Allied warships from incoming enemy aircraft. Aiming an anti-aircraft gun at an attacking airplane was a complicated problem. The ship it was mounted on was moving, and also crashing through waves; the gun doing the shooting was vibrating furiously; and the attacking airplane was moving. The anti-aircraft guns the British and Americans used at the start of the war proved incapable of protecting their ships because gunners couldn't get them aimed fast enough.

Draper's team at MIT, working with Sperry Gyroscope, developed a legendary piece of equipment called the Mark 14 gun sight, which took all the motions into account and automatically smoothed and adjusted the gun, so a gunner sighting incoming aircraft through the Mark 14 was in fact aiming where the plane would be seconds later when the bullets arrived. The Mark 14, said one World War II U.S. Navy gunnery officer, did "four hours . . . of differential calculus in a split second." The gyroscopic gun sight—which was kept secret until just before the war ended—was so innovative and so successful that Sperry made 100,000 of them, which were widely installed on U.S. and British Navy warships and were credited with saving ships, and thousands of lives, from attacking German and Japanese aircraft.[16]

The key elements in the Mark 14 were gyroscopes—sophisticated spinning devices that look like a child's top. One of the people Draper worked with at Sperry Gyroscope during World War II was Jim Webb, the future head of NASA, who was young but already a senior executive at Sperry.[17]

The culture at the MIT Instrumentation Lab was a pure Draper creation: a setting for training graduate students and future engineers, mixing math and physics along with advanced engineering and manufacturing to create new tools that applied advanced science to solving hard problems with real technology. That's what the Mark 14 gun sight had been. That's why Draper had a pilot's license and a plane, so he could go up in the air and see and feel the problems of flight firsthand. Just before World War II, in fact, Draper had spent the summer of 1939 working at Sperry in New York, trying to improve the design, materials, and manufacturing of the gyroscopes he would use so successfully just a couple years later.

In parallel, Draper and his students and colleagues were working on the idea of what Draper called "navigation inside a box": inertial navigation. The idea, the theory, had been around for a couple decades. The German V-2 rockets had used gyroscopes to help with their guidance. (And the creator of those V-2 rockets, Wernher von Braun, had come to the U.S. with his team after World War II.)

But it was Draper and his students and colleagues who figured out how to design and build the sophisticated technology and knit it all together to make inertial navigation a reality. That became Draper's focus after World War II. For him, the science was clear: inertial navigation would work; the equations proved it. What stood in the way was the quality and precision of the instruments necessary to do the measurements: the gyroscopes themselves. It was the perfect example of how the application of advanced science often required pushing forward the state of the art of manufacturing as well.

Draper tapped his months working at Sperry. Dissatisfied with the quality of workmanship he could get, he recruited watchmakers from Waltham Watch, who were skilled at precision work with tiny components. Draper imposed an early form of clean-room techniques in the areas where the gyroscopes were made—workers wearing shoe-covers and head-covers, women forbidden to wear makeup while assembling the gyroscopes—because of the dawning realization that something as small as a hair or a spot of dirt or makeup could ruin the functioning of the sealed gyroscopes, which spun at 12,000 rpm. It was the result of that work that became the huge inertial measurement unit installed in the back of MIT's B-29 bomber for that secret flight in 1953.[18]

The Superfortress took off at 8:25 a.m. on Sunday, February 8, 1953, and flew for 13 hours and six minutes—2,590 miles—across the width of the country, on its own, without any assistance from Draper's chief pilot, Charles "Chip" Collins, until it was time to land at Los Angeles International Airport. Remarkably, when the plane, and the new inertial navigation system, encountered the 100 mph jet stream passing over Lake Mead in Nevada, it adjusted to compensate in a series of maneuvers that the crew found baffling (the jet stream was little understood at the time) but that turned out to be just right.

After more than 13 hours in flight, the plane was just 10 miles off course as it approached LAX—10 miles out of 2,590 flown. As they deplaned, Draper and his colleagues were ecstatic, prompting Collins to tell his boss, "You can celebrate, but I've just lost my job."[19]

Doc Draper was a scientist with a flair for salesmanship and for the theatrical, and this flight was no exception. The technology and the existence of the flight itself were classified. But Draper had timed the trip carefully. He and his colleagues were headed to a secret conference on the state of inertial navigation and whether it would ever live up to its promise, hosted by the federal government and the University of California, which opened in Los Angeles on the Monday morning after the B-29 flight.

Draper and the staff who had been on the flight worked late into Sunday night, making charts and maps showing their path, supplemented by photographs of landmarks across the continent that they had flown over, taken through the nose of the B-29, all mounted in a wide panorama behind the lectern at the conference site.

The event had been set up as a discussion of whether inertial navigation was *possible.* The opening speaker Monday morning: Charles Stark Draper. He had a surprise. "Gentlemen," he said, "we have a system that works. We did it." After the meeting he gave participants tours through the B-29 that had just navigated itself across the continent.[20]

---

The B-29 flight was so significant a milestone in advanced navigation that when the U.S. government finally allowed Draper to talk about it, in April 1957, that one, four-year-old airplane flight made front pages across the nation. "Device Guides Plane across U.S. without Help of Outside Objects," was the *New York Times* headline. The *Washington Post* was more expansive: "New Jamming-Proof Gyro Pilot Guides Any Kind of Craft Anywhere on Earth."

In 1958 Draper reproduced the flight in a B-29 with a modernized inertial navigation unit, half the size of the original, accompanied by a film crew from CBS News and correspondent Eric Sevareid, for the CBS show *Conquest.*[21]

Draper himself described inertial navigation for the ordinary public by way of metaphor. "An inertial system does for geometry," he said, "what a watch does for time."[22]

That technology became the core of the work Draper and the Instrumentation Lab did in the late 1950s, and, quietly, it became the key to the U.S. nuclear strategy and to closing the perceived "missile gap" between the military prowess of the Soviet Union and the U.S.

Draper's—MIT's—navigational prowess was refined and installed in a series of U.S. nuclear missiles: Thor, the first U.S. nuclear missile, deployed in the U.K.; Atlas, the first U.S. nuclear ICBM; and Titan, and then Polaris. Early missiles designed to carry nuclear warheads were guided by radio signals from the ground, but that made them vulnerable to interference from exactly the people they might be aimed at. Inertial guidance made nuclear missiles immune to that kind of interference. It also made their targeting much more accurate, which in the grim world of nuclear strategy made them just a bit less devastating.

The inertial guidance systems were made smaller and more effective, and the Instrumentation Lab produced a version that could be used on ships and, most important, on nuclear submarines, which could cruise for weeks without surfacing (because they made their own air). Submarines were like spaceships under the sea; even if they could cruise without surfacing, they always needed to know where they were, and the way to do that without revealing their location to everyone else was inertial navigation like that on the B-29.

The Instrumentation Lab pioneered not just the technology but a new kind of role as a government supplier. Draper and the Lab did the development work on the technology, figured out the manufacturing challenges, produced working prototypes, and then handed off the designs to defense contractors who manufactured production quantities, typically in close consultation with the staff and engineers of the Instrumentation Lab.[23]

The leap in missiles for the United States was the development of a missile that could be launched from a submarine—a submarine submerged 50 or 60 feet below the surface of the ocean. To do that the U.S. had to design and build the submarine that could carry rockets and launch them from inside its hull, up through the water, into the air, and then on to potential targets.

In the framework of nuclear strategy against the Soviet Union, submarine-launched missiles had stunning advantages: submarines were mobile, so the missiles were mobile, which gave them a certain invulnerability, and also the ability to get close to targets; submarines, especially nuclear submarines, which didn't need to surface, were hidden, and so the missiles weren't just mobile, they were stealthy; and submarine-launched missiles provided certainty for the strategy of "mutually assured destruction": no matter what damage was done to the United States in a first strike from the U.S.S.R., the submarines and their missiles, roaming out at sea, would survive to provide a counterattack. The idea in the sixties and seventies was that the guaranteed submarine counterattack made the world safer, precisely because it made the idea of a first-strike much less likely.

Of course, for a nation struggling to get its routine rockets into the sky from launchpads that were standing still on land, the practical challenges of launching rockets from inside submarines submerged in the ocean were considerable.

The first generation of submarine-based missiles was called Polaris, and the missiles and their submarines were produced with the kind of speed that demonstrates how vital U.S. leaders thought they were. The first submarine that could carry and launch the Polaris missile was called the *George Washington*. Getting Polaris missiles deployed was considered so urgent that the *George Washington* was built using the hull of a sub that was already under construction; the hull was sliced in half, and a new 140-foot section was added between the conning tower and the tail, which contained 16 missile-launching tubes, in lines of 8 on each side of the sub. The initial version of the Polaris missile itself was 28 feet tall and weighed 28,000 pounds. Their devastating mission notwithstanding, just the engineering required to get them into a submarine was astonishing in the late 1950s: first it was necessary to build a sub big enough to carry 16 three-story-tall rockets, upright in launch position, with the technology necessary to use compressed air to blast each missile clear of the surface, where their rocket motors ignited automatically and sent the rockets on their way.

That first Polaris sub, the *George Washington*, was launched in 1959. By 1966 the U.S. had built and put to sea 41 of the huge subs, each longer than a football field. Because of their size and their armament, the subs were nicknamed "boomers."[24]

One thing made the Polaris missiles and subs possible: precision inertial navigation, provided by the MIT Instrumentation Lab. Each submarine had to have an inertial navigation system so it knew where it was. And each missile also had an inertial navigation unit to guide it from its submerged launch platform to its target. The submarine's navigation system fed data to the inertial units in the missiles. Because of the "always ready" nature of nuclear missiles in the heat of the Cold War, the sub's coordinates always needed to be up to date: at any moment the sub might be ordered to launch its missiles, which is to say, the sub might at any moment need to feed those coordinates into the missile's guidance system.

The Instrumentation Lab had a team of more than 100 working on Polaris. Their goal: the missile should have total independence from the ground once it had been launched from the sub. So the onboard inertial navigation system had to include not just the gyroscopes and accelerometers but also a basic computer capable of keeping track of and interpreting that information and doing the math necessary to guide the missile in flight.

One of MIT's principal partners was AC Spark Plug, a division of General Motors. Superficially it seems an odd pairing: the premier technology university in the country and the folks at car giant General Motors who make spark plugs. But by the 1950s AC Spark Plug had long outgrown its name. During World War II its manufacturing capacity had been tapped to help meet the war's industrial demand. In fact it had manufactured tens of millions of spark plugs for the engines in U.S. warplanes. It also made a wild array of products—459 in all—ranging from gas caps and air filters for engines to .50 caliber Browning machine guns that were mounted in warplanes to fully outfitted and sealed control panels for tanks. It made sophisticated bombsights not unlike the gun sights Draper had helped invent. And—under orders from the U.S. government—AC Spark Plug made Sperry Gyroscope autopilots for combat planes because Sperry didn't have the factory capacity to meet the demand. In the Cold War, AC Spark Plug was manufacturing the high-tech guidance and navigation systems that the Instrumentation Lab had designed for the submarines and the missiles, using what it had learned during World War II.[25]

The first underwater Polaris missile was set for launch—not just publicly but with plenty of fanfare—on July 20, 1960. The *George Washington* was cruising in the rocket launch range 30 miles off Cape Canaveral, Florida. A separate navy ship, the *Observation Island*, was on station 1.3 miles away to give reporters and other observers a good view of the launch. As the countdown reached zero, the missile burst from the Atlantic Ocean riding a column of compressed air. Military officers and engineers "gasped," then, after a slight hesitation 50 feet above the surface and, as the *New York Times* defense correspondent Hanson Baldwin put it, "still dripping water, the first stage ignited with voice of thunder, tongue of flame." That first sub-launched Polaris flew 1,150 miles (four times as far as Alan Shepard would fly in the first manned spaceflight 10 months later). The sub itself squeezed 259 people aboard (the regular crew was about 100), including representatives from the Instrumentation Lab and AC Spark Plug. That first Polaris was launched at 1:39 p.m. The *George Washington* launched a second Polaris three hours later. The *New York Times* reported in the fourth paragraph of its front-page story that navy officials described the accuracy of its guidance as "remarkable."[26]

That first B-29 bomber flight from Boston to Los Angeles, proving that inertial navigation would work, had happened just seven years earlier, using an inertial navigation unit that, without a computer, weighed 2,700 pounds and was accurate to within 10 miles out of 2,600. The guidance unit in that Polaris missile launched from a submarine cruising underneath the ocean weighed 225 pounds, including a digital computer, and it likely landed within 2 miles of its target after flying 1,100 miles, crossing that distance in 14 minutes instead of 13 hours. The weight had been cut by 90 percent and the accuracy had been doubled. Two years later the unit would weigh just 140 pounds, and the accuracy would have been improved by a factor of four.[27]

---

Polaris was pursued in the politics of the Cold War, and it was then accelerated in the shadow of the "missile gap," the launch of Sputnik, and the Soviet space successes that followed.

At the Instrumentation Lab the launch of Sputnik in October 1957 had inspired a separate but parallel project. A pair of MIT scientists decided to design a planetary probe to visit Mars, swoop past the planet, snap a single photograph, then come arcing back to Earth, where the canister with the film containing the photo would reenter the Earth's atmosphere and be recovered.

Hal Laning and Milt Trageser teamed up to figure out what a spaceship like that would require, and also what kind of math would be required to get it to Mars and back. Laning and Trageser spent months on the project, working with a small group that eventually included Richard Battin, a legendary MIT mathematician and aerospace scientist who would teach three Apollo astronauts space navigation at MIT.

They envisioned a spacecraft of 330 pounds that was almost completely autonomous from Earth support once it launched: it would have its own onboard digital computer; advanced inertial navigation; a space sextant and telescope for automatically taking star sightings and correcting the inertial system as necessary; four solar panels to provide power; and it would fly in a series of slingshot curves, using the gravity of various planetary bodies to accelerate it on its way to Mars. In part to save fuel and weight, the probe would not enter orbit around Mars after its months-long outbound journey; it was taking just one high-resolution photograph, so it would loop in close and do a U-turn around the planet, using Mars gravity to accelerate itself back toward Earth. Battin, in fact, was one of the originators of that style of interplanetary slingshot navigation that has become the hallmark of sending probes to distant parts of the solar system.

It's hard to appreciate how pioneering every element of the MIT Mars probe effort was. No one knew how to fly to Mars, let alone build the spacecraft to do it. Battin, who went on to be one of the world's leading experts on astrodynamics and space navigation, went looking for some advice on how to get to Mars. In 1958 one of the world's leading observatories, the Smithsonian Astrophysics Observatory, was just a few miles away on the campus of Harvard. "We couldn't even get cooperation from the astronomers," said Battin.

> We talked to people at the Smithsonian Astrophysics Observatory in Cambridge, and they thought we were crazy.
>
> They said, "How are you going to go to Mars? You don't even know where Mars is."
>
> And they were right. At that time with ground-based telescopes, the uncertainty as to [the location of] Mars was about 20,000 miles. You knew that Mars was somewhere in a circle of 20,000 miles diameter.
>
> They never did understand that we were not going to rely on ground-based measurements, that we were going to make those measurements on board the spacecraft.[28]

The MIT Mars study was carefully developed, underwritten by the air force months before NASA was created, when the air force hoped it would become the nation's space agency, and it resulted in a detailed report that ran four volumes. The computer—which was to be a general-purpose programmable computer, capable of handling all the tasks necessary to navigate and run the spacecraft, and doing them simultaneously—was designed with a novel feature: most of its memory would be "fixed," so it would not be possible to change its programming once the computer was built (and once the probe was launched). The programs would be woven into modules, using wires threaded through tiny magnets, because of the density this would provide and also to guarantee that the computer memory would be indestructible during the three years the mission would require. This new style of computer memory was called "core rope" memory.[29]

MIT was dramatically ahead of the curve with its ambitious late 1950s Mars probe, even if it never flew. In 1958 and 1959 the U.S. and the U.S.S.R. would launch 10 robotic spaceships headed to the Moon, all of which would fail at some point. (And Mars is more than 100 times farther away than the Moon.) The Soviets would crash a probe into the Moon in September 1959; the U.S. wouldn't get a probe to the Moon until 1964, after 11 failures (and one partial success). The U.S. was the first to visit Mars, with Mariner 4, in 1965, which took 22 close-up pictures during a fly-by. Mariner 4 radioed the photos back to Earth without having to come back itself and drop off a film canister. The Russians wouldn't make it to Mars until 1971, after nine failed missions.[30]

But given the entrepreneurial spirit of the Instrumentation Lab, and the desire to make things that were useful, perhaps the proposed Mars mission wasn't any further ahead of the curve than that B-29 flight had been. Between Polaris and the Mars project, by the time President Kennedy set the Moon as an urgent goal in May 1961, Draper's Instrumentation Lab had built up significant experience with precision inertial navigation in very demanding settings—inside rockets, in fact—and also experience with thinking through the problems, and the math, of actual interplanetary space travel.

---

On August 9, 1961, Doc Draper received a Western Union telegram, addressed to him at work at the Instrumentation Lab. "Pleased to advise that the National Aeronautics and Space Administration today announced that MIT's Instrumentation Laboratory has been selected to develop the gidance [*sic*] navigation system of the Project Apollo spacecraft. Apollo is capable of carrying three men to the Moon and back. MIT is the first member of the Apollo team to be chosen." The telegram specified that MIT would get $4 million in the first year (more than $30 million in 2019 dollars). It was signed by Senator Leverett Saltonstall of Massachusetts.

It was 76 days after President Kennedy's "go to the Moon" speech. MIT had submitted its formal proposal to run navigation and guidance for the Moon mission just five days before, on August 4. There had been no request for proposals from companies; there had been no competitive bidding.

NASA picked quickly, and picked MIT, for two reasons: no one really knew how to fly to the Moon, and the details of the guidance and navigation system would take time. (The Polaris missile had required four years from start to first underwater launch.) And although there might have been competent companies that would make a strong pitch to run the guidance system, MIT had as much experience and talent as any place in the world. As NASA officials were getting ready to pick MIT, a member of the staff of the Instrumentation Lab was aboard one of the first Polaris-missile submarines, cruising under the Arctic ice pack, monitoring the navigation system. The very week NASA chose MIT, in fact, the Instrumentation Lab had a spacecraft computer up and running, likely the only one of its kind in the world. The Mars probe group hadn't confined itself to blackboards, equations, and reports. The computer folks designed a working model of the Mars computer, and by August 1961 they had it debugged and operating, including doing advanced math while simultaneously controlling a small motor of the kind a spaceship might have.[31]

NASA's confidence in MIT was only underscored by the close personal relationships between Doc Draper and the Instrumentation Lab more broadly and the senior leadership of NASA. NASA chief Jim Webb's experience at Sperry Gyroscope during World War II had taught him to have high regard for Draper. Robert Seamans, the second-in-command at NASA and the man who did much of the day-to-day management of the agency, got his master's and SciD degrees from MIT and taught and did research there from 1940 to 1955, most of that time working for Doc Draper, including working right through World War II with him. In Seamans's autobiography he calls Draper "my mentor": "Of all the people who have had an influence on the way I've thought, apart from my family, Doc Draper is preeminent."[32]

There are a pair of overlapping stories about Draper getting the Apollo contract that have been retold so often—including by Doc Draper himself—that it's hard to pin down the precise details. But they are part of the creation myth of MIT and the Apollo computer and guidance system.

In June, just weeks after Kennedy's speech, Draper had gone to Washington with some senior staff from the Instrumentation Lab to meet with Webb, Seamans, and Webb's other senior deputy and aeronautics expert, Hugh Dryden, and talk about getting to the Moon.[33]

"Doc," asked Webb, "can you design a guidance system that will take men to the moon and back safely?"

"Yes," said Draper.

"When will it be ready?" asked Webb.

"Before you need it," said Draper.

"And how will I know that it'll work?"

"Because I'll go along and run it," said Draper.[34]

Draper didn't lack confidence, and he wasn't kidding about becoming an astronaut, or at least he wasn't kidding about the offer as a way of underscoring that confidence. In the months after MIT won the contract, Draper apparently got wind of doubts about MIT's ability to pull off the system, and so in November he wrote Bob Seamans, his former student and staff member, now the associate administrator of NASA. "I would like to formally volunteer for service as a crew member on the Apollo mission to the moon," Draper wrote. There is nothing whimsical about the letter, which runs on for almost two and a half pages, single-spaced, making the case, including his own 35 years as a pilot, "with much experimental work included," and his "unique" professional experience, more than qualifying him "as the scientific and engineering member of a space craft crew."

"I realize that my age of 60 years is a negative factor in considering my request," wrote Draper, "but General Don Flickinger tells me that this is no sure bar to my selection as a crew member." That apparently passing reference was to Air Force Brigadier General Flickinger, one of five people responsible for having just picked the Mercury 7, the nation's first astronauts, and a name well known to Seamans, Webb, and Dryden. Draper hadn't just consulted a physician about his fitness; he'd consulted *the* doctor who had spent months thinking about who was suitable to fly in space. In closing, Draper asked to be advised "what application blank I should fill out, and what other steps I should take to advance my cause."

The letter is more than just a goof for two reasons. First is the final point Draper makes: "We at the Instrumentation Laboratory are going full throttle on the Apollo guidance work, and I am sure that our endeavors will lead to success. I am also sure that if I am permitted the status of a potential crewmember all our operations will receive a real lift. If I am willing to hang my life on our equipment, the whole project will surely have the strongest possible motivation."

Draper had hit on a point that would resonate through the whole experience of Apollo: the vast army of people working on the project needed to constantly remember that they weren't just building spacecraft and launchpads and sewing spacesuits; they had the aspirations of the nation in their hands, and also the safety of some very specific, very special, very high-profile individuals. That point, which President Kennedy had touched on in his speech to Congress, calling for "every engineer, every serviceman, every technician, contractor and civil servant [to give] his personal pledge" of excellence, became a deep part of the culture of NASA and Apollo. Astronauts tirelessly visited dozens of manufacturing facilities to sustain morale and also to personalize the meticulous, demanding, time-pressed work 400,000 people were doing.

Anticipating exactly the value of that immediacy, Draper wanted to be not just the boss of the Apollo guidance system; he wanted to be the guy willing to risk his life on the work of his staff. And, as Seamans himself points out, Draper wanted to be in the thick of things. To Seamans, Draper had said, "You know, every device that I've ever developed (and there were many of them), I've always wanted to be involved in the first flight, to make sure everything worked properly."

The letter is more than a goof also because of how it landed in the executive suite at NASA. "I took it in to show Jim Webb," says Seamans, "and he got very excited. He said, 'Isn't that wonderful, one of our scientists is interested in going and being directly involved.'" Webb had a quiet, broader mission for NASA than just going to the Moon: he wanted to use Apollo to build competence and skill and interest in science and engineering across the U.S., in schools, in universities, in corporations. Here was a senior scientist and professor at MIT, a man who just 10 months earlier had been one of *Time* magazine's "men of the year," volunteering to be an astronaut.

"I think I'll take it over and show it to President Kennedy tomorrow when I'm with him," Webb told Seamans.

Dryden, the man who had run NACA, NASA's predecessor, jumped in immediately. "Wait a minute, Jim," he said. "Doc Draper is over sixty years old. I'm not sure his health would permit it. We can get in a terrible mess if we start selecting astronauts that way." Dryden and Seamans feared a nationwide astronaut free-for-all, with all kinds of high-profile people publicly volunteering, if word of Draper's lobbying got out. More than that, of course, Draper was much more important at the helm of Instrumentation Lab than he ever would have been in a spacesuit. So his letter never made it to the Oval Office. "And Doc Draper never let me forget it," said Seamans. " 'I was all ready to go to the Moon and you wouldn't let me go.' "[35]

The Instrumentation Lab's assignment was to design and engineer the equipment necessary to handle the navigation and guidance of the spaceships to the Moon. In practice, that meant the inertial navigation units; whatever telescopes, sextants, and other star-sighting equipment might be necessary; the computers to run it all; programming for those computers; and the actual navigation equations necessary to fly from Cape Canaveral to the Moon and back.

Two things were important about that structure and that list. It meant that whatever company or companies actually built the Apollo spacecraft wouldn't get to do the guidance system for that spacecraft. MIT was operating in Apollo much as it had in Polaris—and it was a small world. MIT would do the design and engineering and debugging, but in the end AC Spark Plug would build the inertial units, Sperry would build the accelerometers, Raytheon would build the computers, and a company called Kollsman Instrument would build the star-sighting instruments.[36]

The only things MIT had final responsibility for delivering were the math and the computer software—the other companies would fabricate the actual hardware—but MIT's staff had to design all the other components, including the computer itself.[37]

---

If you're flying to the Moon, accuracy matters.

If you're flying from Washington, D.C., to Boston—400 air miles—if your compass is off by 0.5°, by the time you're approaching Logan International Airport, you'll be 3.5 miles off course. When you're supposed to be over the end of the runway, you'll actually be right over MIT.

But if you're flying from Cape Kennedy to Tranquility Base, on the Moon, and your course is off by 0.5°—by 1/720th of a circle—then you end up 2,100 miles out in space. That's about the diameter of the Moon. In flying to the Moon, if you were just a hair's breadth off course, you would miss the Moon by a space as wide as the Moon itself.

At least while you were headed outbound, you'd have plenty of fuel to correct things. Coming home from the Moon is a lot less forgiving. The heat of reentry, the splashdown targeting into the ocean, and the g-forces piling up on the spaceship and the astronauts inside combine to create a very thin slice of air you need to slide your spaceship into. The command module had just 1° of latitude on reentry. Too shallow an angle, and your space capsule skips off the top of the atmosphere like a flat stone—out into space and a wide orbit around the Earth, from which there was no rescue. Too steep a cut into the atmosphere, and the speed, heat, and g-forces would combine to incinerate your space capsule. And unlike on the way out, on the way back there are no go-arounds. The command module comes in straight from the Moon to the Pacific Ocean; it doesn't go into Earth orbit on the way back. And in those last critical moments, it's flying with just small attitude-control jets to adjust its position. The last thing you do before plunging into the atmosphere is jettison that big engine and its fuel. There's no way to make much of a last-minute fix.

And if you're coming back from the Moon, you're going fast, pulled home with increasing velocity the closer you get to Earth. The accelerating power of gravity, and the way that force magnifies the closer you get to something big, is a principle you learn about in physics class, but the reality is truly astonishing. It takes about 65 hours to fly back to Earth from the Moon; after 63 hours, the command module is going 12,300 mph. Two hours later, because of Earth's pull, the command module is flying 24,600 miles per hour—7 miles a second. You're trying to hit the center of an exit ramp through the atmosphere that's 40 miles wide and if you're distracted for, oh, 7 seconds, you've swerved outside the atmospheric lane lines.[38]

That's why the navigation, guidance, and control of the Apollo spacecraft was a high-stakes game. At the speeds, distances, and tolerances of a flight to the Moon and back, there are no small errors.

Of course, it isn't as if the astronauts came around the back side of the Moon, aimed for home, fired the engine, then covered their eyes for three days and hoped for the best. As the Apollo computers were ultimately designed, they operated all the time. (There had been some very early thought, especially on the part of the astronauts, that the flight computer would be off most of the time and they'd turn it on only when they needed it.[39]) The computers also kept track of where the spaceship was all the time. That's the beauty, and also the method, of inertial navigation. You know where you are because you literally always know where you are. The computer and Mission Control were keeping track of the Apollo spacecraft every minute, and there were many opportunities to make sure you weren't 0.5° off course—at the beginning or at the end.

But that all depended on the computer's ability and reliability and on the accuracy and reliability of the information the computer was gathering as everyone rocketed along through space. And it depended on the care and thoroughness of the programs the computer was running. You didn't need a computer doing pretty good navigation, any more than you needed a spacesuit that was mostly airtight.

Dave Scott was the command module pilot on Apollo 9, a 10-day test flight in Earth orbit, and then the commander of Apollo 15, which lasted 12 days, including three days living, working, and driving around on the Moon. Scott had a doctorate in engineering from MIT, and before becoming an astronaut had taken the legendary astronautics course taught by Dick Battin, who worked on the Mars probe navigation.

"When you come back from the Moon," explained Scott,

> you really have to hit the corridor. If you have a basketball and a baseball [that are] 14 feet apart, where the baseball represents the Moon and the basketball represents the Earth, and you take a piece of paper edgeways—the thinness of the piece of paper would be the corridor that you would have to hit when you come back.
>
> And that's only position. You have to hit with the proper velocity too.

> You have to have a good computer, and when you're approaching the re-entry corridor, you are thinking about that, because you only have one chance.
>
> I remember during the re-entry saying, "Well, here goes."

At key moments, the onboard computer would ask the astronauts to confirm it was the moment for a significant maneuver. "I pushed the PROCEED button [on the computer]," said Scott, "and it was perfect all the way down."[40]

The onboard computers were, in fact, tackling the most elaborate real-life physics problem ever devised. As Armstrong and Aldrin were flying the lunar module *Eagle* down from orbit to the Moon's surface, looking for a place to land, worrying about computer alarms, and watching their fuel dwindle, the computer was not only tracking the LM's position and handling the navigation and thrust of the engine up until Armstrong took control; it was doing all those guidance calculations while also taking account of the fact that the lunar module's mass was changing constantly and dramatically. During the final descent to landing, the lunar module burned off 17,400 pounds of fuel—9 tons of fuel gone in 12 minutes. *Eagle* weighed only 9,500 pounds without fuel. The math was hard, and the mission was to get the hard math into the computer.

The problem was that the world of computers in the late 1950s and early 1960s was anything but reliable. Just the opposite: Computers required whole rooms of space; they required staff people just to operate them and nurse their idiosyncrasies; they required enormous quantities of electricity; and they didn't operate in anything like real time.

It was, said Eldon Hall, "an age when computers were characterized by their inability to operate for more than a few hours without a failure." Hall was a senior engineer at the Instrumentation Lab who was in charge of the computer for Polaris and moved over to be in charge of hardware development for the Apollo computer.[41]

The fussiness, the unreliability seemed baked into the design, components, and operation of computers—almost into the very personality of computers—in those early days, even in the case of computers in critical and high-profile functions.

In 1961, as NASA was ramping up the Mercury and Gemini flight schedules, it was figuring out how to use computers on the ground to run and monitor missions and to run Mission Control. IBM was the only commercial supplier that could provide real-time computing—an astonishing situation that didn't last more than a few more years. And even those IBM machines were only so good. The early Mission Control setup included two full-scale IBM mainframes running in parallel, a primary and a backup. A third identical computer, with identical programming, was installed at NASA's Bermuda tracking station, run offline as a second backup, in case of "a double mainframe failure."

After Alan Shepard's first flight, the next most dramatic event in the U.S. space program was John Glenn's orbital flight, which didn't come until February 20, 1962, almost a year after the Russians had orbited Gagarin. Glenn orbited the Earth three times on that Tuesday, between about 9:45 a.m. and 2:45 p.m. While he was in orbit, the primary computer in Mission Control failed for 3 minutes—just 3 minutes out of 4 hours and 55 minutes. But the IBM and NASA staff couldn't keep the most important computer in the country at that moment running continuously for five straight hours.

When NASA designed the Real-Time Computer Complex for Mission Control in Houston, it specified that the machines needed to operate for 336 hours (14 days) with "up-time" reliability of 99.95 percent, which would have allowed for 10 minutes of downtime in two weeks. In its bid to provide the computers—which was ultimately successful—IBM said it could guarantee only 97.12 percent reliability, meaning nearly 10 hours of downtime during a 14-day spaceflight. (That level of "up-time" wasn't even as good as what had been provided during Glenn's flight.)[42]

Reliability wasn't just a routine concern critical to safety and to being able to finish missions. It was a problem magnified by operating in space with no possibility of support, in zero gravity, and magnified by the sheer complexity of the spacecraft, the number of parts involved that could fail. Reliability was so central, and so challenging, that for nearly three years NASA pursued what seems in retrospect an odd and unwieldy strategy. NASA wanted much of the equipment and electronics on the spacecraft to be repairable by the astronauts while in flight—computers included. Equipment was to be designed and installed with reparability in mind. Tools and spare parts would be stowed alongside oxygen tanks and urine collection bags. One supplier specified that astronauts be equipped with a soldering iron, although how exactly zero-gravity soldering would work, with molten balls of hot metal floating inside the spaceship, wasn't clear.[43]

It was a singular instance of persistent naïveté about spaceflight on the part of NASA. Eventually, for instance, weight reduction efforts for the lunar module were trying to cut 2,500 to 3,000 pounds off the 30,000-pound craft. Every change that could shave 0.1 pound off the lunar module was considered, and a cost threshold was established: Grumman was willing to spend up to $10,000 to cut 1 pound of weight. It's hard to imagine spare parts and tools ever surviving such a scrub.

Thomas J. Kelly, the chief designer and chief engineer for the LM at Grumman, argued forcefully to NASA that in-flight repair made the reliability challenge worse, not better. Making all critical systems accessible to repair, making the mechanical and electronic connections easy to unhook in flight, installing a whole set of fresh sensors and electronics to detect exactly what had failed—all that made any particular component more likely to fail, and also created more possibilities for failure, some of them simply failure warning systems. A carefully built spacecraft, with connections secure and wiring and controls sealed against the hazards of humidity and zero gravity, was, Kelly argued, a much better path to reliability.

The astronauts—almost overwhelmed with what they had to learn as it was—took a dim view of IFM, in-flight maintenance. The topic came up during one of Glenn and Shepard's early visits to the Instrumentation Lab to discuss the computer design. Glenn didn't end up flying any Apollo missions, but Shepard commanded Apollo 14. At MIT that day there was discussion of how to diagnose computer failures and what swapping out computer modules would be like. "Yeah," said Shepard, "and we should all train to be brain surgeons so we can operate on each other."[44]

NASA officially ended the push for zero-gravity repair in 1964. There was a brief effort at MIT to simply include a second identical computer on the command module to provide backup, but that idea didn't survive weight concerns.[45]

In the end, the reliability had to be baked into the Apollo computers—into both the hardware and the software—in exactly the way that the unreliablity seemed wired into those early room-sized machines. That effort would change not just the space program but the perception of computers themselves.

---

The Apollo project was too big for the existing Instrumentation Lab, whose headquarters was in the former Whittemore shoe polish factory. So Apollo got its own building, an old three-story underwear warehouse in Cambridge that was converted to offices. The building sat right on the Charles River and was not air-conditioned, except for the computer rooms.[46]

Much of the leadership and staff of the Polaris missile guidance effort moved quickly to Apollo. Milton Trageser, who worked on Polaris and helped run the Mars probe, became director of MIT's Apollo effort. Ralph Ragan, who had headed Polaris, became operations director of Apollo. David Hoag had been technical director of Polaris and was made technical director of Apollo. John Miller, who had worked on inertial navigation on Polaris, was made head of the inertial navigation hardware for Apollo. Dick Battin was put in charge of the guidance development and software. Eldon Hall, who had been in charge of the Polaris computer hardware, was put in charge of the Apollo computer hardware. In all, 100 people moved from Polaris to Apollo.

The Instrumentation Lab wasn't starting with a room-size computer and trying to miniaturize it while making it better. They were starting with something the size of the Polaris and Mars probe computers and trying to create a computer with more capacity, more reliability, more speed.

That NASA picked MIT so quickly initially posed an interesting challenge. It took four months, until November, for NASA to pick the company to design and build the command module. NASA didn't decide how to fly to the Moon—what kind of spaceship would leave Earth and land on the Moon—until the following summer, 11 months later. And NASA didn't pick Grumman to build the lunar module until November 1962.

Not long after North American Aviation was chosen to build the command module, someone from North American called MIT. "They get on the phone," said Dick Battin, "and they say, 'We understand there's going to be a computer in the command module. How big is it?' "

Battin smiled. "We had no idea how big it was going to be." At that point the computer "was just a bunch of equipment on a rack. So we asked around, What do you think? What should we tell them?

"And I said, oh well, maybe a cubic foot. Let's say a cubic foot." In telling the story, Battin smiled and shrugged. "We were just guessing."[47]

As it happened, North American held Battin and the Instrumentation Lab to that wild early guess, and the Instrumentation Lab made good on it: the volume of the AGCs that flew to the Moon were each 1.04 cubic feet.

Some of the early going went just like that: a certain informality, a certain improvisation that was the hallmark of Draper's Instrumentation Lab. Draper wasn't big on bureaucracy, on carefully thought-out organizational charts and strict reporting lines.

Apollo was bigger than anything the Instrumentation Lab had ever tackled; before long, Apollo would be bigger than the rest of the Instrumentation Lab. But it was also comfortable territory for the people who had gathered around Doc Draper. They were the first people to tackle a set of problems—airplane instruments that were reliable, gun sights that aimed successfully, inertial navigation that could guide a submarine underneath the polar ice cap—and they were used to digging in, figuring out the science, then turning the science into real-world tools.

Apollo was different not just in order of magnitude or profile, though; it was different in the Instrumentation Lab's connection to the rest of Apollo: the computer's successful navigation depended on knowing things like the precise weight and center of mass of the command module and the lunar module, on what would be connected to the computer and relying on it. The successful flights depended on the Instrumentation Lab's ability to supply its equipment and software early enough that Grumman and North American and NASA could test it, could make sure it worked flawlessly with everything else, and could use it in tireless simulations and trainings for the astronauts in Florida and the flight controllers in Houston.

Battin's shrugging guess at the size of the Apollo Guidance Computer—"Let's say a cubic foot"—really was a guess. At that point the working prototype filled four refrigerator-size racks. And when MIT once or twice gingerly inquired about making the flight AGC bigger—for instance, by adding a whole second computer to the command module as a backup—North American said absolutely not. Likewise, during the most serious days of the weight problems with the lunar module, when Grumman was willing to pay $10,000 a pound to lose weight, lunar module designers pressed MIT to make the computer smaller, and MIT resisted.[48]

The conventional wisdom today is that Apollo was possible because it didn't require the kind of dramatic breakthroughs of fundamental science, on a deadline, which the atomic bomb required during the Manhattan Project. But Apollo required literally thousands of "breakthroughs" that pushed the limits of science, engineering, manufacturing, reliability—in the engines, in the spacesuits, in the math of navigation, and in the computers. (Just one example: both the big Saturn V F-1 engine and the relatively tiny lunar module ascent engine—one producing 1.5 million pounds of thrust, the other producing 3,500 pounds of thrust—developed instability problems during tests that, in each case, took two years to figure out.) No one had ever needed a heat shield that could survive 5,000°F before. No one had ever designed an electric car to drive on the surface of the Moon before.

Some work had been done on the kind of navigation Apollo would need. Some reliable inertial navigation technology existed, for use on Earth. Some people were starting to use computers. But for Apollo in 1961, there wasn't anything that could simply be purchased. MIT had to design, and then build, prototypes of the computers, and then the finished models. MIT had to write operating systems and then the programs for navigation and operating the spaceship.

"If we knew then what we learned later," said Eldon Hall, "or if we had had a complete set of specifications [of what the computer would have to do], we would probably have concluded that there was no solution with the technology in the early 1960s."[49]

In fact the spaceships were undesigned, the mission's actual course to the Moon was undecided, and so the IL simply got started on the basics. The job the computer had to do grew in sophistication as the Moon flight itself became clearer, and MIT's confidence grew alongside those demands.

At the time the Instrumentation Lab started working on Apollo, a computer like the AGC would have used transistors (themselves only in commercial production for 10 years), which were relatively compact, low power, and cheap.[50] But the Apollo flight computer was something new: it needed a level of compactness and power-stinginess that even missiles didn't require, and as much performance as could be squeezed into one cubic foot. It also needed to operate for days and days without a hiccup, whereas a Polaris missile needed a computer that could run flawlessly for only 20 minutes. As Apollo was getting started, the "mean time between failures" of aviation guidance systems was 15 hours; meaning that they could be expected to operate for 15 hours without needing repair. NASA wanted MIT to multiply that performance by 10, and then by 10 again, to create systems that could run faultlessly for at least 1,500 hours.[51]

More than a year into the computer design effort, it became clear to Hall that the transistor wouldn't do; the only solution to the conflicting demands of space, weight, power, and performance was the new digital tool, the integrated circuit. Transistors were individual components, reliable and inexpensive. Each integrated circuit could be designed and fabricated to include many transistors in a small space, along with other electronics components, saving weight, complexity, and power. The integrated circuit was the dawn of the computer chip. But integrated circuits were new and their quality couldn't be depended upon.

It was so early in the life of integrated circuit technology that the first samples MIT bought cost $1,000 apiece ($8,000 each in 2018 dollars). It was so early, in fact, that in order to understand the manufacturing, value, and reliability of integrated circuits, Hall visited Texas Instruments to meet with Jack Kilby, who just months earlier had invented the integrated circuit (which would win him a Nobel Prize in physics in 2000). Hall also went to Fairchild Semiconductor and talked to Robert Noyce, credited with co-inventing the integrated circuit independently of Kilby. (Noyce would leave Fairchild before the first Moon landing to cofound Intel.)

"Imagine going to your program manager and telling him you had to buy 4,000 of these"—at $1,000 each—"to build a prototype computer," said Hall. But two things happened. The price started to come down, in part because MIT started buying integrated circuits for NASA.

In 1962 MIT paid $100 per microchip. By 1963, when Hall ordered a single lot of 3,000 chips from Fairchild, the price was $15 a chip.[52]

And Hall turned out to be a persuasive salesman at a key moment. In November 1962 he made a presentation to the chief of the Apollo spacecraft office, Charles Frick, making the case that a computer with integrated circuits would be lighter, smaller, less complicated, less expensive to make, and ultimately cheaper. The margins weren't small: 40 percent reduction in weight and space, while making the computer 2.5 times faster. But the chip would also require redesigning the prototype MIT had been working on. And there was risk: the reliability of integrated circuits was untested, and as of that November there was only one company making the chip Hall and his group thought could do the job—Fairchild.

Doc Draper, himself quite the salesman, was at the presentation. "That was the best sales pitch I've ever heard," he told Hall.[53]

It was a good pitch, but, said Hall, "the choice was far from obvious." Between the existing transistor technology and integrated circuits "there were considerable and sometimes heated debates over the advantages and disadvantages."

NASA sent the Instrumentation Lab a letter three weeks later that was just two sentences long, telling them to go ahead with the new computer design.[54]

Looking back, Hall was wired into the technologies that were coming, in part because of Polaris, in part because of his own curiosity and diligence. He saw the possibilities, and the leap, that microchips could provide. It's also possible that a computer relying only on transistors never would have been powerful enough or robust enough, and that MIT would have realized that, just not as quickly as Hall did. Although Hall predicted weight and space savings of 40 to 50 percent, the computer didn't get smaller—the Instrumentation Lab simply used that space to increase the memory available to its own programmers. And even that turned out not to be enough, precipitating a crisis that almost kept Apollo on Earth.

The real challenge was reliability, and reliability was related to volume: how many chips a company made. The more microchips coming off the line, the higher the quality was likely to be, as companies got better making something new and intricate.

Hall and the Instrumentation Lab were buying thousands and thousands of chips. Through the end of 1963 NASA purchased 60 percent of all integrated circuits made in the U.S. Almost all the rest went to the air force for the guidance system of its Minuteman missile.

It's worth pausing to appreciate how absolutely astonishing that is. Two government customers, working on well-funded cutting-edge programs, were single-handedly creating the market for a new technology.

The key technology that went on to transform nearly every aspect of human life on Earth was not, in any way, an immediate hit. Integrated circuits were expensive compared to the alternative, of uncertain quality, and they simply weren't necessary for most of what people used electronics for. In 1962, as Hall was getting permission to use integrated circuits in spacecraft headed for the Moon, the transistor radio was only seven years old, and the average price was still $29 ($240 in 2018 dollars), making it a novelty or an indulgence.[55]

"Most histories imply that the electronics industry enthusiastically welcomed transistor and integrated-circuit developments," Hall said. "However, the reverse is more historically correct. It took government-sponsored programs like Polaris and Apollo to provide the semiconductor industry with support and motivation."

IBM, battling competitors on all sides to build and hang on to the business computing market, nonetheless didn't have a commercial computer that used integrated circuits until 1970.[56]

In the middle of the race to the Moon, MIT did a study of the microchips it was buying to build the Apollo computer. The paper manages to be blunt while still being couched in the language of academia. The Instrumentation Lab had found that the standard reliability of electronic components for the aviation industry was somewhere between one failure in 1,000 hours and one failure in 10,000 hours. For anyone who owns a dishwasher, that rate seems pretty good. For spaceflight computers that were going to have 5,000 computer chips, it's not nearly good enough. "Today, with the advent of space technology," the Instrumentation Lab engineers wrote, "increased reliability has become necessary and a new term, 'high reliability,' has evolved." Which is to say, MIT was setting a new bar.

The Instrumentation Lab wanted parts with a reliability of one failure in 100,000 operating hours. It had purchased 400,000 chips by the time its own study was done in 1967, and it had accumulated a stunning 330 million hours running those chips, equivalent to running 10,000 of them continuously for four years, watching whether they failed. Some prototype computers were in fact run continuously for two years to see how the integrated circuits MIT had purchased performed.

Two things were at work here: MIT needed to prove the dependability of a technology that was brand new, that was critical to the safety and success of flying to the Moon, and that had a shaky reputation. And MIT was not interested in being the cause of a problem. It wasn't going to buy chips that might fail down the road, build them into its Apollo computer, and then cause a disaster in 1968 or 1969.

The obsessive care was justified. In testing, MIT often found lots of integrated circuits that didn't even meet the lesser standard it was unsatisfied with: chips failing at 20 or 30 or 40 times that ordinary quality. That, mind you, was the Instrumentation Lab's testing of the chips as they came in the door—"acceptance testing" before the real testing began.

The pretest testing was designed to weed out bad chips and to send a message. Chips were centrifuged, x-rayed, vibrated, baked in an oven. These chips were going to the Moon, and it wasn't going to be a gentle ride. They were also tested for leaks. The chips were weighed as they came in, then submerged in a Freon solution. If a chip's weight rose by 0.0005 grams—1/2000th of a gram—that meant the chip wasn't properly sealed, Freon had leaked in, and MIT rejected the whole order as not meeting the Apollo quality standard.[57]

It was all pretty effective. Among the chips that had passed inspection, MIT was able to report on 312 million hours of chips running, installed in prototype computers. That's 10,000 integrated circuits—about two computers' worth, one for the command module, one for the lunar module—running for three and a half years. Total failures: one.

The amazing thing is how vividly right Hall and MIT were about the importance of manufacturing experience and steadiness. There are graphs in the MIT study that show exactly when chip quality plummeted: MIT didn't buy any chips from June to October 1964, and when the manufacturing line started up again, there were lots with 40 and 80 times the ordinary failure rate, before the line settled in and the quality settled down; in June a line moved to a new factory location, and when chip production started at the new place, the same phenomenon occurred, and it took weeks and several shipments for the quality to come back up.[58]

Although MIT would ultimately buy 1 million integrated circuit chips for Apollo, an extraordinary volume in the 1960s, that was still a tiny amount. MIT would build only about 70 computers altogether, and it would build only 20 that would fly with astronauts in space. The million chips were enough for roughly 200 computers. That's why the Instrumentation Lab was simultaneously insistent on a whole new standard of quality and also worried about who would supply the chips.[59]

Hall and the Instrumentation Lab picked Fairchild. That way one company got all the business. They picked a single chip from Fairchild, for simplicity of design and manufacturing in the Apollo computer, and again to give Fairchild maximum output of a single product. They specified that the same chip had to be used in the ground support and testing equipment at MIT and NASA facilities. (Even so, Fairchild actually dropped the particular chip MIT was using, the Micrologic, right in the middle of the race to the Moon because it was becoming outdated. But it was picked up and sustained by Philco Corporation until the Apollo computers were finished.)[60]

Apollo had an indelible effect on semiconductor manufacturing. To meet Apollo's standards, Fairchild had to set up separate manufacturing lines just for MIT. "Asking girls on a standard line to build to Apollo standard is like asking a guy to study in a room where three other guys are having a bull session," Gordon Russell, national sales manager for Fairchild, said in August 1969, after the first Moon landing. "The girls must have a separate facility where they can concentrate and maintain their high standards." Fairchild was very clear about its mission. "The prime consideration of the whole program was reliability," Russell said. "Apollo really taught us a lot about reliability. . . . This thing had to work."[61]

---

Getting a computer designed and built was really just half the challenge—or perhaps a third of the challenge. There were two other big questions: How much would the computer do? And how much would the people flying in the spacecraft with the computer do?

Neither question had a simple answer. In the first few years, the working rule was that the computer would provide navigation and guidance, but there would be a separate, parallel system of electronics and controls that operated the two Apollo spacecraft—the small reaction-control jets, the radar, the big engines.

That's not particularly odd: modern cars have navigation systems built into the dashboard, systems that tell you where you are and will plot a course to where you want to go. But at least for the moment those systems aren't connected to the engine, steering, and brakes of the car. They provide guidance, but they don't provide control; they show you how to drive from Dallas to Houston, but they can't actually drive for you. Apollo was originally designed on the same principle.

And at the beginning the astronauts wanted a maximum role in flying the spaceships, using those controls. The computer was to play an advisory role.[62]

There was a third principle in play in designing how the Apollo guidance computer worked: the Instrumentation Lab believed its guidance systems should be completely independent. That was the Doc Draper way: he proved the power and value of inertial navigation in 1953 by creating a system for the B-29 that was independent of any information from outside the unit itself. The prototype Mars probe was designed to gather navigation data and use math and its onboard computer to guide itself to Mars better than Harvard-based astronomers could have. The same with the Polaris submarine and the Polaris missile. So one of the core ideas of the Instrumentation Lab when it came to Apollo was that the command module and the lunar module should be able to guide themselves to the Moon and back without any help from the ground, if that became necessary. The ships should have all the instruments necessary to take star sightings, to check and recalibrate the inertial navigation units with precise data, and to calculate how to get where they were going with the accuracy necessary to take care of the mission and the astronauts—the course, the angles, the timing and duration of rocket burns.[63]

Within a few years all three of these core ideas were transformed, one completely upended, the other two compromised in ways that would prove invaluable.

First, NASA gave up on the idea of a computer that provided only guidance but not control. It turned out that a whole separate system of electronics and controls to fly the command module and the lunar module was clumsy, duplicative, and hard to execute. The Apollo computer was stitched right into the control of both spacecraft. Part of the programming for each, in fact, was an autopilot program: the Apollo computer was, in theory, capable of flying the entire mission, right to the gritty dust on the Moon, all by itself. The Apollo guidance computer became an Apollo guidance and *control* computer. That meant the computer had more work to do at every moment, and it ended up with an astonishing 200 inputs and outputs—200 connections to other systems on the spacecraft, a whole network of information it needed to absorb every second, coming in from sensors, radar, gyroscopes, and also a whole network of instructions going out, to antennas, gimbals, thrusters and engines, and Mission Control. Giving the computer the ability to actually control the flight of the spaceships made the work of MIT's programmers more demanding, but it greatly simplified the spacecraft themselves, and also the work of the astronauts. They only had to learn to operate the ship from the computer, not from a second system of controls as well.

Once the onboard computer was given control responsibilities, it became a different machine. It couldn't be kept in some kind of "sleep" mode, to be woken when the astronauts wanted a little guidance (which was unlikely to have happened anyway). The Apollo computer became an always-on sentinel. Its role was so indispensable as time went on that for some at the Instrumentation Lab and NASA the Apollo computer became known as "the fourth crew member."[64]

But even as its capabilities expanded, the Apollo computer's autonomy in some measure was diminished. One of the quiet triumphs of the Apollo era for NASA was the construction of a worldwide space tracking network with astonishing capacity and resilience; it had 14 tracking stations on land, two satellites in geosynchronous orbit, four tracking ships at sea during missions, and, during reentry, eight planes in the air. Around the world there were 30-foot antennas, 85-foot antennas, and a pair of 230-foot antennas that proved crucial. The network, which cost an astonishing $1 billion to build and operate, 5 percent of the spending on Apollo, maintained continuous real-time voice and data communication with the spacecraft, except when they were behind the Moon. It could track both spacecraft, including the lunar module from orbit down to the Moon's surface. On Earth the network of tracking stations was connected by the world's first dedicated high-speed data network; NASA had 2 million miles of land-based and undersea cable linking Houston to the tracking stations around the world. The system cost $70 million to $80 million a year to run during the Moon landings, and had a dedicated worldwide staff of 2,300 people.[65]

The network's navigational fixes were so accurate, and could be calculated so quickly, that at most points the best navigation data for the astronauts didn't come from their own ships and onboard inertial navigation instruments but from the ground. In lunar orbit, 240,000 miles from Earth, the NASA tracking network could pinpoint the command module and the lunar module to within 30 feet (0.006 miles) and could clock their speed to within 1 mph. An intergalactic radar gun.[66]

That kind of accuracy eliminated the need for Apollo's onboard computers and equipment to be the primary source of navigation data and calculations, which was deflating for MIT. But when the Instrumentation Lab's computer programs were wildly too large for the space available for them—creating the possibility that MIT would delay the Moon landing—it turned out to be a blessing. Some of the first programs to be pruned were those dealing with precise, real-time onboard navigation.

The tug-of-war between crew control and automation was part of the culture of the NASA astronaut corps from the Mercury 7 on—the challenge of hiring high-performance military pilots, often combat, carrier, and test pilots, and putting them in spaceships that, by their very nature, needed a very different kind of piloting.

The computer itself ended up with a fairly slick and sophisticated interface, the DSKY, among the first examples of what we would come to call a "user interface," a way for people to communicate with a computer. It was designed completely from scratch, with a particular group of users in mind, and a particular kind of work in mind. The Apollo computer was for flying to the Moon, and so the DSKY too was for flying to the Moon. So while the DSKY had a keyboard, for instance, the keyboard had no letters, only numbers. It was eight inches square and seven inches deep. The keyboard, along the bottom, had early versions of function keys that would become familiar to everyone 10 or 20 years later: ENTR, RSET, CLR, PRO (proceed), and KEY REL (keyboard release). The keyboard also had two wholly original keys labeled VERB and NOUN.

Above the keyboard were two kinds of displays. On the left a set of 10 lights, just like on a jet control panel; these were status and warning lights: UPLINK ACTY, STBY, and OPR ERR (operator error). On the right, five lines of digital displays that could display numbers in the classic calculator format, as segments of the number 8.

The first two lines of number displays were dedicated; the first, PROG, told the astronauts what program was running. The second provided space to display VERB and NOUN. Below that were three open lines that could be used to display all kinds of data, although with some interesting quirks. Each line had space for five numbers and a plus or minus sign. But there were no decimal points and no indicators of what was being displayed. Just numbers. In that sense, the displays were akin to what you got from a slide rule. The astronaut—the user—had to know what kind of data he asked for, what kind of data he was looking at, what the units were, how many digits were being provided, and where the decimal point went.[67]

Sometimes the top line was velocity. Sometimes the top line was hours. Sometimes the bottom line was time, in hundredths of seconds. Sometimes the bottom line was altitude from the surface of the Moon. The astronauts entered data or program instructions using numbers on the keyboard.

The VERB and NOUN keys and the VERB and NOUN number displays were the key to understanding the computer and to using it. This was the syntax—the operating technique—of the Apollo computer. VERB and NOUN were how the astronaut told the computer what to do.

When the astronauts wanted the computer to do something, they punched the VERB key, then the two-digit code for the function they wanted. Then they punched the NOUN key, and the two-digit code for what they wanted the computer to do that function to. VERB 16 NOUN 36 ENTER meant "Display" the "ground-elapsed time"—the length of the mission since launch. The time would appear as lines of numbers.

If the astronauts wanted the computer to execute a program, they used VERB 37 and then the number of the program to be executed; 64, for instance, was the guidance program to take the lunar module from orbit on the first phase of its flight down to the lunar surface.

For a skilled and practiced user, the computer had tremendous flexibility and power. Some VERB-NOUN combinations just displayed information. Some set the spacecraft to doing a whole series of maneuvers in space, controlled in the background by the computer itself.

And while the computer was running those programs, it used the number lines to display data back to the astronauts, and sometimes to request information from them or to request permission to proceed to the next step—hence the PRO key.

When the computer wanted attention from the astronauts, it flashed the indicator lights or the register displays or both, depending on what kind of information it needed. And the computer also maintained a whole set of functions in the background—monitoring position, for instance, and sending and receiving data to the ground.

Ramon Alonso, an Instrumentation Lab engineer who worked on the Mars probe computer, had come up with the VERB and NOUN idea as a way of giving the computer instructions and the actual terms to put on the keyboard buttons. The idea came to him, he said, simply from realizing that what the astronauts would want the computer to do matched basic sentence structure: "display velocity," "fire engine."[68]

MIT's Apollo computer didn't make beautiful displays on a screen—that technology and processing power didn't exist yet—but it was more sophisticated in its routine operations than the laptops we use every day 50 years later. It knew how to do things on its own, and it was connected to the equipment to do those things—to rocket engines and radar antennas and gyroscopes. The computer knew how to ask for information, wait for it, and then use that information to handle sophisticated tasks like navigating from orbit to the surface of the Moon. And of course, the Apollo computer could function in what we think of as the more traditional user-machine mode, accepting requests and providing real-time data.

It's worth underscoring that the Instrumentation Lab created all this from scratch, well before the folks in Silicon Valley came up with the mouse and the graphical user interface. The VERB-NOUN combo was a simple, clear, easy-to-understand way of running the computer. You thought, "I want the computer to do this," and then you used your list of two-digit codes to tell it to do exactly that. Putting aside the simplicity of the display, it was a much more intuitive and easy-to-use system than the DOS command line, with the C> prompt, that Microsoft would first offer computer users 12 years after Apollo 11.[69]

Dave Scott was one of the main points of contact between the astronauts and MIT, and he developed great affection for the Apollo computer, and also the ability to add his own programs to it, which he did as commander of Apollo 15. "How do you take a pilot, put him in a spaceship, and have him talk to a computer? That's not easy, in real time," Scott said. "[The onboard computer] was, with its computational ability, a joy to operate. It was just a tremendous machine. It was so simple and straightforward that even pilots could learn how to use it."[70]

Using the Apollo computer did involve a lot of button pushing, something like 13,000 keystrokes for a weeklong Apollo mission, but the astronauts who really absorbed the computer's personality and logic could, as NASA's Jack Garman said, learn it "like playing a piano—you don't have to see your fingers to know where they are."[71]

Designing and building the Apollo computer grew into a large undertaking, especially considering that in all fewer than 100 computers, and fewer than 150 DSKYs, were built, and just 20 computers flew with astronauts in space. At the Instrumentation Lab, the number of people working full time on the hardware peaked in 1965 at 600. Raytheon, which had assembled the much simpler computer for Polaris, expanded its Waltham operation from 800 to 2,000 employees to tackle the Apollo guidance computer, with much of the work, especially the actual "weaving" of software, done by women, and done by hand, one wire at a time.

But as challenging and cutting-edge as the design and demands on the hardware for the Apollo computer were, it was the software that would create the real drama. By the mid-1960s NASA had begun to fear that Apollo might miss Kennedy's "before this decade is out" deadline to land on the Moon because the brilliant staff at Draper's Instrumentation Lab weren't going to be able to get the software finished in time.

# 5

# The Man Who Saved Apollo

You sit at the very center of the success or failure of this extremely important program. You're behind. Get it through your head: You are fucking this thing up.

> **Bill Tindall**
> *the NASA official put in charge of Apollo software, to a group of MIT engineers*[1]

In the dark on Sunday morning, July 22, 1962, NASA launched the first-ever U.S. interplanetary space probe. It was Mariner 1, headed for Venus, Earth's neighbor in the solar system, next closest to the Sun.

Mariner 1 was launched atop a 103-foot-tall Atlas-Agena rocket at 5:21 a.m. EDT, and for 3 minutes and 32 seconds it rose perfectly, accelerating to the edge of space, nearly 100 miles up.

But at 3 minutes and 32 seconds into flight, Mariner 1 started to veer in odd, unplanned maneuvers, first aiming northwest, then pointing nose down. Mariner 1 was off course, out of control, and headed for the shipping lanes of the North Atlantic. At 4 minutes and 50 seconds into flight, a range safety officer at Cape Canaveral flipped two switches, and explosives in the Atlas blew the rocket apart in a spectacular cascade of fireworks visible back in Florida, to prevent it from hitting people or land. Just six seconds later, the first stage would have dropped away from the spacecraft, and it wouldn't have been possible to destroy it. The Mariner 1 probe itself was blown free of the debris, and its radio transponder continued to ping flight control for another 67 seconds, until it fell into the Atlantic Ocean.[2]

The disappointment—just in 1962, NASA had launched two probes to the Moon and this one to Venus, and all three had failed—was softened by the fact that a second, identical Mariner spacecraft, and an identical Atlas-Agena rocket, were already in hangers at the Cape, ready to be prepared for launch. Mariner 2 was launched successfully a month later and reached Venus on December 14, 1962, where it discovered that the temperature was 797°F and that the planet rotated in the opposite direction of Earth and Mars, so the Sun on Venus rises in the west.[3]

It was possible to launch Mariner 1's twin just 36 days after the crash because it took scientists at NASA's Jet Propulsion Laboratory only five days to figure out what had gone wrong. In handwritten computer coding instructions, in dozens and dozens of lines of guidance equations, a single letter had been written incorrectly, probably forgetfully.

In a critical spot the equations contained an "R" symbol (for "radius"). The "R" was supposed to have a bar over it, indicating a "smoothing" function; it told the guidance computer to average the data it was receiving and to ignore what was likely to be spurious data. But as written and then coded onto punch cards and into the guidance computer, the "R" didn't have a bar over it. The "R-bar" became simply "R."

As it happened, on launch, Mariner 1 briefly lost guidance-lock with the ground, which was not uncommon. It was supposed to follow its course until guidance-lock was re-achieved, unless it received instructions from the ground computer. But without the R-bar, the ground computer got confused about Mariner 1's performance, thought it was off course, and started sending signals to the rocket to "correct" its course, instructions that weren't necessary, that weren't correct; "phantom erratic behavior" became "actual erratic behavior," as one analyst wrote. In the minute or so that controllers waited, the rocket and the guidance computer on the ground were never able to get themselves sorted out because the "averaging" function that would have kept the rocket on course wasn't actually programmed into the computer. And so the range safety officer did his job.[4]

A single handwritten line, the length of a hyphen, doomed the most elaborate spaceship the U.S. had until then designed, and its launch rocket, at a cost of $18.5 million ($154 million in 2018 dollars). Or rather, the absence of that bar. In the popular press, for simplicity, the missing bar became a hyphen. The *New York Times* front-page headline was "For Want of a Hyphen Venus Rocket Is Lost." The *Los Angeles Times* headline: "'Hyphen' Blows Up Rocket." The science fiction writer Arthur C. Clarke, in his 1968 book *The Promise of Space*, called it "the most expensive hyphen in history."[5]

For NASA's computer programmers, it was a lesson in care, caution, and testing that ended up steeped into their bones. It was, in fact, an arresting vulnerability of the new Space Age. A single missing bolt in a B-29 Superfortress wasn't going to bring down the plane, but a single inattentive moment in computer programming—of the sort anyone can imagine having—could have a cascade of consequences.

Dick Battin, for whom the software programmers on Apollo worked, knew the story of Mariner 1 well. He became friends with George Mueller, who was NASA's associate administrator for manned spaceflight from 1963 to 1969. Just before that Mueller had been an executive at Space Technology Laboratories, which had responsibility for writing the guidance equations for Mariner 1, including the equation with the missing bar.

Mueller, says Battin, took an interest in MIT's performance at writing code. "He was particularly concerned because he had a flight software problem that destroyed . . . the Mariner mission. Somebody had left a hyphen out of the code, and the missing hyphen caused the Mariner to be aborted and the system lost. And this made George Mueller particularly interested in flight software.

"In fact, he had a hyphen framed, hanging behind his desk, to remind people that it just takes one little thing like that to ruin a program and to abort a mission."[6]

---

By the spring of 1966, President Kennedy's challenge to NASA to take America to the Moon, to do it before the Russians, to do it before the 1960s were over, had given the American space program a sense of focus, urgency, and purpose. Not to mention success.

The one-man Mercury missions were history, six men launched into space and returned home safely, four after orbiting the Earth. In 1965

U.S. manned spaceflight seemed to grow up. No more cannonballs into the Atlantic, no more tiny capsules that looked like they would fit in a Ford pickup. Gemini launched three missions in 1965 in quick succession—in March, June, and August—each with two astronauts. The June mission lasted four days and included the first U.S. spacewalk, Ed White opening the hatch of the spacecraft and floating out in space. The August mission lasted eight days—more than a week in space. These weren't space shots; these were missions. Then came December: two Gemini capsules launched into orbit at the same time, four astronauts, and the two spaceships maneuvered within 12 inches of each other (they didn't have docking capability), while talking to each other by radio and traveling at 17,500 miles an hour around Earth. The combined Gemini 6 and Gemini 7 flight was space travel. The crew of Gemini 7 lived and worked in space for two weeks, longer than any previous flight by either the U.S. or Russia, and longer than any of the Moon landing missions would be. Some of those Gemini missions developed serious problems, from which the astronauts and NASA recovered with calm and quick thinking, imagination and good training.

By the spring of 1966 the U.S. was a spacefaring nation. NASA had flown a dozen manned space missions in a row, carrying 15 different astronauts to space (and three astronauts had flown twice). The "flopnik" era was over. NASA was competent and confident, but without a hint of cockiness. Apollo, the missions that would fly to the Moon, was just over the horizon; the first manned test flight of the Apollo command module was set for February 1967. But to start flying Apollo spaceships, they needed Apollo flight computers, and those computers needed finished programs.

NASA scripted missions second-by-second so that both flight controllers and astronauts knew what was happening, exactly when, and what each maneuver or event required. That kind of planning also helped set the agenda for training and simulations, for astronauts and ground personnel, and it was the framework for all kinds of "alternative scenario" planning and practice as well. If things didn't go as expected, what were the options, what was the timing of those options, what role did each person need to play—all that was gamed out for every part of every mission.

In the end, all that planning produced a bible for each flight, a set of mission rules, committed to paper, that everyone worked off. Even when there were surprises, how you handled them would not be a surprise. The mission rules for Apollo 11 were 330 pages long.[7]

The era of spaceflight had dawned just a little before the era of computers. Early flight trajectories for solo-astronaut Mercury launches were calculated using primitive computers, the results then laboriously plotted by hand on wide expanses of paper. And if the hand-drawn curves didn't match the expectations, you went back to your original formulas, riffled through your computer punch cards, puzzled over your plot, trying to figure out where the error was. That era didn't last long.[8]

Gene Kranz, the flight director who watched over much of mission planning, described the transition from handcrafted trajectories to computer-designed trajectories as a revelation for spaceflight planners. The engineers, mathematicians, and programmers, Kranz said,

> started planning every aspect of the launches, the rendezvous, and reentry. They provided us with options that just months before we did not know existed. We had no choice but to believe in the data and methodology they came up with, so our trust in their work was absolute. They designed the mission, then loaded their software in the computers in the spacecraft and in [Mission Control]. Their work had to be perfect—and it was, thanks to increasing computer capacity, speed, and availability.[9]

*They came up with options we didn't know existed.*

Spaceflight simply couldn't get very far without computers. Rockets travel too fast and too far, tracing trajectories through three-dimensional space that require too much math for people to figure in their heads or react to fast enough in real time. Rocket guidance requires data transmitted at the speed of light, decisions made at the speed of electronic circuits, and instructions issued at the speed of light.

You had to use computers, and you had to use computer programs, and that means you had to use a whole new class of professionals—computer programmers—in whom you had to place enormous faith. Because computer programming wasn't just advanced math applied to advanced engineering; computer programming was a whole new set of languages and protocols that, without training and experience, was anything but accessible.

As Gemini wound down and mission planning for Apollo started in detail, the flight planners in Houston started to get nervous about the state of the software for the computers from MIT's Instrumentation Lab. Without mission software, it was impossible to fully test the spacecraft. Without mission software, it was impossible to fully train astronauts and flight controllers sitting at consoles.

Chris Kraft, who invented and then built Mission Control in Houston, had all the NASA-side mission planners working for him. "George Low called me in to talk about mission software, particularly for the Moon flights," said Kraft. Low was second-in-command in Houston. "We can't get the software moving out of MIT," Low told him. "So I'm giving you responsibility for making it happen."

Whether by instinct or design, Kraft made an inspired choice. "I sent Bill Tindall to the MIT Instrumentation Laboratory . . . to find out what was wrong. The legendary electronics expert Stark Draper himself was running the place and welcomed Bill with open arms."[10]

Bill Tindall knew the theory, the math, and the reality of space navigation as well as anyone. He'd worked in the radar room of the destroyer USS *Frank Knox* in the Pacific during World War II, joining the navy straight out of high school, and developed a fascination with math and engineering. He got an engineering degree from Brown University after the war and went to work the month after graduation for NASA's predecessor, NACA, in 1948.

In 1960 Tindall provided the orbital calculations for the world's first communications satellite. Echo 1 was an almost fanciful creation of the Eisenhower era of the space race; it was launched as a small payload into a 1,000-mile-high orbit, where it unfolded and then inflated itself into a vast gleaming silver balloon 100 feet across. It was, quite literally, a satellite as big as a 10-story building, but it was made entirely of filmy Mylar thinner than a single sheet of plastic wrap (the same Mylar we use today for "Happy Birthday" balloons filled with helium). It was positioned so that it spent weeks at a time in direct sunlight, even when it was over nighttime areas on Earth (when it was in the Earth's shadow, it actually deflated), and it was so large that it was visible with the naked eye to almost every person on Earth, a kind of glittering answer to Sputnik. Echo 1 was passive: to use it, you bounced radio signals off its shiny surface, from one location to another. And it worked perfectly. NASA issued detailed information about its orbit and location, and President Eisenhower, whose voice was the first to bounce off Echo 1, from NASA's Jet Propulsion Lab in California to Bell Labs in New Jersey, issued an invitation to any nation on the planet to make use of it. Eisenhower's picture was also transmitted via Echo 1. Of course, to bounce your signals off Echo 1, you had to have Bill Tindall's orbital calculations.

Tindall moved his family from Langley to Houston as the manned space program ramped up, and went on to guide the trajectory and orbital calculations for much of the Mercury and Gemini missions. His group did the elaborate math necessary to allow Gemini 6 and Gemini 7 to rendezvous in space during their December 1965 mission, which, like Echo 1, was a worldwide first. The U.S. had beaten the Russians to being able to rendezvous two spacecraft in orbit.

On that occasion Tindall's mother observed that her son had come a long way. In high school, she said, "he had some ability in math. But . . . he was not an outstanding student."[11]

By the mid-1960s Tindall was widely regarded at NASA as a genius of orbital mechanics, the master of the math necessary for rendezvous, one of the people, in fact, who helped invent it, who could untangle the equations necessary to navigate in three dimensions and help his staff do the same.

Tindall, who was 41 years old when he was dispatched to Cambridge from the Manned Spacecraft Center, was gracious and funny, a good listener but also discerning and decisive. He knew when the discussion was over, and he knew how to make tough decisions, of which there were many in a high-pressure, high-stakes, tight-schedule engineering program.

By 1966 Tindall had had years of management experience; one engineer who worked for him said Tindall liked remaining the deputy in the divisions where he worked because it gave him more actual ability to get things done, more maneuvering room, and considerably less bureaucratic hassle. Said his wife, Jane, "He was the opposite of the Peter Principle."[12] Tindall had the ability and experience to absorb, understand, and sort out serious technical problems, and that ability earned him the respect of his colleagues, even when they didn't get the decision they wanted.

At MIT's Instrumentation Lab, Tindall found a mess.

---

Here's how bad it was. The software MIT had written for the Apollo command module was 15,000 words over the computer's capacity of 36,000 words. MIT had written programs that took up 42 percent more space than would fit in the computer. The software for the lunar module was either 4,500 or 6,000 words over the capacity of 36,000 words—between 13 and 17 percent too large for the computer. And that was true even though MIT's own hardware staff had in the previous four years doubled the memory available, from 12,000 words to 24,000, and then increased it again by 50 percent to 36,000.

The more Tindall dug in, the clearer it was how the core program for flying to the Moon had ended up 40 percent bigger than anyone could use.

The Instrumentation Lab had no real sense of urgency about figuring out how to fit the programs into the computer; about getting them finished so they could be tested and used for training, not to mention used in actual flights. The documentation of the programs themselves—records of the software's development and the changes as things were fixed and adjusted—was spotty and unenthusiastic, on a project where NASA wanted every ounce of metal to be traceable back to the mine it was dug out of.

The Instrumentation Lab had no thorough testing program, and it had no management organization that created accountability. A lot of talented people were creating ingenious computer programs to solve space travel problems, but there was no one worrying about whether the whole thing would work, or even be practical, for actually flying Apollo space vehicles to the Moon.[13]

Whether or not Doc Draper, the head of the Instrumentation Lab, welcomed Bill Tindall with open arms—there's no record of Draper's reaction to Tindall's arrival—the staff of the Instrumentation Lab reacted with irritation and dismissiveness to this guy flying up from Houston, from the customer, for goodness' sake, to figure out what was going on.

"He started off . . . as an object of real derision," said Fred Martin, who was about 31 when Tindall arrived, had been at the Instrumentation Lab for a decade, and was working in Battin's software group. "Because we were rolling along, doing all of this fantastic work building software, doing this, doing that, when NASA somehow woke up and decided that these guys were totally out of control. Their documentation wasn't any good. They didn't have schedules. They weren't doing this. They weren't doing that.

"So he started to show up at [MIT] week by week by week, punching and pushing us into shape . . . to a great deal of discomfort at the Lab. We really started to bellyache about it, and a lot of people started to talk about how he's totally unreasonable. NASA is being totally unreasonable. . . . They were the customer, but that was beside the point."[14]

It was an interesting moment in the world of software, as well as the world of Apollo. "Software" as an idea—and as a word synonymous with "programming"—was so new that it was often put into quotes when it was used in newspaper stories. In fact the idea of software was still so novel that the spellings "softwear" and "software" were often used interchangeably, in newspaper stories, in headlines, even in advertisements for jobs from computer companies as well established as Control Data Corporation.[15]

NASA wrangled throughout the 1960s with the project management elements of Apollo—the balance between headquarters and the NASA centers, the balance between NASA authority and the vast army of contractors, and the need to stay both on schedule and on budget. NASA's chief James Webb said during and after the race to the Moon that the real achievement of Apollo was twofold: that the management system created to go to the Moon was as valuable as the Moon landing itself, and that Apollo showed the world that democracy could be successfully combined with management of complicated, large-scale projects, that, as Webb put it, democracy could "out-manage" authoritarianism.[16]

There were well-established protocols for managing engineering projects: the U.S. built Hoover Dam in four years, the Pentagon in 16 months, and the Empire State Building rose at a rate of one story a day and was completely finished in 14 months. Not to mention the execution of the Manhattan Project, just part of the vast logistical effort of

World War II. But there were no practices or management systems for big software projects; indeed, when Doc Draper's staff got started on the Apollo software in 1961, there hadn't been any software development projects of the scale and complexity of Apollo. It was the first of a whole new kind of engineering project.

Instrumentation Lab software engineer Margaret Hamilton, who graduated from college in 1958, joined the MIT Apollo project in 1963, and by 1969, just 11 years out of college, was overseeing software for the command module, and is often credited with popularizing the phrase "software engineering." "Software during the early days of this project was treated like a stepchild and not taken as seriously as other engineering disciplines, such as hardware engineering," Hamilton said. "It was regarded as an art and as magic, not a science. . . . I began to use the term 'software engineering' to distinguish it from hardware and other kinds of engineering." It didn't immediately catch on. "When I first started using this phrase, it was considered to be quite amusing. It was an ongoing joke for a long time." (The phrase started to appear in computer job advertisements in 1966.)[17]

In managing Apollo software, that lack of experience, or even a real model, was made worse by the evolving nature of the task. Every time the spacecraft changed, the software needed to change. More to the point, it wasn't possible to write mission-specific software until the folks at the Instrumentation Lab knew what the specifics of a mission were.[18]

MIT had gotten accustomed to a certain latitude. If NASA didn't quite know what it wanted, the Instrumentation Lab would tell them. "We were all very self-confident, egotistical guys," said Malcolm Johnston, who wrote algorithms and software for flight dynamics, for keeping track of what was happening while the engines were firing. "We didn't manage things well. We had too much software trying to squeeze into too little computer."[19]

That spring of 1966 Tindall was going to MIT from Houston every week for two or three days to try to get a sense of the dimensions of the problems and how to get both the software and the Instrumentation Lab itself on track.

The immediate focus was for a planned mission more than a year off, August 1967, when NASA hoped to launch both the command module and the lunar module, on separate rockets, into orbit together;

have the astronauts rendezvous with the lunar module; and then test out both spacecraft. Even with 17 months to go, the situation seemed grim.

"He'd come up, and we'd have lunch," said Ed Copps, one of the MIT software engineers who had joined Apollo from Polaris. "We used to have beer—somebody would go out for subs and the beer . . . and people would sit around." Tindall was a good-humored man, his temperament described as sunny. But at a lunch that spring, his frustration boiled over.

"One day," said Copps, "Tindall just gave us hell, you know? He really beat us up. 'How can you possibly do this? Here you sit at the very center of the success or failure of this extremely important program. You're behind. Get it through your head: You are fucking this thing up.'

"And he was right in many ways. I don't think there are many people who we could have taken that from, because we were pretty snotty and pretty arrogant people."[20]

In fact, before Tindall was able to get his arms around the problems, and before he had won the confidence of the MIT crowd, there was a brief rebellion against his oversight. "I can recall a meeting where we all got together and actually complained about how difficult it was to work under this environment," said Fred Martin. The gathering was in the office of Ralph Ragan, one of the senior-most managers on Apollo at MIT. "There were about 20 of us in there. . . . And Doc Draper came to that meeting, and listened to all of these childish complaints about what was happening." Draper not only wasn't one for "org charts"; he didn't personally manage people or projects closely. He wanted talent, and he wanted results. Despite his prominent role in getting the Apollo project, and his personal vow to Jim Webb that MIT would deliver, Draper didn't in any way manage the program directly.

And so Draper's reaction to Tindall's arrival, to Tindall's diagnosis of the problems—essentially, that everything was wrong except the quality of the programs themselves—and to the burst of anger from the IL staff was, in some ways, another masterstroke. "Listen," he said, "if you guys don't want to do the (Apollo) program, we'll get out of the program. It's really up to you. If the environment is no good for you to work in . . ."

Draper knew the crowd and the culture at MIT, and he also understood the demands of a program like Apollo. And at that meeting, in the three sentences Martin recalls, Draper picked sides. He wasn't siding with the whiners.[21]

Tindall's spring visits to MIT culminated in an all-hands meeting on Friday, May 13, 1966, that came to be known in the Instrumentation Lab as Black Friday.

"On May 13 and 14, 1966, a flock of [NASA] people met with MIT people in Boston to discuss the spacecraft computer program requirements," Tindall opened his account of the Black Friday meeting, which spilled into Saturday. "My main purpose is to describe the situation as it exists on these important programs; it is not altogether a happy one."

Tindall had gathered everyone who understood the Apollo flight and the computers, and their mission was clear: cut enough software so what was left would fit in the memory available but also still fly America's spaceships to the Moon.

"It was evident from the start that there were very few programs which could be easily deleted. In fact, it was a very painful process," Tindall wrote. Everything that was deleted "could only be dropped at some cost in probability of mission success or by putting a greater workload on the crew or reliance on ground support."

They did it, though. And Tindall pushed even deeper. "We identified the next computer routines which would be deleted in the event storage was ultimately exceeded, forcing the removal of more routines."

Tindall went on to list some of what had been cut. The ability for the onboard computers to take over guiding the launch of the Saturn rocket, if its own computer should fail, even though "it has been directed by NASA Headquarters that this [backup] capability be provided." Programs to let the computer run attitude maneuvers on the spacecraft: "[T]he pilot could do the job instead of the computer, although probably at some extra cost in our precious RCS [reaction control system] fuel."

"It is evident," Tindall continued, that what was cut "would be extremely valuable . . . and the necessity of deleting these programs is probably the best indication of how critical the computer storage problem is."[22]

That memo—three densely typed single-space pages—had a distribution list that included more than three dozen names (including Shepard and Kranz and Kraft) and a couple additional distribution lists.

It was just the beginning of Tindall's blunt and public accounting of the Apollo computer software—public not in terms of the general public but in terms of the widespread distribution of his memos inside NASA.

Two weeks later, in a memo he labeled a "newsletter"—"the first of a series"—Tindall was pessimistic: "There are a number of us who feel that the computer programs for the Apollo spacecraft will soon become the most pacing item for the Apollo flights." By "most pacing," he meant that the software could well be holding things up. A year hence, with spacecraft and rockets arriving at Cape Canaveral ready to launch, "we working on the computer program development will still be 'sloshing through the mud.' "

That very month, Tindall said, the Instrumentation Lab had reorganized its work, although even so, he confessed, "I still do not have a clear understanding [of how the work is getting done]."

In an oddly inverted moment of triumph, MIT—the contractor—had agreed, under Tindall's pressure, to hire more people to accelerate the pace. (MIT resisted on the theory that new people would need weeks of training, which would slow the work of the experienced staff, not speed it up.) And just to help move things along, in what may well have been a move considerably more galling than his own presence, Tindall had on his most recent visit brought with him an IBM executive to brief the folks at MIT on how NASA and IBM had organized IBM's work on the Mission Control computers. "I hope and expect [MIT] will draw heavily on this experience in setting up a similar system," Tindall wrote.

Of the software itself, Tindall wrote in the "newletter" memo, "I am still very concerned about unnecessary sophistication in the program and the effect of this 'frosting on the cake' on schedule and storage.

"It is our intention to go through the entire program, eliminating as much of this sort of thing as possible. I am talking about complete routines, such as 'Computer Self-checks,' as well as little features, such as including the third and fourth harmonics of the earth's oblateness and drag in programs for the lunar missions."[23]

MIT had done impressive work, but they'd spent a little too much time, and way too much computer memory, "frosting the cake."

It was a perfect Tindall moment. He knew the details of flying to the Moon, he had absorbed the details of the computer programs, and he knew you could fly to the Moon without the charming but excessive flourishes MIT programmers had built in. And he'd come up with the perfectly memorable phrase to describe this MIT tendency: "frosting on the cake."

Just two days later there was a fresh update. This time Tindall was into management details. MIT was analyzing the programs themselves to identify those that were most behind and—revealingly—to give the programmers working on those "top priority" for computer access. Among the problems that were slowly dawning on Tindall: the Massachusetts Institute of Technology did not have enough technology to handle the work. Progress on the computers was being slowed to a crawl by lack of computers. (This problem did not get fixed quickly: by February 1967 the Apollo group's work computers were so backed up that programmers waited one to four *days* to get test runs back.)[24]

"I'd like to make one final observation regarding the overall situation," Tindall wrote. "It's probably terrible; I really don't know yet. But it's my feeling that everything that can be done to help has been done. We are reacting to the problem areas as fast as possible; MIT has reorganized in what seems to be the best possible way, and they appear to be getting things on a businesslike basis, which up to now has probably been our worst problem." As pessimism goes, Tindall was pretty optimistic.[25]

In addition to his other abilities, Bill Tindall was a great writer—funny, literate, capable of cutting through technical complications to explain problems and choices with clarity. We know this because in May 1966 he started a series of memos—dispatches from the front lines of, first, fixing MIT, and then the larger mission of getting Apollo to the Moon—and those memos run into 1970, nearly four years. These memos quickly acquired a nickname: Tindallgrams. They started out documenting the MIT efforts but gradually expanded—as Tindall's own authority and responsibility expanded—to become a four-year account of a whole range of Apollo decisions and problems. A year after Tindall was put in charge of wrangling MIT, he was put in charge of the much larger project of fine-tuning every navigational and procedural detail of landing on the Moon. To do this he held meetings in Houston where NASA officials, astronauts, and staff from contractors sat around a U-shaped table, with Tindall presiding at the base of the U, and walked through every minute—really every second—of the lunar landings and every decision that had to be made, in the spacecraft and in Mission Control, to get to the Moon and back. The meetings weren't informational. Tindall decided things, step-by-step: heard the arguments, sometimes sent people off in a side meeting to figure something out, weighed the information, and decided. That meant if some part of your work was at stake, you couldn't afford to miss the meetings. So many people came that they lined the walls. "Each of the astronauts was on six or seven different panels," said Malcolm Johnston, one of the MIT engineers assigned to staff Tindall's larger mission planning meetings in Houston. "And a lot of those they didn't show up for. They just didn't have time. But they always showed up for Bill's meetings."[26]

The Tindallgrams range from the serious and astonishing (NASA wasn't sure how to figure out where the lunar module and the astronauts were on the Moon once they landed, because the Moon was unmapped) to the quietly revealing (Tindall was constantly worried that the astronauts had too many tasks to do during the last hour as they were landing on the Moon) to the easily overlooked but vital. Tindall wrote a memo firmly establishing how everyone at NASA would count how many times a spaceship had orbited the Moon: "This may seem like a trivial matter—however, before any confusion arises let's firmly establish the means of identifying revolutions in lunar orbit by number. . . . Revolutions will be started and ended at the 180° lunar longitude, i.e., the back of the moon near the point of lunar orbit insertion. . . . The first revolution in lunar orbit shall be, appropriately, called number one (1)."[27]

The Tindallgrams shaped the course of the flights to the Moon; by one account Tindall dispatched 1,100 over six years, which comes to three a week; during the critical year 1968, there are 184 surviving Tindallgrams, 15 a month, but the NASA numbering scheme on the memos indicates that there might have been more than 200. He kept writing memos after the initial Moon landings (which continued to be called Tindallgrams regardless of the topic). Tindall was capable of demanding attention for the big issues—how could MIT, of all places, not have enough computers?—and also spotting the smallest quirks of spaceflight that might seem silly but could have unanticipated consequences.

Just weeks before Apollo 10 blasted off, headed for a full dress rehearsal of a Moon landing, flying to within nine miles of the Moon's surface but not actually touching down, Tindall wrote a memo titled "Let's have no unscheduled water dumps on the F mission." (Internally the Apollo missions were lettered: Apollo 10 was F, Apollo 11 was G.)

At a recent meeting, Tindall wrote, "we were informed that the [command and service module] has some sort of automatic water dump system. It was even rumored that it might be enabled on [Apollo 10] while the crew is sleeping. . . . This memo is to inform everyone that an unscheduled water dump can really screw up . . . orbit determination. Accordingly, if we have a vote, this automatic capability, if it exists, should be inhibited and water dumps should only be performed as scheduled by [Mission Control]."

Tindall had worked for NASA since the day of its founding, but he had discovered something new about the Apollo command module in February 1969, after two Apollo missions had already flown: it ejected waste water into space automatically. Tindall, the orbital trajectory expert, knew that even the minor "thrust" provided by water being expelled from the command module could alter the orbit of the command module and the lunar module around the Moon in ways no one had accounted for, and in ways that might screw up navigation to the Moon's surface. Message of Tindallgram 69-PA-T-31A: Only Mission Control can authorize an overboard water dump.[28]

This kind of venting from the spacecraft was one of Tindall's minor obsessions—a small, apparently innocuous act of routine maintenance, the kind of thing an astronaut or Mission Control could do almost without thinking that could send the spacecraft instantly off course. In another Tindallgram, sent just eight days after Apollo 11 returned to Earth, Tindall explained how to get Apollo 12 onto the Moon with pinpoint accuracy; the goal was to land the second lunar module next to Surveyor 3, an unmanned probe that had landed on the Moon in 1967. Tindall's expert opinion was that Apollo 12 lunar module *Intrepid* would be lucky to set down within a mile of Surveyor 3, but he issued a five-page Tindallgram that included instructions in a "9-step program" to increase the chances of that pinpoint landing. Point #3: "Absolutely no venting or dumping allowed!! For heaven's sake, will all spacecraft system people please take note of this. What seems insignificant to you is a nightmare to orbit determination people." Tindall's "program" worked: Apollo 12 commander Pete Conrad put *Intrepid* down just 535 feet from Surveyor 3, a short stroll even in spacesuits, and there are vivid photos of the astronauts visiting the robotic probe, with their lunar module as the backdrop.[29]

The Tindallgrams are a remarkable window into a reality it was hard for ordinary people to understand at the time, and is even harder to appreciate from half a century away: flying to the Moon in the 1960s was hard, it was dangerous, it was filled with uncertainty, because space really is a whole different world, and a whole new world, and the technology to fly there had just been created, and it too was unfamiliar, and at least to start, it was untried in space. The image of three spacesuit-wearing astronauts, holding their portable air units, striding off confidently to board the Saturn V is real and true, but also very misleading.

Depending on how you count, there were 14 flights of Apollo spacecraft to space. NASA built 15 Saturn V rockets, 18 flight-ready command modules, 13 finished lunar modules—total.

In the end, 11 Apollo missions flew with astronauts on them; two stayed in Earth orbit; three orbited the Moon; six landed on the Moon.

It took 410,000 people to design, build, and deliver the spacecraft for those 11 flights. That, in itself, is the measure of complexity—410,000 people supporting just 33 crewmembers, 12,000 people on the ground for each person flying in space. But all those people had been working for years and years to make the flights possible.

Apollo spacecraft flew in space with astronauts for 2,502 hours—about 100 days over 11 missions. On Earth there had been a decade of work, 2.8 billion work-hours. Every hour of Apollo spaceflight required 1 million hours of work on the ground.[30]

The Tindallgrams are a journey into the intensity of that work and that complexity. And not just that: they document, in fascinating and accessible detail, legitimate disputes about how to get to the Moon, disputes that time has smoothed away but that are a reminder that even in a program as high-stakes and high-profile as flying to the Moon, where almost everyone's motivation and mission are unquestioned, building spaceships is still work, and there will be arguments, and also serious mistakes.

Among the most avid readers of the Tindallgrams were the astronauts, who realized that Tindall was chasing important questions until he got the answers on which their lives would depend.

Ken Mattingly was the command module pilot for Apollo 16, the astronaut who stayed in lunar orbit. He recalled becoming immersed in both Tindall's Houston meetings and the Tindallgrams:

> I remember going home one night . . . [thinking] it's like no one had ever thought about going to the Moon. We've been in this program for how many years, and yet people are asking questions that are almost like, "Does anyone know where the Moon is and how to find it?"
>
> . . . There were so many questions, and every one of them needed an answer. . . . Bill started having these meetings. . . . That kind of put some sanity and sense to it. . . . Because Bill Tindall would listen.
>
> These meetings would go on sometimes two days, and they would be eight in the morning until eight in the evening, whatever it took. Room filled with people. Not always a lot of decorum. Bill was after answers. It was nowhere near as collegial an environment as you see in some organizations today. But they were after what was right, and everybody was passionate. Everybody was young so they were kind of brash and there wasn't a lot of patience anywhere. So some of those meetings were very, very colorful. Some of the characters were colorful. At the end of this, you were just inundated with all of this stuff you've heard. And now what?
>
> And the next day you would get this two-, maybe three-page memorandum from Bill Tindall written in a folksy style, saying, "You know, we had this meeting yesterday. We were trying to ask this. If I heard you right, here's what I think you said and here's what I think we should do." And he could summarize these complex technical and human issues and put it down in a readable

style. I mean, people waited for the next Tindallgram. That was like waiting for the newspaper in the morning. They looked forward to it.[31]

---

Bill Tindall wrote his slightly optimistic Tindallgram about MIT—"Everything that can be done to help has been done"—on June 2, 1966.

June 13, 1966:

> I just got back from MIT with my weekly quota of new ulcers, which I thought might interest you. . . .
>
> The first estimate was that the program tapes could be released for . . . manufacture on about November 15, which is exactly three months too late. Rather an interesting proposal, I thought, since it is so obviously so unacceptable. After recovering from our complete shock, we started looking into the alternatives. . . .
>
> The program paring must be done, I feel, solely for schedule reasons, which is really kind of weird when you think about how long the programs have been under development. It will mean that we fly to the Moon with a system which does not minimize fuel expenditure nor provides the close guidance tolerances which are ultimately within its capability. . . .
>
> I certainly don't want anyone to think that we feel that situation is any better than barely tolerable.[32]

July 1, 1966:

> I would like to emphasize that [NASA] has bent so far over backwards in an attempt to reduce requirements on the . . . program that we all look like a double pretzel. . . . We have adopted a course of action which seriously perturbs other interfacing activities, which is annoying, to say the least, if not on the verge of being unacceptable. . . . It is galling that the "good guys," who have

> really been doing the job right, are forced into a position of seeing their efforts go right down the drain [because of MIT], and then to be forced into a crash effort to make up for deficiencies in another system.[33]

October 11, 1966:

> At present MIT has two [Honeywell] 1800 digital computers on which all program development and verification is carried out. These machines have been and are currently completely saturated. There are no other facilities in the entire universe, to our knowledge, of proper configuration to relieve this situation completely. This is identified as a major problem area particularly during the months of November and December. However, an IBM 360 is to be installed at MIT very soon and it is currently estimated that it will be online no later than February 1st.[34]

November 3, 1966:

> This program has gone together very nicely. Dan Lickly and his team of [AC Spark Plug] and MIT people are to be commended for the professional manner in which they handled this job. . . . This program has no known bugs or deficiencies at this time. If development of all the [Apollo] programs went like this, we'd be out of a job.[35]

That November 3 Tindallgram was making reference to a single specific flight program, rope weaving, that got finished and released for manufacture on time. It was hardly the last of the serious problems at MIT, but it was an indication that most of a year of attention by Tindall had given the Instrumentation Lab's software work the focus and discipline it desperately needed.

Interwoven with the Tindallgrams about MIT's herky-jerky progress was a steady thread of memos on other issues Tindall was fretting about. One of his favorite sources of irritation was the Apollo computer's "self-check" software, which ostensibly allowed the computer to check its own operations—particularly ironic in light of MIT's somewhat lackadaisical early attitude toward testing the software thoroughly while it was still on Earth.

On September 21, 1966, Tindall reported that "the self-check programs are still in," but he added, "I would like to make sure that this program really provides a useful operational function . . . before we decide to carry it to the Moon at the exclusion of some other program someone wants."[36]

Then, on January 25, 1967, he devoted an entire memo to self-check, beginning, "If they ever have a contest to select the piece of Apollo with the funniest history, I would like to enter 'computer self check.'" MIT, it turned out, had deactivated the self-check routine in a test flight because of its own doubts about self-check's reliability. "I guess we should be happy they discover these problems before the flights instead of during them. Apparently if the system were left as it is now, it has the potential of bombing out the system irrecoverably. I assume, or at least hope, that if it did that, it would light the little red light."[37]

The memos had a conversational tone—Tindall often addressed readers as "you," as if he were talking directly to them—but they were galvanizing. By the end of 1966, dozens and dozens of people were reading each one, so even the kind of memo that was mocking and funny had a serious purpose. Self-check did not, in fact, fly to the Moon.

Tindall often raised problems that were known but weren't getting enough attention. "I think this will amuse you," begins a Tindallgram from November 25, 1968.

> As you know, there is a light on the LM dashboard that comes on when there is about two minutes worth of propellant remaining. . . . This is to give the crew an indication of how much time they have left to perform the landing or to abort out of there. It complements the propellant gauges. The present LM weight and descent trajectory is such that this light will always come on prior to touchdown. This signal, it turns out, is connected to

> the master alarm—how about that! In other words, just at the most critical time in the most critical operation of a perfectly nominal lunar landing mission, the master alarm with all its lights, bells, and whistles will go off. This sounds right lousy to me. In fact, Pete Conrad tells me he labeled it completely unacceptable four or five years ago, but he was probably just an Ensign at the time and apparently no one paid any attention. If this is not fixed, I predict the first words uttered by the first astronaut to land on the moon will be, "Gee whiz, that master alarm certainly startled me."

Tindall knew enough about how the low-fuel light was wired into the master alarm that the Tindallgram goes on to suggest how to rewire the lunar module control panel. The low-fuel light was not, in the end, allowed to trigger the LM master alarm.[38]

The Tindallgrams often deal with the kinds of issues that people without math and engineering training can only dimly understand, except that they so clearly show how complicated flying in space is. Fuel sloshing in the fuel tanks of the descent stage of the lunar module could interfere with navigating (but only when the fuel got low).[39]

Tindall's memos are meticulous; in hundreds of densely typed, single-space pages, misspellings and typos number just half a dozen. This accuracy may be due, in part, to the fact that he dictated most of the Tindallgrams to his longtime secretary, Patsy Sauer.[40] His style is so distinctive you can tell within the first sentence—often within a few words or just by reading the title—that he is the author. In the subject line to an early Tindallgram about MIT, he refers to their computer programs as "a bucket of worms." He dismissed something unlikely to be useful with "Holy waste of time, Batman!" (The TV series starring Adam West and Burt Ward had debuted on ABC just 10 months earlier.) He devoted one memo to the concerns of senior MIT manager Ed Copps ("Some Things Ed Copps Is Worried About"), which was serious, but concluded this way: "This whole business apparently scares the hell out of Ed Copps, and I guess if I knew enough I would be frightened too." One memo mounting a blunt and deadly serious campaign against removing a radar unit from the lunar module is framed with these two sentences, the first and the last: "A rather unbelievable proposal has been bouncing around lately. . . . Please see if you can stop this if it's real."[41]

George Low, the senior NASA official in Houston whose worries about MIT originally prompted the dispatch of Tindall to Cambridge, was so devoted to the Tindallgrams that he refused to let his secretaries summarize them for him and insisted on reading the originals.[42]

Tindallgrams never dealt with trivia—even when Tindall himself suggested he was tackling something trivial—but routinely dealt with the most urgent of subjects.

In April 1967 a serious problem was discovered on the lunar module. Somehow Grumman, the designer and builder of the LM, and MIT had gotten their signals crossed. When the LM computer was instructed to aim the LM descent engine one way, the engine aimed the opposite way. (The technical term for this is gimbaling the engine; the exhaust bell on the LM engine could move to aim the thrust in the desired direction.) In addition to the seriousness of the problem itself, Grumman blamed MIT, whose computer programming fumbles were widely known across Apollo.

Tindall waded in directly and bluntly in an April 21, 1967, Tindallgram:

> A serious misunderstanding between MIT and Grumman resulted in a situation which would have been catastrophic . . . if it had not been discovered. Specifically, the direction the LM descent engine gimbals move when commanded by the spacecraft computer was opposite to the way they were supposed to.
>
> Upon discovery of this, Grumman immediately [placed] . . . the fault for this inconsistency on MIT. Since it was easier to make the necessary modification in the software than the hardware, we chose to do that—thus giving further weight to the idea that the MIT work was faulty, which I simply do not believe to be the case. I am distributing the attached letter from MIT which explains the situation in some detail in order to dispel the erroneous accusations you may have heard and believed. . . . I would like to emphasize that a number of

> positive steps have been taken . . . to make sure a vehicle is never flown with anything as fouled up as this. Also, it is worth noting that this discrepancy was detected well in advance of the mission and in time that something could be done about it fairly easily.[43]

Tindall said often that most of what the job required of him and his mission planners wasn't just to plan for how to fly to the Moon; it was to figure out all the things that could go wrong while flying to the Moon, and figure out a plan for resolving every one of those as well. He said that "80 to 90 percent" of his efforts went into that kind of contingency planning.[44]

The most vivid and prescient example of that is found in a Tindall-gram from July 1968. Subject line: "LM propulsion of the LM/CSM configuration as an SPS backup technique." Translated from NASA's compressed lingo, the question was: Can you use the lunar module's engine to fly the whole docked Apollo spacecraft if the main engine for the command and service module (SPS) should fail for some reason? Could the lunar module rescue the mission and the astronauts?

Tindall's note is really a means of passing on a detailed assessment of this issue from the legendary NASA engineer Max Faget, the man who came up with the idea for space capsules in the shape we're familiar with—their backsides broad and blunt, to dissipate the heat of reentry. Faget's memo on using the lunar module to power the mission back to Earth starts with a full-page index of a dozen earlier NASA studies of this possibility—going back to 1963. While hardly a key concern, Tindall had several times—in September 1966, in August 1967—discussed with MIT what kind of computer programming would be necessary to navigate using the lunar module's engine and thought it was important enough to squeeze into the overburdened computers. He wrote in the 1968 memo that Faget's information had been widely disseminated throughout Houston.

In April 1970 this somewhat remote possibility would become the focus of all of NASA's efforts and the whole world's attention, as the explosion of an oxygen tank on Apollo 13 crippled the service module and disabled the command module, and the lunar module *Aquarius* was used in just this way, as a lifeboat for the Apollo 13 astronauts who powered down the command module and retreated, as a group, to the cramped lunar module for the flight home. When that crisis overtook Apollo 13 and Mission Control, they had at hand a small reference library of guidance about how to use the lunar module, its supplies and engine, to navigate just such an emergency.[45]

---

It worked. Bill Tindall rescued MIT, and MIT rescued the Apollo computer software.

The cutting of carefully crafted programs—really subroutines that did specific actions—never stopped being painful and contentious for MIT. Norm Sears, a senior MIT manager for Apollo software, spent days going over every subroutine of the programs at meetings with Tindall, and had some of his own work cut. "Bill would listen. He would argue with you. There's a lot of give and take if you are trying to hold on to some cherished design, to find some other way. A lot of those meetings got tense." In the end Sears lost many of those battles because a lot of software had to be cut. "I knew his job was tough. He commanded enough respect that even though we didn't agree with him, we would go with his judgment."

In his assessment of the Apollo computers—part of a sweeping history of NASA's early use of computers on the ground and in flight—computer scientist James E. Tomayko wrote, "No one doubted the quality of the software eventually produced by MIT nor the dedication and ability of the programmers and managers at the Instrumentation Lab. It was the *process* used in the software development that caused great concern, and NASA helped to improve it."[46]

What saved the process from disaster was Tindall himself—his technical proficiency, his personal modesty, his personality, which came through in those meetings as it did in the Tindallgrams. "Bill had this sense of humor, which he used very effectively," said Sears. "It was a gift. Another person could have made the same decisions and come out with a totally different reaction."[47]

About a year after starting his visits, in March 1967, Tindall could write, "It is my feeling that no major problem exists any longer. . . . MIT has an organization and facilities geared up to handle the workload in an orderly, professional, unharried manner. High quality flight programs should be available well ahead of their need. . . . Your comments shall be received with relish." As if to underscore his enthusiasm for the change in fortunes, the distribution list for that Tindallgram was 64 people.[48]

At MIT there was a universal sense that Tindall had been their savior. David Hoag, who ended up as program manager for the whole Apollo effort at MIT, called Tindall "one of my biggest heroes for the whole program." Ed Copps, one of MIT's senior managers, and one of those lunching with Tindall when he bluntly asserted MIT was "fucking up," said, "I met a lot of people on Apollo at NASA and everywhere and I would say that certainly, he was one of the really giant figures." Malcolm Johnston, who worked with Tindall both at MIT and at the big planning meetings in Houston, pointed out the feeling for Tindall wasn't just grudging respect among the MIT crowd. "Everybody Tindall argued with, everyone whose arms he twisted—they ended up agreeing with him," Johnston said. "In many respects, Bill saved our ass."[49]

The Instrumentation Lab had reorganized the work and the lines of responsibility. Tindall had lit a fuse of urgency in Cambridge; MIT did not want to be the one—out of 20,000 contractors—preventing Apollo from reaching the Moon by Kennedy's deadline, or even holding up a single launch.

But something else had happened, a devastating event that served as a reminder of the stakes of Apollo, right on the doorstep of the first flight. On January 27, 1967, three Apollo astronauts suffocated in their capsule when a fire swept through the interior of the pressurized spacecraft as they were doing a test on the pad at Cape Kennedy. The deaths of Gus Grissom, Ed White, and Roger Chaffee were horrifying for NASA and for the nation. But the disaster did two absolutely indispensable things for NASA. First, it imprinted the can-do spirit in NASA at the most elemental level—that what the dead astronauts would most want would be for the space agency to investigate the fire, fix what was wrong, then go fly in space. Second, while the agency took the months required to do those things, whole swaths of the space program got a year to pause, reassess their own work, and get a lot of things right that were being done too hastily or too sloppily.

The staff of the MIT Instrumentation Lab knew that the Apollo fire stand-down was a moment not to exhale, but to redouble their own efforts.

Tindall's memos, at least the ones we have, make no mention of some of the biggest events in the space program during the decade. There are three Tindallgrams from January 30 and another from January 31, 1967, just three and four days after the fire that briefly froze the nation and NASA in astonishment and grief. None of the four makes mention of the fire. In the memo from March 24 praising MIT's progress and newfound maturity in software development, Tindall opens with an important but oblique reference to the fire: "It is possible to take advantage of the stretchout of the Apollo flight schedule in the manner in which we develop the spacecraft computer programs at MIT." The "stretchout"—no U.S. astronauts would fly in space at all in 1967, and not for the first nine months of 1968—was solely because of the fire. (There is, likewise, a Tindallgram from August 1, 1969, just eight days after the triumphant splashdown of Apollo 11, that makes no mention of that flight and its successes.)[50]

"It was about a year and a half before anything flew again," said Dick Battin, who had been in charge of the software operation at the Instrumentation Lab, "and that was the time that we needed to get our software act together and to get things moving along. . . . Without that . . . year and a half, we would have always been the late ones. They'd be ready to go and they'd say, 'MIT, where's the software?' 'Well, we're not ready yet.' And we never wanted to be in that position.

"Then we got the time, but I would just as soon have gotten it some other way."[51]

It's easy to be critical of Doc Draper's lab, but as the history of software projects large and small has shown, half a century after the first big project software engineering remains muddled, messy, and often—perhaps even usually—badly mismanaged. Not enough planning, not enough people, not enough attention to either documentation or the schedule. The problems of today's big projects are identical to the problems of Apollo. The effort programming takes makes people optimistic. David Hoag was one of the widely respected senior managers of MIT's Apollo effort, and in the history he wrote for the *Journal of Guidance, Control, and Dynamics* a decade after the events, he assessed the software operation with a single sentence of understatement: "The effort needed for the software design turned out to be grossly underestimated."[52]

You can see the transformation in the number of people working on the flight software. When Tindall first started visiting MIT, the software group had about 130 people. By the time Tindall would write, 10 months later, that MIT was finally tackling software in an "orderly, professional, and unharried manner," the software staff had doubled to 260. During most of 1968 it would be more than 350.[53]

And the computer itself had an impact on every element of Apollo, big and small. The flight computer was the way the astronauts flew their ships; it was at the center of navigating and piloting both the command module and the lunar module. The astronauts were in no way merely symbols; the early fear of the original Mercury 7 astronauts that they would be, in Tom Wolfe's memorable phrase, "Spam in a can," turned out to be the least of their worries. The Apollo spacecraft, and the Apollo computers in particular, required a lot of flying, and Tindall, for one, was constantly watching to make sure they weren't overworked.

But as the rest of society would discover over the next 50 years, digital technology has a subtle tendency to reshape life around it in unexpected ways. Apollo was no exception. The original plan for the control panels of both the command module and the lunar module was to have standard analog clocks, aviation style. Both Mercury's and Gemini's control panels had clocks: black faces with white numbers, hands and minute lines, and a sweep second hand.[54]

Apollo's computer had its own internal clock, of course, and many instruments on the spacecraft required precise timing signals, to the millisecond, which the computer supplied. The autopilot received instructions 10 times a second; the computer downlinked data to Mission Control 50 times a second.[55]

But in the mid-sixties, said astronaut Dave Scott, digital clocks didn't exist. "Everybody had analog clocks and watches. A computer naturally expresses its time digitally. It was quite a consideration on what kind of clocks to have. I think the influence of the digital computer ultimately showed the advantages, especially in the business of traveling in space, of a digital clock. The initial Apollo design had three analog clocks on the panel, and ultimately we ended up with digital clocks. In fact, the whole control center in Houston ended up with digital clocks."[56]

As pioneering as the design of the computer itself was—with its reliance on the integrated circuit, its real-time speed, its use by the actual user, its insistence on reliability—the software was just as inventive, tackling problems that had never been tackled before. The astronauts were unquestionably young and fearless, but as one of the MIT programmers, Margaret Hamilton, observed, the software engineers were young and fearless too.[57]

The two key software innovations—the Apollo computer's ability to make decisions about what work to do, and the ability to recover gracefully and quickly from being overloaded or from failures or bugs—are so distinctive they can be traced to specific people. J. Halcomb Laning, the mathematician and computer pioneer, was one of the two people at the Instrumentation Lab inspired by Sputnik to create the Mars mission project that laid the groundwork for the Lab getting Apollo. Laning was one of the elder statesmen at the Instrumentation Lab by the 1960s. Born in 1920, he had started at MIT as an undergraduate in 1938, got his degree in chemical engineering, a PhD in applied math, and started working for Doc Draper in 1945. By 1965 he was 45 years old, and the IL had staff members young enough to be his kids.

Laning loved the pre-Apollo world of the Instrumentation Lab, small projects, the ability to tackle problems, often by himself. In 1953 he came up with the first compiler for a computer, the forerunner of FORTRAN and every other "higher level" computer language we rely on today. His compiler was used for advanced math; he created a program that instantly translated mathematical equations into an assembly language that the computer could understand, so people using the computer could enter their problems in the terms they were used to and not spend time laboriously translating the math terms into machine language. "The effect of our program is to create a computer within a computer," Laning wrote. The compiler, which he named George, ran in real time on MIT's early vacuum-tube computer, Whirlwind.[58]

It was Laning who came up with the idea of giving the Apollo computer executive function, in which every subroutine and task the computer had to do would be given a priority, in advance; tasks like managing a rocket firing or using thrusters to keep the spacecraft stable could "cut in line" in front of tasks the computer might be actively doing, like updating the displays for the astronauts or sending data to Mission Control. For the executive to work, the computer's software had to be designed so lower priority tasks could be paused midstream, but also so that everything those tasks were in the middle of could be stored temporarily, to be resumed when the higher priority task was done.

"He basically made it up out of whole cloth," said Don Eyles, one of the young programmers on Apollo. "But it was brilliant."[59]

Giving computers that kind of priority-based decision-making skill was an essential insight, and also an essential milestone for the development of modern, high-value computing. The Apollo flight computer—any computer juggling many tasks from routine background calculations to life-critical functions—would need the ability to do the essential, time-critical tasks the moment they needed to be done, regardless of what else was going on. All this happened in the space of fractions of a second, of course, giving a sense that not only did the Apollo computer operate in real time but that it could do more than one thing at once.

Laning, who was Battin's boss and then his colleague and tennis partner, was revered at MIT. "Hal was the most brilliant person we ever had the chance to work with," said Dan Lickly, who led the group writing guidance programs for reentry to the Earth's atmosphere.

But he let others handle Apollo. "When we got the Apollo job," Battin said, "he told me, 'Dick, I'd like to help out, but I do not want to be a manager. The endless meetings and trying to explain things to people who don't understand them—I can't do that.' "[60]

If the idea of the executive came early to the programming efforts and shaped the work itself, the idea of creating a computer that could be interrupted—that could suffer a fault, a software bug, or a power failure—then instantly restart itself came late to the process. The restart ability was a considerable hassle to weave into programming that already existed, and its value, even its actual workability, was controversial.

Ed Copps, the Instrumentation Lab software manager mentioned occasionally by Tindall, found the restart ability baffling, frustrating, and distracting to an effort already under pressure. The Apollo computer, he said, "among its other attributes, would periodically fail to work. But, some brilliant person invented this way to write programs, so that if it didn't work for a while, you could start it back where it was last known to be working, and hope that it would work again. That was actually the right thing to do, there's no question about it, but it really made things . . . a lot more complicated." Among the problems: How do you test a computer's ability to recover from unplanned failures, so you know what actually happens as it restarts?—which is the critical question. The software, installed on simulators or on flight-ready computers, might run for months *without* failing. "It was always something that kept me up at night," said Copps.[61]

The problem that restarts were initially designed to solve, most simply, was what happened if there was a brief power interruption in either the command module or the lunar module during a critical flight maneuver. What if the spaceship suffered the equivalent of the lights in your house flickering just as the lunar module was hovering for a Moon landing, and you suddenly lost the ability to guide, stabilize, and land the lunar module because you suddenly lost your computer?

The restart system was designed by a guy named Charley Muntz, who began working at the Instrumentation Lab one summer as a janitor while he was an MIT undergraduate. "Trundling from office to office emptying waste baskets gave him a taste for the sort of projects Doc Draper's lab was tackling," said his colleague Don Eyles. Muntz graduated from MIT in 1962 and immediately went to work at MIT on Apollo; he did critical Apollo coding before his 25th birthday. Restart protection was ultimately coded into the computer for hardware problems, for software glitches, and also for occasions when the computer's processing capability itself—its decision-making capability—was overloaded, to give the computer a way out of having too many programs needing to be paused at once.[62]

Restart does come up in the Tindallgrams, as something that is complicated and space-consuming, although desirable. Copps decided that the only way he could really understand it was to write a paper about it: "I thought I should know how it worked, so I figured it out."

MIT's approach was so novel that Copps opened by saying, "The intent of the paper is to explain what restart protection is, why it is done, what things might have been done instead and implications, large and small, of various design decisions." One of the paper's sections is titled, simply, "Will Restart Work?" Another is titled "The Cost of Restart Protection," in which Copps detailed how much space the restart coding takes up: 4 percent of the total computer memory available on Apollo.

Copps wrote his paper in August 1968, months before the first Apollo flights with astronauts, when the flight computer had flown only in unmanned tests. He almost seems at pains to convey skepticism about the value of restart, except for a single prophetic sentence, the last sentence of the paper. In the closing paragraph, Copps writes:

> The establishment of restart protection in a computer such as the AGC presents somewhat of an enigma. In a total of over 25 hours of space flight, the computer has yet to have a transient failure from which the restart feature could be called on to demonstrate its worth. This could well be the experience for the whole Apollo program. We have seen that the provision for restart in the computer program complicates the generation and test of [the] program. We have seen that there is a significant class of transient failure events which restart will probably fail to cure. And yet only one successful recovery by restart might save a mission.[63]

---

The Apollo spacecraft computers contained memory composed of 1s and 0s, like all modern computer memory. For Apollo, the fixed memory that contained the programs was composed of exactly 589,824 1s and 0s.[64]

Inside the memory portion of that one-cubic-foot box that was the Apollo Guidance Computer, that meant there were 589,824 wires, and each wire was either a 1 or a 0. If the wire was threaded through the center of a tiny ring magnet, that wire was a 1. If it was threaded just to the outside of the tiny ring magnet, it was a 0.

And every one of those 589,824 wires was threaded by hand, in a factory in Waltham, Massachusetts, in a process that looks like very painstaking weaving, because that's exactly what it was: using a needle and a piece of wire to literally weave the software code that the programmers at the Instrumentation Lab had written.

In the mid-1960s "producing" software for Apollo spacecraft was different from just writing it, testing it, checking it, documenting it, and loading it into a computer, as we do now. At MIT, at Mission Control, at Cape Kennedy, at IBM, computer programs were stored and moved around using stacks of punch cards and huge reels of magnetic tape. A single reel of tape could hold 40 megabytes, an unimaginably large amount of data for the era. But the Apollo flight computer didn't have card or tape readers; they were too bulky, impossible to use on a spaceflight. Once a mission's programs were finished, they were loaded onto tapes and cards and taken to the factory where the computers themselves were manufactured, that Raytheon facility in Waltham. There, in a blending of the 19th-century horse-and-buggy era and the 20th-century Space Age, the programs for what was then the smallest and most innovative computer ever conceived, the programs with instructions for how to fly to the Moon, were stitched into the computer's memory by dozens of women wearing blue smocks, sitting at carefully designed looms.

It was a mark of the times—although, as we'll see, no sign of disrespect—that the memory became known not just as "core rope memory" but as "LOL memory": constructed by "little old ladies." The weavers were women, although based on video and press accounts, they weren't particularly old.[65] Because of the density of braided wiring, the programs were called "ropes," and at MIT, each flight had a slightly different version of the software, and each version had a manager watching over it, called a "rope mother" (although almost all of them were men).[66]

The core rope memory was awkward to create, and it also presented an awkward dilemma for MIT: because the manufacturing took so long, programs for a flight had to be finalized—quite literally locked down—8 to 12 weeks before the flight was scheduled. Early in Apollo, among MIT's programmers, says Ramon Alonso, "The reaction was, What? I can't walk up to the launch pad and change whatever the program is like?" The programmers had to finish, and their work had to be perfect.[67]

Then the women went to work, and their work too had to be perfect. In one stage of the process, two women sat across from each other, between them a matrix in which were mounted a grid of dozens of the tiny round magnets. The women would pass a needle about eight inches long threaded with wire back and forth through a magnet, or around it, repeating the process over and over. At another stage of the process, a woman working alone would use the same kind of needle to thread the wire through or around the magnets in a grid, based on a tiny, computer-controlled loop, that moved after each threading to show the weaver where to put the needle next.

The work was both demanding and painstaking. The women who did it came from a Waltham community deeply experienced in the textile industry; many were hired from Waltham Watch Company. "They were the only ones I ever saw with that much patience," said Raytheon supervisor David Bates. "There was a bit of tender loving care in that too," said Eldon Hall of MIT.

Creating the rope cores for a single computer took about eight weeks; just one 12-inch-long module contained a half-mile of wire, and all the elements of the computer had to be wired to each other with the same level of precision and care. The weaving part of the manufacturing looked relatively simple, but it required absolute perfection; a single misrouted wire meant that some aspect of the computer wouldn't work right. The quality and consistency were related to both skill and experience, which Raytheon quickly appreciated. During a strike at Raytheon in the mid-1960s, supervisors and managers sat down to do the weaving. "Everything they made was scrap," said Ed Blondin, who worked for AC Spark Plug, which was overseeing Raytheon's Apollo work.[68]

Raytheon created a dedicated group at the Waltham factory, which also produced guidance systems for Polaris missiles, just to do the Apollo work. "The girls who worked on ropes, that's all they could do because they had got good at doing ropes and we didn't want them to go off doing a bunch of other things and not be able to get them when you needed ropes," said Jack Poundstone, who was technical director of Apollo at Raytheon. "We paid those women to sit there and wait until the deck of [punch] cards or the tape came out. And they would be sitting there knitting for, you know, two or three weeks. And then a deck of cards would come out, or the program would come out, and then they'd just go like hell."[69]

When the manufacturing and assembly were done, the memory and processing modules were mounted on computerized equipment and tested relentlessly before being sent south to Cape Kennedy to be installed on the spacecraft.

In its own way, the need to finalize the software three months in advance served the folks at the Instrumentation Lab. "I think that that helped a great deal to get the discipline necessary to make sure that this thing worked as advertised," said MIT's Alonso, "or as close to it as humanly possible."[70]

And yet, when errors were discovered, the artisanal nature of the software was arresting. Tindall documented MIT's discovery of a small but significant navigation error in early versions of the flight software (before the Apollo 1 fire), after that software had been sent to Raytheon for manufacture. The error totaled 8 words of the 36,864 in the program, all in a single memory module. "It is currently our consensus that we would be wise to manufacture a single new module to be substituted in the spacecraft when it's available. It will cost about $15,000 and will take about 30 days to make starting after delivery of those now in process. The cost in effort and treasure is justifiably small to procure the insurance the new rope would provide."[71]

It was a crazy way to create software, to weave it by hand. But it was also a particular moment in time. The needs of a spaceship computer were just two or three years ahead of the sophisticated technology necessary to make it easily. For Apollo, rope-core memory had powerful advantages. In the mid-1960s it was the densest memory available—which is to say, the most memory to be had for the weight and space, between 10 and 100 times more efficient than other kinds of computer memory. On spaceships that needed as much computer memory in as little space as possible, that was vital.[72]

Rope-core memory was also, essentially, indestructible. It couldn't be accidentally erased by a mistaken computer command; it couldn't be damaged by an electrical surge or radiation. In fact, in a near disaster, Apollo 12 was launched on November 14, 1969, into low, dark storm clouds and was struck by lightning twice as it raced for Earth orbit, once at 36 seconds after launch, and again at 52 seconds. The command module's main electrical power was knocked off-line, the control panel in the command module lit up with "so many [warning lights], we couldn't read them," as Commander Pete Conrad told Mission Control, and the CM guidance system was also knocked off-line; its inertial platform lost its lock on where the rocket was, and the AGC's erasable memory was wiped clean. Untouched and undamaged: the computer's hard-wired rope-core memory. (The Apollo 12 astronauts were able to reset most of the electrical system while still rocketing to orbit, and restored the computer's guidance system, using manual star sightings, once in orbit.)[73]

Rope core was indispensable to Apollo, but it apparently wasn't used for any other significant major computing projects, being quickly supplanted by the use of silicon chips for memory, which were orders of magnitude more efficient, of course, and ushered in the era when software was 1s and 0s stored deep inside those chips—weightless, and also instantly changeable the way we think of it today. For Apollo, though, the software was hardware.

The Waltham factory was in no sense an obscure corner of the Apollo operation. The astronauts routinely visited MIT to learn how to use the computer and how to apply their guidance and navigation skills to it. They also visited the Raytheon factory.

"NASA was smart. They brought these astronauts around not just to meet management. They brought them down to meet everybody that was on the floor," said David Bates of Raytheon. "These little old ladies adopted every one of those astronauts."

The astronauts, said Herb Briss, also at Raytheon, "were brought in under the guise of wanting to see how the components are made. But . . . they talked to the ladies like they were [talking to] their mothers."

Those visits connected that relentless threading of wires to the real people depending on them. At one point, said Bates, a set of rope cores had passed its acceptance testing at the factory, but the women themselves rejected it. Bates went to see them. "It passed everything," he said. "It costs $75,000 and you've scrapped the thing. Why can't we use it? . . . [One of the ladies] looks up at me with this face, and she says, 'You know, I built that and it passed. But I don't think it's too good. So you wouldn't want me to pass something that I thought wasn't too good . . . on to one of our boys.' . . . This isn't a missile that we fire. . . . Somebody's going to be out there having to count on this thing." The rope core got scrapped.[74]

---

In 1969 the world's computer skills weren't *primitive*—a condescending adjective that does injustice to the genius and insight of the people who invented modern computing and drove it forward. Those early machines were the opposite: given the electronic and manufacturing technology available at the time, they were marvels of ingenuity and determination. You didn't build a machine out of vacuum tubes to do some basic differential equations because you couldn't do them yourself; you built that machine because you knew what it might ultimately do.

The computers of 1969 were, however, undeniably basic.

But here's the amazing thing. We didn't take our basic computer skills and computing tools and try something simple like running the elevators in a skyscraper or the inventory control system for a factory or the scoreboard at a football stadium. We took those very limited skills and we used them—the folks at MIT and NASA used them—to do the very hardest thing that had ever been done: fly to the Moon.

There would be two decades of step-wise growth in computing after Apollo until the World Wide Web was invented in 1989, but the first step was in fact a giant leap. No one needed more inspiration for what a computer could help us do, or what we could depend on it to do, than "Houston, Tranquility Base, here. The *Eagle* has landed."

The Apollo guidance computer was a hero of Apollo. Several times.

For many of the men and women at MIT, that first flight to the Moon, Apollo 8, was the emotional climax of what was then seven years of work. After everything that had happened, Apollo 8 proved that MIT computers could fly to the Moon. For the rest of us, flying the command module to the Moon without the lunar module, orbiting 10 times on Christmas Eve, then flying safely home without landing was wonderful; it was emotional and emotionally satisfying. American astronauts reading from Genesis while in orbit around another heavenly body was riveting and cathartic after the devastating year of 1968. The United States had *made it to the Moon* before the Soviet Union.

No one in the press or the U.S. political leadership cast it that way, however. Indeed the just-named acting NASA administrator Thomas Paine did just the opposite. "A hundred thousand miles from Earth there is no room for a space race, no place for Russian-American competition," Paine said. "This is something for all mankind." But inside NASA, where senior officials had been briefed by U.S. intelligence agencies, there was real fear that the Russians could still beat the U.S. to the Moon, if not to the Moon's surface. On December 6, 1968, *Time* magazine had a cover story, "Race for the Moon," featuring an American astronaut and a Soviet cosmonaut, in spacesuits, matching strides in a race for the Moon. "Lassoing the Moon" with Apollo 8 didn't fulfill President Kennedy's promise, but it prevented the sense of defeat that the Russians orbiting the Moon first would have meant. As important, it kept Apollo on track for a 1969 Moon landing, even though the lunar module remained badly behind schedule.[75]

For the Apollo project staff at the Instrumentation Lab, the triumph was big and it was personal. Apollo 7, the first flight with astronauts in almost two years, lasted 10 days, but it was really a test of the Saturn V and the command module. There was no lunar module (it was not finished), and while the astronauts were using the Apollo computer to navigate and fly the command module, there wasn't much navigation involved in orbiting the Earth for 10 days. Apollo 8 was completely different. It was going to fly the 240,000 miles to the Moon, and it was going to have to do the very precise rocket burns necessary to enter Moon orbit and the equally demanding rocket burns to leave Moon orbit and return to Earth. (To slide into lunar orbit, you turned your command module around so you were traveling engine first, and fired it to slow down and let the Moon's gravity grab hold.)

Most important, Apollo 8 was going to do something no spacecraft with astronauts had ever done before: it was going around the back of the Moon (every time it orbited, in fact), and so Apollo 8 would be flying 100 percent autonomously, dependent exclusively on the technology and skills aboard the spacecraft to get the flying right. In fact, because of the realities of orbital mechanics—the orbital mechanics of orbiting the Moon, in this case—the two most crucial rocket firings happened while Apollo 8 was blocked from assistance, or even being monitored, by Mission Control. The burn to put Apollo 8 into lunar orbit, and the burn to pull it back out of orbit and aim it for home, both happened while the spacecraft was behind the Moon.

Everything depended on the precision and reliability of the Apollo onboard computer and on the skill of the astronauts using it.

Chris Kraft, in charge of Mission Control and flight operations during Apollo 8, recalled Bill Tindall sitting him down to explain, as only Tindall could,

> all the things that could happen when you fire that engine on the back side of the Moon. If the attitude control systems did not work perfectly, when the engine stopped burning you could be going into the lunar surface. Or you could be going out into deep space and never [seen] again, if it cut off at the wrong time in the wrong attitude. . . . It was like kicking the bird out of the nest when you were on the back side of the Moon for 30 minutes. . . . It was dangerous. It was risky.[76]

NASA and MIT had done the precise calculations of when the deep space telecommunications network would lose contact with Apollo 8 and when the network would pick their radio signals back up—to the second, according to Kraft. In Cambridge, as in Houston, this was the critical test of the Apollo computer: MIT had fought for the computer to have all the capability it needed to navigate and operate the spacecraft without guidance from Mission Control.

In the room they had set up to monitor missions in their old underwear warehouse, dozens of MIT staffers waited to hear the voices of the Apollo 8 astronauts—Frank Borman, Jim Lovell, Bill Anders—on their real-time squawk box, as the command module emerged from behind the Moon. CapCom Jerry Carr started hailing Apollo 8 by radio 47 seconds before the command module was expected to be back in contact. From CapCom: "Apollo 8, Houston. Over."

Four times, Carr radioed Apollo 8, and by the fourth call, Apollo 8 was 19 seconds overdue for the calculated moment of radio contact being restored. But, in fact, within one second of the calculated moment, the spacecraft itself had begun radioing back telemetry data on its own condition to Mission Control. It was just voice communication that seemed delayed.

On the sixth hailing, this reply from Jim Lovell came over the speakers in both Mission Control and MIT: "Go ahead, Houston. Apollo 8. Burn complete. Our orbit is 169.1 by 60.5. 169.1 by 60.5."

From CapCom: "Apollo 8, this is Houston. Roger, 169.1 by 60.5. . . . Good to hear your voice."

Those orbital parameters—the shape of the orbit in miles—weren't trivial. The flight calculations called for the burn to put Apollo 8 into an orbit of 60 miles by 170 miles. Apollo 8 had flown 240,000 miles, and its computer had fired behind the Moon with such precision that, in the end, the orbit was off by 2,500 feet in one direction and 4,700 feet in the other.

Seconds later the computer itself radioed back its calculation of Apollo 8's precise position in space, which matched exactly what NASA's ground network had calculated.[77]

In Cambridge the room erupted in cheers of joy and relief.

"It brought tears to the eyes," said Alex Kosmala, an MIT programmer. " 'We did that'—it was a staggering thought."

"There was a tremendous amount of relief in the room," said Fred Martin, "a lot of cheering, a lot of satisfaction that what we had produced actually worked."

Said Malcolm Johnston, "Apollo 8 was the most exciting flight, period, end of story."[78]

There was one unabashed moment of patriotism, right at the end of Apollo 8. The astronauts splashed down in the Pacific about 40 minutes before dawn on December 27, and while helicopters from the aircraft carrier USS *Yorktown* hovered over the capsule as it bobbed in the ocean, navy rules said the rescue swimmers couldn't be dropped into the ocean and begin recovering the astronauts until sunrise.

The pilot of one helicopter, Commander Donald S. Jones, talked with the astronauts in their capsule by radio while waiting to put his swimmers in the ocean. He asked whether the Moon was actually made of green cheese. "It's not made of green cheese at all," replied Borman. "It's made out of American cheese."[79]

Six Apollo missions later, on Apollo 14, there was a problem at once so minor and so serious that it perfectly captures the reasons for NASA's obsessive caution about cleanliness and precision manufacturing—a problem not unlike the missing hyphen in the software coding that doomed the Venus probe Mariner 1. Except in the case of Apollo 14, it came out the opposite way: the Apollo computer's nimbleness and some quick and ingenious thinking on the part of an Instrumentation Lab staffer rescued the mission from failure.

The lunar module for Apollo 14 was named *Antares* (after the star), with America's original astronaut, Alan Shepard, aboard as mission commander and Ed Mitchell as LM pilot. After they had undocked from CM *Kitty Hawk*, and just before they were to begin their descent from lunar orbit to the Moon's surface, a NASA engineer in Houston noticed an errant red warning light pop up on his panel: the light indicated that one of the astronauts had pressed the red "abort" button on the lunar module control panel. Nothing had happened because the abort button was a way of returning to orbit, immediately, in case of trouble while the lunar module was landing. Pressing it had all kinds of instantaneous and irreversible consequences: it meant you weren't landing on the Moon. But it only worked during the landing. The Apollo computer was programmed to ignore signals from the button except when the lunar module was descending to the surface.

The code that signaled the computer to abort was a single computer bit. If the bit was set to 0, everything was normal; if the bit was set to 1, the computer was being signaled to abort the descent and fire the ascent engine to return to orbit immediately.

On the computer screen, and then in the raw data coming down from *Antares*, NASA engineers in the backroom of Mission Control could clearly see that specific bit set to the number 1. After a round of consultations, an engineer suggested they have the crew rap on the control panel near the abort button.

First, CapCom Fred Haise, one of the astronauts who survived Apollo 13, had the astronauts check to make sure they hadn't pressed the abort button by accident (an unlikely event), and then press it purposefully and reset it.

The computer bit did not change back from 1 to 0, an ominous sign that something was wrong in the lunar module wiring or the switch itself.

Haise said to Mitchell, the lunar module pilot, "Ed, could you tap the panel around the abort pushbutton and see if we can shake something loose?" Mitchell reached over with a flashlight and tapped the control panel next to the bright red button.

It worked. "You sure tap nicely," Haise said.

The crew resumed preparing for the trip from orbit to the Moon's surface.

But no one was particularly satisfied with that. If the button threw its abort electrical signal while the lunar module was descending, the flight computer would immediately throttle up the engine or separate the ascent stage from the descent stage and fire that engine and send the astronauts right back into orbit.

Thirty-six minutes later, Haise radioed up, "Okay, Ed. That bit just showed up again. Wonder if you could try tapping the panel there by the abort switch again." The tapping worked again, but that was just a sign of how glitchy and dangerous the switch was.

The immediate thought in Houston was that a loose piece of metal solder inside the switch was floating around in zero-g, sometimes closing the contact on the button, as if it had been pushed, sometimes drifting away and reopening the contact.[80]

There was an open line between Houston and Cambridge, and at the Instrumentation Lab, although it was 1:00 in the morning, staff was on hand for the landing maneuvers, including a young programmer named Don Eyles, the same Don Eyles who had written the descent program that guided Apollo 11 (and all the lunar modules) from orbit to the surface of the Moon.

Eyles had stumbled into working for MIT programming the Moon spacecraft quite late. He had graduated from Boston University in 1966 with a degree in math and was on the way back to his apartment from another job interview one day that summer. "I passed by the Instrumentation Lab, and I thought, why don't I go in and investigate?" At that stage, Bill Tindall had just started spending two or three days a week there, the pressure was on, and the IL had begun the hiring that would quickly double the number of people working on software. That very day, Eyles filled out an application and was interviewed by the Intrumentation Lab's personnel officer. The Lab offered him a choice of two jobs: programming a ground-based computer or programming the lunar module's onboard computer.

Eyles picked the lunar module computer, and started working on July 1, 1966. He was 23 years old. While he was a math major, he had never programmed a computer before. He'd been asked at a previous job interview how he would program a computer to alphabetize a list of names. "I pulled a blank," he said. But it didn't matter. "There weren't so many people who already knew how to program a computer. I was going to learn to do that at the same time as I learned how to go to the Moon."

The math and the computer programming made sense to Eyles, and he found the loose work culture of the Instrumentation Lab appealing: demanding in terms of results, open in terms of work style. He ended up writing the 2,000 lines of AGC code that flew the lunar module for about 15 minutes, from low orbit to the Moon landing itself, precisely the code that was running when those alarms went off during Apollo 11's first landing. From never having written a line of code in July 1966, by July 1969 he had written arguably the most important stretch of code in the most important computer in the world.

At about 1:30 in the morning on that Friday, February 5, 1971, with Apollo 14 in orbit around the Moon, a colleague came to Eyles's office to report the trouble aboard *Antares*—that the abort bit on the lunar module was coming and going, and that it could be "deactivated" by tapping on the control panel. Was there a way to bypass the abort switch?

As it happened, Eyles, by now 27, had also written the computer code that monitored the abort switch. He pulled a printout of the code off the shelf and went to work. "It was my own code that I had to defeat," he said.[81]

There was a distinct window for allowing the lunar module to land. Between Mission Control and Cambridge, they had two hours to come up with a solution, test it, get it approved, get it to the astronauts, and allow them to fire their engine to head for a landing. If they missed that window, there would be no landing. In the back of everyone's mind was the knowledge that the previous Apollo mission, Apollo 13, had been a near disaster turned into a fantastic rescue—but without accomplishing its mission. Another failure to land on the Moon might deflate national enthusiasm for doing something that had already been done twice.

In Houston, in Cambridge, and onboard *Antares*, it was clear how dangerous this little intermittent short-circuit was. You simply couldn't start the descent flight to the Moon without knowing if the abort program would be triggered at any moment by the jiggling, flying lunar module. It would have been not only foolish but dangerous.

Without an explicit discussion of the stakes, Shepard and Mitchell knew their machine and their computer well enough to know something bad was afoot. Ninety seconds after being asked to rap on the control panel a second time, Mitchell asked Mission Control, "Do you think we're going to come up with something on this problem with the abort button?"

> ***CapCom:*** Roger. We're working it right now and also MIT's working it. Needless to say, we're busy here, but we think we got a solution.
>
> ***Mitchell:*** Good enough. Something—is it something like a solder ball?
>
> ***CapCom:*** Well, we don't know yet.

The astronauts had figured out what was likely going on inside the errant switch.[82]

In the backroom of Mission Control, Jack Garman was part of the group working on the solution on the NASA side; it was Garman who had given the go for Apollo 11 to land. At this point, he was 26 years old.

"We're animated and talking, 'What about this? What about that?' talking to the Instrumentation Lab. . . . I looked around, and standing behind me were about ten people," said Garman. "Every icon of the space program was standing behind me. I mean all of them." Bob Gilruth, head of the Manned Spacecraft Center. Chris Kraft, head of Mission Control and flight operations. Bill Tindall. "It's like a private turning around and seeing all the four-stars standing behind them or something. It scared the you-know-what out of me, because I woke up that we were in serious trouble at this point."[83]

In Cambridge, Eyles had a flash of insight: the way to fix the problem was to tell the computer that an abort was *already in process,* so if the switch sent its signal, the computer would, effectively, shrug and say, *Thanks, we've already aborted.*

There were three key issues. Because of the slightly quirky way the computer was programmed, explained Eyles, this fix "could be done without actually touching off an abort." But it meant that several of the normal functions of flying to the Moon's surface—throttling the engine, running the right guidance equations, incorporating data from the landing radar—would have to be triggered manually. One brief set of instructions to tell the flight computer an abort had already happened, then, as Eyles put it, several sets of instructions to clean up after that instruction, so *Antares* would actually land on the Moon.

Finally, disabling the abort button deep inside the computer left the astronauts without the use of it; if they needed to abort, they would have to enter the abort program instructions manually into the computer's DSKY, and do so under emergency circumstances, which is why there was a single red button in the first place.

Eyles wrote out the procedures and handed them off to a group of MIT colleagues who had gathered to help. They ran up one floor to try them out on MIT's real-time lunar module simulator. That first time, said Eyles, the lunar module "crashed into the Moon at 600 feet per second"—410 miles per hour. He had forgotten to trigger the descent guidance program. "I added the keystrokes to turn guidance on." On the second run on the IL simulator, the procedure worked perfectly.[84]

Meanwhile *Antares* had slipped behind the Moon on what would be the last orbit before they would either have to land or lose their chance. There was an hour remaining.

Eyles's solution was read to the folks at Mission Control. "As soon as he identified it," Garman said, "everybody went, 'Yep. That's it.'" Houston ran it on their simulator as well.[85]

The procedure was 61 keystrokes on the DSKY, in several distinct groupings, which had to be entered perfectly, at just the right moment.

When *Antares* came around from behind the Moon, CapCom Fred Haise read the procedure up to Mitchell and Shepard, and Mitchell transcribed it by hand.

The astronauts knew enough about the computer and the situation to have confidence in the set of commands, even though they had never used them before. Indeed Mitchell had a doctorate in aeronautics and astronautics from MIT and had taken Battin's course, Astronautical Guidance, in 1964.[86]

Eyles's solution worked. Perfectly. Between disabling the abort switch, reengaging the landing programs, then flying to their landing spot, there wasn't time for much chatter. But six minutes after landing, from their base at Fra Mauro on the Moon, Mitchell told CapCom, "That was really great work you did on that abort problem. . . . We sure appreciate that."

Shepard added immediately, "It really saved the mission."[87]

The success was clear to everyone back on Earth as well. The computer hack made the front page of the *New York Times*. The *Boston Globe* published two front-page stories about the MIT rescue, one from the perspective of the astronauts, the other from the perspective of the Instrumentation Lab staff, headlined, "Apollo Team at MIT Lab Saves Day."[88]

It was a remarkable moment for the Instrumentation Lab and also for computing. The Apollo guidance computer was not only capable of flying a pinpoint-perfect landing to the Moon (for the third time); it was also adaptable and capable enough to solve problems caused by failures in other parts of the spacecraft.

Eyles—longhaired, peace-protesting, with John Lennon wire glasses—was both hippie and skilled mathematician coder. He was quietly asked if he would accept a congratulatory invitation to visit the White House, but after spending a night imagining a polite anti–Vietnam War speech he might deliver while shaking Richard Nixon's hand, he declined. Not long after Apollo 14 returned to Earth, *Rolling Stone* magazine sent Tim Crouse to profile Eyles. He's pictured with his hand on the abort button on MIT's lunar module simulator. The headline: "EXTRA! Weird-Looking Freak Saves Apollo 14!"[89]

---

On each piece of navigation and guidance equipment that flew the Apollo spacecraft to the Moon, a small silver-and-black identification tag was riveted into place. It had blanks to be filled in to identify the type of equipment, the part number, the serial number, the manufacturer. Along the top it said, "Apollo G&N System." Two other things were etched on every tag. The largest item was the original *Jetsons*-style 1959 NASA logo. And in small type along the bottom appeared this line: "Designed by MIT Instrumentation Lab."[90]

You don't find design credits on most industrial equipment. The MIT Instrumentation Lab was justifiably proud of its contribution to Apollo, and no matter what happens to the details of the written history, the Instrumentation Lab's role is permanently etched onto the artifacts themselves. One hundred years from now, or 250 years from now, a historian or curator turning over the equipment from that first leap of human beings from the Earth to another planetary body will see the evidence of MIT's role quite clearly.

The Apollo guidance computer never failed during the 11 Apollo flights with astronauts. In 2,502 hours of spaceflight—104 days and 6 hours—there was not a single computer hardware failure. There was not a single software glitch. As David Hoag's 1983 history puts it, the AGC performed better than "any other computer designed then or since for aerospace application. Such near perfect reliability was achieved at considerable effort [and] attention to design."[91]

And achieved, also, with the full knowledge of the ghost of Mariner 1's missing hyphen—that a single small error, even an unintentional slip, could trigger a cascade of consequences. "There was no second chance," said Margaret Hamilton. "We knew that."[92] The floating ball of solder inside the otherwise ordinary switch in the control panel of lunar module *Antares* was exactly that: a tiny flaw in manufacturing and quality control that could have thwarted an entire Moon mission.

It was more than that too, of course. MIT's Apollo computer and guidance system got the astronauts to the Moon and back with grace and precision. On Launchpad 39-A at Cape Kennedy, the full-up Apollo stack—Saturn V rocket, lunar module, command module, astronauts, and a full load of fuel—weighed 6.5 million pounds, as much as a U.S. World War II destroyer. In that 6.5 million pounds of weight, the Apollo computer weighed 70 pounds, the DSKY weighed 17.5 pounds. With two computers and three DSKYs, computer equipment accounted for 192.5 pounds (as it happens, about what a beefy fourth crew member might have weighed). And yet Apollo and America flew to the Moon as much on MIT's rope-core memories and computer chips as it did on the Saturn V's enormous engines. And Apollo couldn't have landed on the Moon, and returned safely, without the computer, any more than it could have flown to the Moon without the rocket engines.

George Low, who ran the Apollo Spacecraft Project Office in Houston before becoming deputy administrator at NASA right after the first two Moon landings, said, "If you had to single out one subsystem as being most important, most complex, and yet most demanding in performance and precision, it would be guidance and navigation."[93]

The astronauts who had been so skeptical of the computer at the very beginning came to have respect, and even a certain affection, for the computer as they learned how sophisticated it was, and also how underpowered for the job it did. "By the time we got around to the missions, there just wasn't enough space in that computer," said Apollo 14's Ed Mitchell. "It was a masterful programming job to get all the functions that had to be done—the guidance and control, orienting telescopes and the [gyroscopes]—and still having the descent programs and everything we had. . . . Absolutely a masterful job. I don't know how the guys at Instrumentation Lab ever made it work."[94]

On that first pioneering flight, Apollo 8's Christmas journey to the Moon and back, the computer's performance was so good it provoked a memo from Tindall and a change of procedure. After Apollo 8's return, Tindall wrote Jerry Hammack, NASA's head of recovery operations:

> Jerry, I've done a lot of joking about the spacecraft hitting the aircraft carrier, but the more I think about it the less I feel it is a joke. There are reports that the [Apollo 8] command module came down right over the aircraft carrier, and drifted on its chutes to land only 4,572 meters [2.8 miles] away. This really strikes me as being too close. The consequence of the spacecraft hitting the carrier is truly catastrophic. I seriously recommend relocating the recovery force at least 5 to 10 miles from the target point.[95]

Apollo 8 had flown 504,006 miles and landed just 1.6 miles from its target point in the Pacific Ocean,[96] so close that the man in charge of all that navigation wanted the U.S. Navy to stand off its aircraft carrier so that future spaceships didn't accidentally land right on the flight deck.

Despite his public anonymity—at his death, on November 20, 1995, not one newspaper in the country ran an obituary on Bill Tindall—those inside NASA were fully aware that Tindall had contributed as much, perhaps, as Doc Draper himself to the ultimate success of MIT's computer, and to getting to the Moon. Said Chris Kraft, who by the end of Apollo was the head of the Manned Spacecraft Center and who first sent Tindall to MIT, "It would be difficult for me to find anyone who contributed more individually to the success of Apollo than Bill Tindall."[97] Gene Kranz, who became director of flight operations and was the flight director for Apollo 11, a legendary and commanding figure, said of Tindall, "He was one of the great pioneers of manned spaceflight."

Tindall was happy in the background. When he was running a meeting, when he was writing a pointed Tindallgram, he was happy to be the pivot, the center of the action, but not the center of attention. For that first Moon landing, except for Neil Armstrong, Kranz was the commander of the operation, including the final descent to the Moon's surface. Although Tindall had planned every second and knew the mission as well as anyone, he would not normally have been inside Mission Control. He planned missions, but he didn't run them. For Apollo 11's final ride to the Moon, Kranz wanted Tindall with him, a quiet but vivid acknowledgment of Tindall's role. "I asked him to sit next to me at the console for the lunar landing," said Kranz. Tindall declined. "But I persisted in my request and he finally agreed."[98]

And so as the drama of Apollo 11's landing unfolded, Tindall was at Kranz's elbow, witnessing every moment of what he'd help make possible.

As the *Eagle* headed for the surface, as the lunar module's master alarm triggered off, as Mission Control relied on the guidance provided by the 25-year-old Jack Garman, the *Eagle*'s computer was being overloaded with tasks, dumping the work it concluded was low priority, and then was restarting itself and resuming flying the lunar module to the surface of the Moon. Twice the astronauts' computer display went blank as the computer cycled itself. One of the reasons Garman, Kranz, and the rest of Mission Control were confident in moving forward is that despite the tumult, the *Eagle* stayed perfectly on course.

That Kranz and the team in Mission Control were able to make decisions to press forward, with confidence, is a tribute to their own training and experience. The folks in Cambridge were listening on the edge of their seats, but the *Eagle* was flying down too fast and too close to the surface for them to contribute anything.

What had caused the alarms—which NASA and MIT were able

to figure out over the course of just the next half-day so Aldrin and Armstrong could take off with confidence—was complicated and subtle. The lunar module's rendezvous radar had been sending streams of errant signals to the computer, even though it wasn't used while landing; it was turned on when leaving the Moon, to find the command module and navigate to it. But because of a quirky and uncorrected electrical problem deep inside the wiring of the lunar module, the rendezvous radar was pouring signals into the computer as the *Eagle* headed for landing, overloading it. The computer knew those signals were irrelevant to landing on the Moon. It dumped them, wiped itself clean, started again, and sounded the alarm. The alarms, in fact, were informational: the computer wasn't in trouble; it was just telling the astronauts and Mission Control what it was doing (and raising a flag that some other part of the lunar module was doing something odd). That's why Garman kept saying with such confidence, *GO! Same alarm! GO!*

Even the blank displays, as unsettling as they would prove to be in retrospect, were a sign of the same thing. The computer missed a couple cycles of "refreshing" the display because in a computer as tight for power and memory as the Apollo computer, the display was a lot less important than the lunar module's engine and guidance equations. All of which worked flawlessly, in their automated mode, and as Armstrong himself took control.[99]

One thing is certain: any other computer, at that moment in that era, would have choked. Indeed even a computer designed for landing on the Moon, but designed with less forethought and resilience, would have choked. The 12-minute flight from 50,000 feet in lunar orbit to landing was the most demanding moment for the Apollo computer in the whole journey from Launchpad 39-A to the Pacific Ocean. And in that very moment, on that very first Moon landing, a small flaw in the wiring harness of the lunar module overloaded the computer that was controlling that landing, and the computer was able to perfectly pick the work that absolutely had to be done, and also restart itself so it didn't freeze. Both of the Apollo computer's most innovative qualities—the decision-making ability and the ability to restart itself in the middle of its work and resume working without a stumble—turned out to be absolutely essential and to work perfectly.

Jack Garman, Gene Kranz, Buzz Aldrin, and Neil Armstrong made decisions based on what the Apollo flight computer was telling them—decisions that held the fate of Aldrin and Armstrong, and also the fate of this most dramatic moment in human history, with half a billion people watching live. And the Apollo computer didn't hesitate, and it didn't disappoint.

# 6

# JFK's Secret Space Tapes

Anybody who would spend $40 billion in a race to the Moon for national prestige is nuts.

> **Former president Dwight D. Eisenhower**
> *who created NASA, on the Apollo program, June 12, 1963*

I'm not that interested in space.

> **President John Kennedy**
> *in a Cabinet Room meeting with NASA officials, November 21, 1962*[1]

The Apollo 11 spaceship that carried Michael Collins, Buzz Aldrin, and Neil Armstrong from the Earth to the Moon was big: the command and service module and the lunar module, docked nose-to-nose as they flew to the Moon, were 53 feet long, from the business end of the service module's engine to the round footpads at the bottom of each of the lunar module's four legs. When Collins fired the service module engine to settle into orbit around the Moon—the big engine ran for 357.5 seconds to slow the ship enough to get into orbit, six long minutes—there was already another spaceship in orbit around the Moon waiting for them. It had arrived two days earlier, from the Soviet Union.

Luna 15 was a Russian robotic craft that was at the Moon on a mysterious mission. It was certainly no coincidence that at the moment when the United States was getting ready to land people on the Moon's surface, with the whole world watching, the Russians had decided they too ought to have a spacecraft at the Moon. Luna 15 had been launched on Sunday, July 13, before the Wednesday launch of Apollo 11, and the Russians said it was simply going to "conduct further scientific exploration of the Moon and space near the Moon."

But from the moment of Luna 15's launch, U.S. space scientists and NASA officials speculated that it was a "scooping" mission, designed to land on the Moon, extend a robotic arm, scoop up some soil and rocks, and deposit them in a compartment on the spacecraft, which would then zoom back to Earth, bringing home Moon rocks just like Apollo 11 would, and maybe, just maybe, arriving back on Russian soil with its Moon rock cargo before the Apollo 11 astronauts could make it home.[2]

Frank Borman, the commander of the Apollo 8 mission that had orbited the Moon, had just returned from a nine-day goodwill tour of Russia—the first visit by a U.S. astronaut to the Soviet Union—and appeared on the NBC news show *Meet the Press* the morning of Luna 15's launch. "I would guess it's probably an effort" to bring back a soil sample, Borman said. "I heard references to that effect [in Russia]."[3]

NASA, at least publicly, was mostly concerned that Russian communications with Luna 15 might interfere with Apollo 11. In an unprecedented move, Chris Kraft asked Borman to call Soviet contacts from his just-finished trip and see if they would supply data on Luna 15. The Russians promptly sent a telegram—one copy to the White House, one copy to Borman's home near the Manned Spacecraft Center—with details of Luna 15's orbit and assurances that if the spacecraft changed orbits, fresh telegrams would follow. It was the first time in the 12 years of space travel that the world's two space programs had communicated directly with each other about spaceflights while they were being flown. At a press conference, Kraft said Luna 15 and the Apollo spacecraft would not come anywhere near each other. He observed that Armstrong, Aldrin, and Collins would have neither opportunity nor time to look out the window in search of their fellow spacecraft.

Luna 15, at least to start, succeeded in making sure the Soviet Union's space program wasn't overlooked while Apollo 11 dominated the news worldwide. The Luna 15 mission made the front pages of newspapers around the world. On July 19, 1969, the third day of the Apollo 11 mission, the *New York Times* wrote four stories about Luna 15 and published the full text of the telegram from the Russians; two of those stories were on the front page, including the lead story for the day: "Moscow Says That Luna 15 Won't Be in Apollo's Way." (That day there were only four other stories about Apollo.)[4]

At the time, NASA and the public never did find out what Luna 15 was up to. Now we know it was a well-planned effort to upstage Apollo 11, or at least be onstage alongside the U.S. Moon landing, according to documents released and research done since the breakup of the Soviet Union and thanks to the rich and detailed history of the Soviet space program written by historian Asif Siddiqi, *Challenge to Apollo*.

By 1969 the Russian manned space program had fallen well behind the determined, even relentless effort of the United States. But the Soviet robotic program remained ambitious. Pricked in part by the worldwide success and acclaim of Apollo 8 and by the daunting expectation that the Americans would land astronauts on the Moon by the summer of 1969, Russian space scientists had assembled five identical Moon probes. They were designed to fly to the Moon, land, then drill a foot into the Moon's surface to obtain a sample of soil uncontaminated by the lander. This would be sent back to Earth in a small upper stage that would blast off from the Moon, delivering its sample to the ground back home by parachute. No one in Russia was under the impression this would match a landing of American astronauts, but if they could do it before Apollo 11, what had for half a decade been the world's premier space program would be able to retain a certain pride, and also lay claim to the scientific breakthrough of having brought samples of the Moon back to Earth first. As Siddiqi points out, it was not lost on the Russians that their lander would have all the more potency if, for some reason, Apollo 11 didn't succeed.

The first Russian Moon scooper was launched, without public announcement, on June 14, 1969; the fourth stage of its booster rocket failed to ignite, and the probe landed in the Pacific Ocean.

The second attempt—like the first, its timing dictated not just by competition but by the fixed launch windows related to getting a spaceship to the Moon from the Soviet Union—was Luna 15.

When Luna 15 arrived in lunar orbit on July 17, two days ahead of Apollo 11, Siddiqi says, Russian space officials were surprised "by the ruggedness of the lunar terrain" where it was headed, and that the craft's altimeter "showed wildly varying readings for the projected landing area." As Armstrong and Aldrin stepped out onto the lunar surface, Luna 15 was still swooping around the Moon, her engineers back in the Soviet Union trying to find a landing site they had confidence in.

Two hours before the *Eagle*, with Armstrong and Aldrin aboard, blasted off the Moon, Luna 15 fired its retrorockets and aimed for touchdown. The British radio telescope at Jodrell Bank Observatory, presided over by Sir Bernard Lovell, was listening in real time to the transmissions of both Apollo 11 and Luna 15. And Jodrell Bank was the first to report the fate of Luna 15. Its radio signals ended abruptly. "If we don't get any more signals," said Lovell, "we will assume it crash-landed." Luna 15 was aiming for a site in the Sea of Crises, about 540 miles northeast of *Eagle*'s spot in the Sea of Tranquility, the distance from Atlanta to Richmond.

The Soviet news agency Tass reported that Luna 15 had fired its retrorockets and "left orbit and reached the Moon's surface in the preset area." Its "program of research . . . was completed."

Despite taking almost a whole extra day to figure out the terrain issues, Soviet space scientists apparently missed a mountain in the Sea of Crises. On its way to the "preset area," Luna 15 slammed into the side of that lunar mountain, going 300 miles per hour.[5]

Wrote Siddiqi, "There was one small irony to the whole mission. Even if there had not been a critical eighteen-hour delay in attempting a landing, and even if Luna 15 had landed, collected a soil sample, and safely returned to Earth, its small return capsule would have touched down on Soviet territory two hours and four minutes *after* the splashdown of Apollo 11. The race had, in fact, been over before it had begun."[6]

At about 1:15 p.m. ET Tuesday, the Apollo astronauts woke from a 10-hour rest period; Armstrong and Aldrin had blasted off from the Moon about 24 hours earlier, and the three Apollo 11 astronauts were 12 hours into their 60-hour ride home. As they got started on their day, CapCom Bruce McCandless radioed, "Apollo 11, this is Houston. If you're not busy now, I can read you up the morning news." Replied Aldrin, "Okay, we're all listening."

A lot of the news was about Apollo 11. Reported McCandless, "Things have been relatively quiet recently in Vietnam. GIs on patrol were observed carrying transistor radios tuned to your flight." About one-third of the way through McCandless's space newscast, slipped in between telling the astronauts that President Nixon would head to Romania after meeting them onboard their recovery aircraft carrier and the Vietnam news, McCandless reported, "Luna 15 is believed to have crashed into the Sea of Crises yesterday after orbiting the Moon 52 times."[7]

If ever there was a moment that captured the crushing reversal in the performance of the world's two space programs of the previous decade, that was it: Mission Control, Houston, matter-of-factly reporting the crash-landing of the Soviet Union's somewhat flailing attempt to use a robotic probe to bring home Moon rocks, to the three astronauts who were flying home from the first landing on the Moon, with 47.5 pounds of Moon rocks.[8]

---

Apollo became a narrative thread, a dramatic story line—the astronauts, the Saturn V, the lunar module, astronaut wives and astronaut food—braided into the other narratives of the 1960s: the Vietnam War, the civil rights movement, the assassinations of Kennedy and King and the political unrest that followed, the Cold War, rock 'n' roll, the sexual revolution, the decision of LBJ not to run for president again, the victory of Richard Nixon, who had been president just six months when Armstrong and Aldrin walked on the Moon and made a call from the Oval Office right into their spacesuit helmets. The decade had begun with the cultural hangover of the 1950s, but with Jack and Jackie Kennedy in the White House. It ended with a revolution in every part of American popular culture, but with Richard and Pat Nixon in the White House.

Going to the Moon had a momentum all its own, especially in that last dash to make President Kennedy's end-of-the-decade deadline. From the moment on Christmas Eve when the Apollo 8 astronauts read to the world from the book of Genesis while orbiting the Moon, to the moment of Armstrong's one small step, just seven months passed. America flew three nearly flawless Apollo missions in that time: Apollo 9, the first test of the lunar module with astronauts, flying in space, just in Earth orbit; Apollo 10, whose lunar module (call sign: *Snoopy*) flew to within 47,000 feet of the Moon's surface before returning; and then Apollo 11. Between Apollo 8 and Apollo 9, 66 days. Between Apollo 9 and Apollo 10, 66 days. Between Apollo 10 and Apollo 11, 51 days. The pace was breathtaking. And before 1969 was out, Apollo 12 would also land on the Moon and return safely.

In some ways Apollo separated itself from the rivalry that inspired it. People listening to Genesis read aloud from space weren't thinking about the Cold War, and neither were people watching Armstrong and Aldrin bouncing around on the Moon in front of their lunar module.

But as Luna 15's desperate leap to the Sea of Crises reminds us, politics shadowed Apollo right through the 1960s to the Sea of Tranquility.

The excitement of Shepard's first flight, the stirring call to reach for the Moon from President Kennedy, proved surprisingly gossamery. With both a clear mission and a deeply experienced boss, NASA's performance stabilized and regularized. Eleven weeks after Shepard's cannonball arc from Cape Canaveral into the Atlantic, Gus Grissom flew the same path, 16 minutes of rocket flight, just edging into space for 5 minutes of weightlessness. But there was no White House meeting for Grissom, no convertible ride up Pennsylvania Avenue to the Capitol through a crowd of 250,000 people. Grissom got his Distinguished Service Medal from NASA administrator Jim Webb, not from President Kennedy. Grissom was, after all, the second U.S. astronaut into space.[9]

When it came to public events, Americans' attention spans were no longer in the 1960s than they are today. We were no more inclined toward the virtues of slow-and-steady progress, no more capable of delayed gratification. Just over a year into the enterprise that would take at least six or seven years, with the billions of dollars just starting to flow, without any particular space spectaculars on the horizon, the thrill of spaceflight, adventure, and conquering the next frontier had faded. More than that, there were prominent public voices stoking skepticism and dissent.

The week after Grissom's flight, U.S. Senator Paul H. Douglas released his own poll, not of the American people but of U.S. space scientists. The question: Was sending astronauts to the Moon, "at the earliest feasible moment," of great scientific value? Douglas had arranged to poll the membership of the American Astronomical Society, and received 361 written replies from astronomers and space scientists. Of those, 36 percent said a manned Moon mission had "great scientific value," and 35 percent said it had "little scientific value." And unmanned, robotic missions to the Moon? Sixty-six percent of space scientists said they would have "great scientific value." Douglas was a member of Kennedy's own party, a liberal Democrat, and he had gone to some trouble to establish that America's actual space scientists judged that the race to the Moon wasn't worth it. "If the astronomers are not competent [to decide]," asked Douglas, "who is?"[10]

In January 1962 the *New York Times* published an editorial pointing out that "the grand total for the Moon excursion would reproduce from 75 to 120 universities about the size of Harvard, with some [money] left over"—a Moon landing, or a Harvard University for every state? In August 1962 the *Times* pointed out in another editorial ("To the Moon") that while "many will be filled with wonder and admiration for the boldness" of Project Apollo, "it is not curiosity that has set the timetable, and hence the annual cost, of Apollo." Not curiosity, but the Cold War.

In September 1962 the *Times* asked somewhat churlishly, "In view of the country's late start, is it wise to stake so much of the nation's prestige on the proposition that an American will reach the Moon first? . . . There are manned landings on Mars and Venus, for example, to shoot at too." And returning to the theme of building a Harvard in every state, the *Times* noted, "This country still has tremendous need for more schools, more hospitals, more housing. . . . The American people must never lose sight of the fact that, while ardently competing with the Soviet Union in space, a concurrent goal is to create a better world here on Earth."[11]

Norbert Wiener, a professor and highly regarded mathematician at MIT, dismissed Apollo in a late 1961 interview as a "moondoggle," a word the press and NASA critics loved; through the end of 1961 and into 1962, "moondoggle" started to pop up regularly in coverage of the space program, particularly in stories about spending and in editorials.

(By 1964 a young sociology professor at Columbia University, Amitai Etzioni, had written a book called *The Moon-Doggle*, which was reviewed by the space correspondent of the *New York Times.*)[12]

In 1962 John F. Kennedy was unbending on the critical importance of the race to the Moon—at least in public. At a half dozen press conferences where space came up, he combined his tireless agreement that the United States was indeed in second place in space with his insistence that we were catching up. In February he said, "We are making maximum effort." In June he said, "I think the American people have supported the effort in space, realizing its significance, and also that it involves a great many possibilities in the future which are still unknown to us. . . . I don't think that the United States is first yet in space, but I think a major effort is being made."

In August 1962 the Russians launched two cosmonauts, in separate spaceships, within 24 hours of each other, the double mission totaling seven days in space at a moment when the total for all four American spaceflights was 11 hours. Kennedy was asked why Americans shouldn't be pessimistic since they were not just second to the Soviets but "now a poor second." "We are behind and we are going to be behind for a while," he replied. "But I believe that before the end of this decade is out, the United States will be ahead. . . . This year we submitted a space budget which was greater than the combined eight space budgets of the previous eight years. So this country is making a vast effort."[13]

The press conference comments were defensive and reflexive—a hockey goalie swatting the puck away from the net. There was no eloquence about space in them, the responses more dutiful than enthusiastic.

In the fall of 1962, though, Kennedy did a two-day tour of space facilities to see for himself how the Moon program was taking shape. It was a brisk visit—just two and a half hours total at Cape Canaveral, for instance—but it was immersive. Kennedy took Vice President Johnson with him, and for security, they flew on separate planes. The entourage included NASA chief Webb, NASA second-in-command Bob Seamans, Defense Secretary Robert McNamara, budget chief David Bell, and a handful of key senators and House members from budget and space committees. A third plane carried the press.

Huntsville, Alabama, home to Wernher von Braun's rocket team, was the first stop. Von Braun showed the president a model of the Saturn rocket that would eventually launch astronauts to the Moon. (The Saturn V and its engines were still under development.) "This is the vehicle which is designed to fulfill your promise to put a man on the Moon by the end of the decade," von Braun told Kennedy. He paused, then added, "By God, we'll do it!"

Von Braun took Kennedy to the firing of a Saturn C-1 rocket as a demonstration of the coming power of American rocketry. The test—eight engines firing simultaneously, roaring red-orange rocket thrust out of a test stand, with Kennedy, von Braun, and the visiting party in a viewing bunker less than a half-mile away—shook the ground and sent shockwaves across the Alabama test facility. When the engines stilled, Kennedy turned with a wide grin to von Braun and grabbed his hand in congratulations. The president was apparently so captivated by von Braun's running commentary that he took the rocket scientist—the biggest U.S. space personality outside the astronauts themselves—on the plane with him to Cape Canaveral.

At the Cape, Kennedy visited four launchpads, including one where he got a guided tour from Wally Schirra of the Atlas rocket and Mercury capsule Schirra was set to ride into orbit in about two weeks.[14]

The three planes ended the day in Houston, where Kennedy's popularity was on vivid display. The city's police chief said 200,000 people—more than one in every five residents of Houston at the time—had come out to see the president, who rode in an open car from the airport to his hotel. Kennedy spent part of the next day at NASA's temporary Houston facilities—the space center itself was under construction—including seeing a very early mock-up of the lunar module, then called "the bug." But the emotional and political climax of Kennedy's tour came Wednesday morning at the Rice University football stadium. In the blazing early-morning Texas heat—already 89° at 10 a.m., with Kennedy and his party wearing dress shirts, coats, and ties—the president gave a speech designed to lift the space program up out of the political squabbles and budget bickering that was starting to beset it, as if going to the Moon were just like any other government program.[15]

Kennedy's speech that morning was just 18 minutes long, but it started with a tour through all of human development, emphasizing just how recently humans had figured out how to invent technology to make their lives better, at least compared to the long stretch of history. He began with humans learning to use the skins of animals for clothing, and then in half a minute he swept from the wheel, the printing press, and the steam engine to penicillin and nuclear power.[16]

"This is a breathtaking pace," Kennedy said. "So it is not surprising that some would have us stay where we are a little longer to rest, to wait."

Kennedy was having none of it, and he was certain the audience wasn't either. "The United States was not built by those who waited and rested," he said. "This country was conquered by those who moved forward—and so will space."

That morning Kennedy was connecting his vision of space to three powerful forces of human nature and American character, delivered with a particularly impressive demonstration of his rhetorical power. First was the irresistible tide of discovery and new human knowledge. If his tour through invention showed anything, "it is that man, in his quest for knowledge and progress, is determined and cannot be deterred. The exploration of space will go ahead, whether we join in it or not, and it is one of the great adventures of all time."

The second force at work, no less irresistible, was the very character of Americans. "Those who came before us made certain that this country rode the first waves of the industrial revolutions, the first waves of modern invention, and the first wave of nuclear power—and this generation does not intend to founder in the backwash of the coming age of space. We mean to be a part of it—we mean to lead it."

Since its own creation, Kennedy was saying, the United States had led the world in innovation—not least the innovation that secured victory in World War II, the cracking of the atom. (Kennedy mentions nuclear science three times as an example of great technological innovation.) He did not intend to preside over a generation in which that trait of the American character would fade, and he was certain Americans today, Americans right now, would not want to be the first generation to falter.

Space was the frontier of science, technology, discovery, and adventure. The exploration of space would be driven by innovation and discovery, and would in turn create innovation and discovery. Space was America's manifest destiny, just as the continent itself had been. Space technology was just in its infancy, Kennedy said, but the U.S. space effort would be driven, in part, from Houston—so much so, Kennedy predicted, that "Houston" would become synonymous with space travel. "What was once the furthest outpost of the old frontier of the West," Kennedy told the crowd of 40,000 Texans, "will be the furthest outpost of the new frontier of science and space."

Kennedy did talk about costs—when it came to space, money was always on his mind—and also dangers. NASA's budget had tripled just from 1961 to 1962, "a staggering sum, though somewhat less than we pay for cigarettes and cigars every year." If we can afford our Marlboros and Parliaments, we can afford our rockets and spaceships.

Space was not just about human progress and American destiny. There was an urgency about the project, because space was also about human freedom, or human tragedy. The choice would be ours. "For space science, like nuclear science and all technology, has no conscience of its own. Whether it will become a force for good or ill depends on man, and only if the United States occupies a position of pre-eminence can we help decide whether this new ocean will be a sea of peace or a new terrifying theater of war." America had vowed that space would not be "governed by a hostile flag of conquest, but by a banner of freedom and peace"; space would not be "filled with weapons of mass destruction, but with instruments of knowledge and understanding." Those vows, Kennedy said, "can only be fulfilled if we in this nation are first, and therefore we intend to be first."

We must do it, and we must do it first, and to do that, Kennedy said, "we must be bold."

Inventive and courageous, determined and free and bold. It sounded like the speech could have been written while watching TV images of America's first astronauts, in their gleaming spacesuits, striding to their rockets, waving and grinning through their clear space helmets.

Space didn't just create the opportunity for these things, Kennedy said—the opportunity for knowledge and adventure, the opportunity for American destiny and American values. It created an obligation to do them, the responsibility to reach for the Moon, and to reach beyond.

That is the point of the most famous passage of the Rice University speech: "We choose to go to the Moon. We choose to go to the Moon. . . . We choose to go to the Moon, in this decade, and do the other things, not because they are easy, but because they are hard, because that goal will serve to organize and measure the best of our energies and skills, because that challenge is one that we are willing to accept, one we are unwilling to postpone, and one which we intend to win, and the others, too."

We choose to go to the Moon because we are up to the challenge.

We choose to go to the Moon for the knowledge that exploration will unleash, the knowledge that will create the future, as those bold enough to bequeath us the printing press and the airplane created the world in which we live.

We choose to go to the Moon because only America can lead the world into space in the name of both freedom and peace.

Kennedy said "We choose to go to the Moon" three times; he was fortuitously interrupted by applause after the first time because of a well-received joke about Texas sports teams he had just delivered, and he clearly felt the need to repeat himself to take advantage of the momentum from the audience and to pull their attention back for his most important and most eloquent moment.

But in saying it three times, what Kennedy was really saying was this: We have no choice about whether to go to the Moon. We must go to the Moon. It is who we are—it is who you, hearing this speech, are. Reaching for the Moon isn't just the thing Americans are doing; it is the American thing to do.

"It will be done," Kennedy declared in closing, stepping back from the lectern, looking from one side of the audience to the other and slapping one hand against the other so there was no mistaking his determination. "It will be done before the end of this decade."

Kennedy makes only one passing reference at Rice to "the coming age of space." The Space Age had started that Friday night five years earlier, on October 4, 1957, its arrival signaled by the beep-beep-beeping of Sputnik streaking through the sky overhead. But if the Russians managed to set the Space Age in motion, it was Kennedy who defined the Space Age that day in Houston, unfurling an almost irresistible vision of the value and power of space exploration. Riding rockets was an inevitable next step in human development, a necessary next step in human society, as inevitable and necessary as the discovery of fire.

---

The Rice speech took place on September 12, 1962. Ten weeks later, on November 21, in the Cabinet Room, Kennedy presided over a meeting about America's space program with a very different tone. It was fractious and frustrating, driven by the president's own impatience. He didn't like the pace the U.S. space program was moving at; he didn't like what it was costing; and he didn't like the answers he was getting from the people gathered around the table with him, including the three most senior people at NASA, along with the man then charged with managing Apollo specifically.

Ostensibly the occasion for the meeting was to hash out whether NASA and Kennedy were going to push Congress for an extra $400 million for Apollo before the next budget cycle. Not even the NASA people agreed about the wisdom of that.

But Kennedy used the meeting to press hard on Webb and his lieutenants about the schedule and about their sense of urgency, and it quickly became clear that the president and the people running his space program didn't see things the same way.

The poetry of the Rice speech, even the vision of a space future from the speech, is nowhere to be found in the Cabinet Room that Wednesday. We know this because Kennedy had a secret taping system installed in the White House, as FDR had, as LBJ would, as Nixon, most famously, would. Kennedy's system, installed by the Secret Service in July 1962, was quite elaborate. There were microphones in both the Oval Office and the Cabinet Room. Unlike Nixon's system, which was voice-activated, the Kennedy system had to be turned on; Kennedy himself decided which meetings and phone calls to record. In the Oval Office there were three concealed buttons for the system: on Kennedy's desk, in a bookend near his favorite chair, and in a coffee table. In the Cabinet Room there was one button on the table near the chair Kennedy customarily took. The buttons turned on a red light at the desk of Kennedy's secretary, Evelyn Lincoln; when Kennedy turned on the red light, she turned on the taping system: reel-to-reel recorders for the meetings, a Dictaphone for the phone calls. In the course of 16 months, Kennedy recorded 325 meetings and 275 phone calls, about 9 a week. Lincoln said Kennedy recorded the meetings with a sense of history in mind. "He never once went back and listened to one," she told the *Washington Post* at the time details of the tapes were revealed. The system was removed by the Secret Service the day Kennedy was killed.

Knowledge of the taping system in the Kennedy White House was so closely held that when the stunning news first broke that Nixon had recorded hundreds of Oval Office conversations, close aides to Kennedy were among those expressing outrage. His appointments secretary David Powers insisted he would know if any such system had existed in the JFK White House, and the historian and presidential aide Arthur Schlesinger called the idea that Kennedy might have had a similar system "inconceivable." It was those comments, in part, that convinced the Kennedy Presidential Library to come forward and reveal the existence of Kennedy's system right after the revelation about the Nixon tapes. The Kennedy tapes themselves were not released for decades.[17]

In those hundreds of meetings were two high-level conversations about space that reveal a very different Kennedy attitude about the race to the Moon.

Just 10 weeks after his Rice University speech, Kennedy spent 30 minutes asking questions about NASA's budget and spending, trying to get to the bottom of the schedule. "Gemini has slipped how much?" he asked.[18]

To much laughter—there were nine people in the meeting besides the president, four of them space agency people all too familiar with countdowns and launches that frequently slipped—Webb responded, "This word 'slip' is the wrong word." To which Kennedy says, "I'm sorry, I'll pick another word."

At that point NASA chief Jim Webb had been telling Kennedy that a Moon landing was possible in late 1967, but was more likely in 1968. Kennedy wanted it sooner. How do you move it back into 1967? Would the $400 million they were there to discuss do that? How about early 1967? What would that take? Kennedy seemed puzzled that more money wouldn't necessarily make it happen sooner.

There is a long exchange in which Kennedy tries to understand why getting $400 million extra right now would help Gemini but wasn't likely to move Apollo any sooner. He didn't understand the details of staged technology development, that you have to build and fly Gemini in part to help you make the right decisions about Apollo. Four months here or there over four years is hard to nail down.

Kennedy was impatient with both the conversation and the inability to get to the Moon quicker. Bob Seamans, one of Webb's two deputies, sensed Kennedy's misunderstanding and stepped into the conversation.

"You have to understand what these dates really are," he said. "These are dates for the internal management of the projects. They have to be dates that people believe are realistic. I mean, you have to have a fighting chance to achieve these dates, but they're by no means dates that you can absolutely guarantee at this time." Apollo "is a development program, and you are learning as you go along, and if you crank up too much of a crash program and you start running into trouble, it can take more time to unsort the difficulties than if it is a better paced program."

With something as complicated and untested as space travel, the best way to go fast, in fact, is to go slow—or at least methodically.

Thirty minutes into the conversation, the president takes a step back. "Do you think this program is the top-priority program of the agency?" Kennedy asked Webb.

"No sir, I do not," Webb answered without hesitation.[19] "I think it is *one* of the top priority programs, but I think it's very important to recognize here—" Webb started to explain the importance of some of NASA's non-Moon programs. Kennedy lowered his voice and simply stepped into Webb's conversational stream.

"Jim, I think it is the top priority. I think we ought to have that very clear.

"Some of these other programs can slip six months, or nine months, and nothing in particular is going to happen. . . .

"This is, whether we like it or not, in a sense *a race*. If we get second to the Moon, it's nice, but it's like being second any time. So that if we're second by six months, because we didn't give it the kind of priority—then of course that would be very serious.

"So I think we have to take the view, this is the top priority."

Webb, utterly unpersuaded, jumped right back in. "But the environment of space is where you're going to operate the Apollo, and where you're going to do the landing." Webb seemed baffled by Kennedy's insistence that going to the Moon—alone—was the most important thing NASA should be doing.

"But I know all the other things," Kennedy said, interrupting Webb again, "the satellites, the communications, and weather and all—they are desirable but they can wait."

At this point, neither the president nor the NASA chief was listening to the other. Webb tried again: "I am talking now about the scientific program to understand the space environment within which you got to fly Apollo and make a landing on the Moon." What he means is, you can't fly to the Moon without figuring out how to fly to the Moon. That's not some ancillary program—it's at the heart of getting to the Moon. But Webb was also talking about the "environment" of space, which clearly sounded to Kennedy like some kind of basic science research that didn't have to do with aiming a rocket at the Moon.

Kennedy had started out calm, but at this point he was irritated. He tried again: "Wait a minute. Is that saying that the lunar program to land a man on the Moon is the top priority of the Agency? Is it?"

There was a moment of silence in the room.

An unidentified voice jumped in: "And the science that goes with it."

Another voice seconded that: "The science that is necessary—"

For the fourth time the president became insistent: "Going to the Moon is the top priority project. Now, there are a lot of related scientific information and developments that will come from that which are important. But the whole thrust of the Agency, in my opinion, is the *lunar* program. The rest of it can wait six or nine months."

It's really not clear why Webb, or Seamans, didn't simply agree with the president. He didn't expect NASA astronauts to click their spacesuit boots together and magically transport to the Moon.

In fact Webb, who seemed to be seriously misreading Kennedy's mood and intention, then launched into a soliloquy on the virtues of a wide-ranging space program, including the idea of discerning "the laws of nature that operate out there."

Kennedy's science advisor, Jerome Wiesner, jumped in to try to rescue things. He pointed out that, for instance, "we don't know a damn thing about the surface of the Moon," but if you were going to land on it, you needed to find out about it. "The scientific programs that find us that information have the highest priority. But they are associated with the lunar program. The scientific programs that aren't associated with the lunar program can have any priority we please to give 'em."

Then the space people really lost their president.

"Why are we spending $7 million on getting fresh water from salt water, when we're spending $7 billion to find out about space?" Kennedy asked. "Obviously, you wouldn't put it on that priority, except for the defense implications. And the second point is that the fact that the Soviet Union has made this a test of the system. So that's why we're doing it. . . . We've got to take the view that this is the key program."

Kennedy was being as clear as he possibly could: the U.S. focus on the Moon was motivated by the Russians. It was fine to fly to the Moon, but the point of going to the Moon with this sense of urgency—NASA's growth from the federal agency with the tenth largest budget to the one with the third largest—was to get to the Moon before the Russians. It didn't seem clear to the people in the White House Cabinet Room that day, but the only reason they were there at all was because Kennedy needed to beat the Russians. Not because he needed to fly to the Moon.[20]

At the same time, the space people were doing a particularly poor job of helping the president understand that they weren't off doing space research because it was interesting and they suddenly had a lot of money. Wiesner, who didn't much support NASA or human spaceflight, compared to robotic exploration, did a better job of making that point than anyone. Webb, his brilliance as a manager notwithstanding, utterly failed in this setting to do what needed to be done. In the end, the president and the Moon people talked right past each other.

The frankness of the exchanges, even the lack of understanding, can make one quietly grateful for the secret tapes. Although the stakes were nothing like those for the Cuban missile crisis—many of those meetings were also taped—it might be hard to credit an account of this meeting without actually hearing it.

In a way, you have to admire Jim Webb's courage—or his incaution. He did not give up. "But you see," he said, "when you talk about this, it's very hard to draw a line between what—"

Kennedy interrupted: "Everything we do ought to be really tied into getting onto the Moon ahead of the Russians."[21] He didn't have any trouble drawing a line. If you need a line, here it is: Does it help us beat the Russians to the Moon?

Said Webb, his voice rising, "Why can't it be tied to preeminence in space, which are your own words? . . ."

"Because, by God, we've been telling everybody we're preeminent in space for five years," said Kennedy, "and nobody believes it."

When the president of the United States underscores his point to you for the sixth time with the words "By God!" it's probably time to stop talking. Webb, God bless him, took one more run at the president.

But the president had had enough. He said the $400 million budget question wasn't going to be settled that morning. The Moon landing, he said again, firmly, "is *the* top priority program of the Agency, and one of the two things—except for Defense—the top priority of the United States government. . . .

"Otherwise, we shouldn't be spending this kind of money, because I'm not that interested in space. I think it's good. I think we ought to know about it. We're ready to spend reasonable amounts of money. But we're talking about these *fantastic* expenditures which wreck our budget."[22]

And hard as it is to believe, Webb continued to push back, responding, "I'd like to have more time to talk about that." Once more. Twice more. Finally, the president asked Webb to write him a letter, outlining his views on paper. And suggested that he would write back. By then Kennedy had resumed a calm tone. And then the president got up and left.

The conversation continued well after Kennedy's departure. But no one took up, or even commented on, those arresting words, which must have been quite stunning to the space people in the room: *I'm not that interested in space*. The man who launched the United States to the Moon wasn't that interested in space travel, "the greatest adventure on which man has ever embarked," as he described it at Rice University, the task that will measure the best of us. He just—by God!—wanted to get to the Moon before the Russians.

---

Kennedy's frustration ultimately inspired him to tell the truth: "I'm not that interested in space."

We know that during the campaign, Kennedy used Eisenhower's

unhurried response to the Soviet space successes against Nixon, with wit and precision: the first creatures into space weren't named Fido or Rover. Where was the American vitality?

But we also know that, going back to that meal at Locke-Ober in 1958, before he was running for president, when he and his brother Bobby met Doc Draper from MIT, Kennedy was immune to the charms of space travel. Draper, a legendary persuader in his own right, had utterly failed to entrance the brothers with the value of space travel, or its romance.

We know that his NASA chief was the last cabinet-level official named to the Kennedy administration.

And we know that Kennedy had used exactly the same contrary example on the campaign trail as he used with Webb and Seamans and the rest of the space crew the morning of November 21, 1962, in the Cabinet Room. In a speech in Valley Forge, Pennsylvania, on October 29, 1960—a week before Election Day—Kennedy said, "The conversion of salt water to fresh water—a project widely neglected in recent years—could end forever the domestic squabbles between the states of this nation and the peoples of this Earth and, if we develop it first, mean more to our prestige than all the Soviet moon-rockets combined in those underdeveloped nations where great deserts border great oceans."[23] Kennedy had a fascination with the possibility of cheap desalination to transform the lives of the poor and hungry and the economic prospects of developing nations. Want to win hearts and minds and political loyalty in the developing world? Forget Moon rockets: give those nations unlimited fresh water.

The Rice speech wasn't an act, and it wasn't insincere. It was politics. Kennedy needed Apollo to succeed; to succeed, Apollo needed money; to get the money from Congress, it needed popular support. The eloquence and persuasive power of the speech were authentic. For people who believe in the value of the space program, Kennedy's eloquence defined the Space Age itself. For Kennedy, the speech wasn't a defining moment of either his presidency or the Space Age; it was a skirmish in the Cold War.

And that's why that meeting in the Cabinet Room should have set off alarms for Webb and his senior staff. Kennedy wasn't all-in on going to the Moon by 1970. He was all-in on going to the Moon by 1970 *if he had to.* The president had hired, and then inspired, a staff of true believers, and they in turn were building an army of true believers in factories and research facilities across every state of the country. But the real message of what Kennedy said that day in private in the fall of 1962 was something else. Kennedy was flexible. We're going to the Moon to beat the Russians. Simple and direct. If the politics between the U.S. and the U.S.S.R. changed, the race to the Moon might change too. Kennedy wanted to do a lot of things in this world; he wanted America to do a lot of things in this world; and he clearly saw the race to the Moon as instrumental. But also as a project that was wrecking—his word, "wrecking"—his ability to get a lot of other important things done.

In 1963 the politics of going to the Moon got even more challenging than they were in 1962.

Webb was worried about the scientific community, many of whom felt that a space program that sent humans into space would consume huge amounts of federal money that could be used for scientific research with more immediate value on Earth. At the November meeting, in one of his rejoinders to Kennedy, Webb told the president that simply labeling the Moon landing as the No. 1 priority, just talking about it as the No. 1 priority, "would lose an important element of support for your program and for your administration."

"By who?" the president asked. "Who? What people?"

"Particularly the brainy people in industry and the universities," Webb replied.[24]

This exchange came right before Kennedy excused himself from the meeting.

In April, in an editorial in the prestigious journal *Science*, the magazine's editor Philip Abelson provided precisely the cerebral, almost disdainful critique Webb had been hearing in his conversations with scientists.

Abelson walked through each justification—military value, technological innovation, scientific discovery, and the propaganda value of beating the Russians—and dismissed each in turn. "Military applications seem remote," he wrote. The technological innovations "have not been impressive." If actual science was a goal—and no scientist was on any imagined Moon landing crew yet—"most of the interesting questions about the Moon can be studied by electronic devices," at about 1 percent of the cost of using astronauts.

As for the worldwide prestige, "the lasting propaganda value of placing a man on the Moon has been vastly overestimated. The first lunar landing will be a great occasion; subsequent boredom is inevitable."

An editorial in *Science* was hardly a popular uprising against going to the Moon. But the next morning Abelson was on NBC's *Today* show to make the same case. The *New York Times* covered the editorial. The *Washington Post* also wrote a story about it, and then did a three-part series on the question of whether the Moon race was scientifically justified, asking if going to the Moon were merely "a stunt." The *Post* series was titled "Moon Madness?"[25]

And then on June 10 Abelson was one of a group of scientists called to testify before the Senate Committee on Aeronautical and Space Sciences about the future of Apollo. Abelson, a physicist and a key contributor to the creation of the atomic bomb, told the senators, "[The] diversion of talent to the space program is having and will have direct and indirect damaging effects on almost every area of science, technology and medicine. I believe that [Apollo] may delay the conquest of cancer and mental illness. I don't see anything magical about this decade. The Moon has been there a long time, and will continue to be there a long time."[26]

Abelson's Senate testimony came on a Monday. Two days later, on June 12, former president Dwight Eisenhower spoke to a breakfast gathering of Republican members of Congress in Washington, where he was sharply critical of Kennedy's spending plans overall. Asked specifically about the space budget, Eisenhower replied, "Anybody who would spend $40 billion in a race to the Moon for national prestige is nuts." The line drew sustained applause from the 160 Republican congressmen at the event.

Leave aside that Eisenhower was going with the most extreme estimate of the Moon cost (one that didn't come close to the truth in reality, even nine years later). That was the immediate past president of the United States calling the current president of the United States "nuts" (and doing so, remarkably, on a day when Eisenhower went from that breakfast to an afternoon meeting with Kennedy at the White House to talk about supporting the administration's civil rights efforts).[27]

Headline writers from one side of America to the other loved the story, which made the front pages of dozens of newspapers with some variation of the headline "Ike Calls Moon Race 'Nuts.' "

The very day Eisenhower called the Moon race "nuts," NASA announced the end of the Mercury program, the small capsules with just a single astronaut. Next up, the much more sophisticated, and much more ambitious, missions of Gemini. But the last Mercury flight was May 1963, and the first manned Gemini flight wouldn't come until March 1965—a long time between "space spectaculars" to fire the public imagination, and enough time for an entire presidential and congressional election to play out without a single spaceflight.

In Congress, which was also thinking about elections coming the following year, NASA had gone from receiving near-unanimous support after Kennedy's initial "go to the Moon" speech to being viewed as an agency where money might be harvested for other purposes.

The world had changed dramatically between the middle of 1961 and the middle of 1963. The civil rights movement, which was just beginning to gather force when Kennedy took office, had become a dominant issue, impossible to ignore in the South and in the nation's capital. Kennedy did not mention segregation or civil rights—or even allude to them in passing—in his inaugural address, but in the 1962 State of the Union, civil rights got its own section. In October 1962 James Meredith became the first African American to enroll in the University of Mississippi, his safety and the safety of the university secured with federal troops. The protests in Birmingham, Alabama, in which Martin Luther King was arrested and jailed—and during which he composed his essay "Letter from a Birmingham Jail"—took place in April 1963.

On June 11 Kennedy gave a major address on civil rights to the nation during prime time, carried live by all three news networks. The speech came the day that federal troops in Tuscaloosa, Alabama, under his direction, secured the safe admission of two black students to the University of Alabama, vanquishing the protests of Governor George Wallace. Also that day, 300 federal troops were encamped in Oxford, Mississippi, to guarantee the safety of black students enrolling at the University of Mississippi.

Kennedy's speech was brief—13 minutes—but had a new emotional force that asked Americans to reconsider not just their laws but their own behavior. "We face . . . a moral crisis as a country and a people," he said, because of the second-class treatment of blacks in America. "Law alone cannot make men see right. We are confronted primarily with a moral issue. It is as old as the Scriptures and is as clear as the American Constitution. . . . I hope that every American, regardless of where he lives, will stop and examine his conscience." The *New York Times* story on the speech described it as "the broadest appeal on civil rights ever addressed to the nation by a President."

"A great change is at hand," Kennedy said, "and our task, our obligation, is to make that revolution, that change, peaceful and constructive for all."[28]

The night of that speech, civil rights leader Medgar Evers was shot by a sniper while walking across the front lawn of his home in Jackson, Mississippi. Hit by a single rifle bullet through the heart, Evers was rushed to a hospital in Jackson, where, despite being mortally wounded, he was at first refused treatment because he was black.[29]

Kennedy did in fact propose a sweeping package of civil rights reforms that summer, within weeks of his speech and Evers's assassination. And that August came the March on Washington and Martin Luther King's "I Have a Dream" speech.

The day before his civil rights address, Kennedy had given an equally dramatic and equally pioneering speech on foreign policy at the commencement of American University, proposing a pause in the Cold War between the U.S. and the Soviet Union, a reassessment of the relationship between the two great powers, and announcing as a sign of goodwill an immediate halt by the United States in atmospheric testing of nuclear weapons. "Let us re-examine our attitude toward the Soviet Union," he said in the 30-minute address. "We are both caught up in a vicious and dangerous cycle with suspicion on one side breeding suspicion on the other."

Just the previous October, the U.S. and the Soviets had faced off over Soviet nuclear missiles installed in Cuba, when the young, somewhat underestimated American president successfully blockaded the island and forced the removal of the missiles (in quiet exchange for removal of U.S. nuclear missiles that were in place in Turkey, within quick-strike distance of Moscow).

In any war between the U.S. and the U.S.S.R, Kennedy said at American University, "all we have built, all we have worked for, would be destroyed in the first 24 hours." In addition to proposing a reassessment of the Cold War, he announced that within weeks, the U.S., Britain, and Russia would begin nuclear test-ban talks at a meeting in Moscow.

The speech was considered such a departure for U.S.-Soviet relations, and for a U.S. president, that its complete unedited text was printed two days later for Russians to read themselves in the government newspaper *Izvestia*.[30]

In the space of two days—on a Monday morning at American University and on a Tuesday evening from the Oval Office—Kennedy sought to change the course of the Cold War that put the whole world at risk and sought to revolutionize the approach to the civil rights movement convulsing his own nation.

Within two months, in fact, Kennedy and Khrushchev had signed a nuclear test ban treaty, as dramatic a development in the Cold War as the Berlin Wall and the Cuban missile crisis, but in the opposite direction: a calming of the tensions and the expensive weapons rivalry. In that context, Eisenhower's referring to the Moon race as "nuts" had a different resonance than it would have had two years earlier.

As if to underscore the shift in public attitude, on September 13 the *Saturday Evening Post*, one of the widest circulation weekly magazines in the country, published a story titled "Are We Wasting Billions in Space?" On the cover the headline was just "Billions Wasted in Space," without the question mark, a crisper summary of the story's point. "The space program stands accused today as a monstrous boondoggle—a circus intended just as much to keep American prestige afloat throughout the world as it is to exploit space." Assuming the role of well-meaning everyman, the writer observed, "The layman can only ask in his small voice, 'Is this trip necessary?' " The Moon race wasn't a serious endeavor anymore; it was, according to a chief rival of *Life* magazine, a "boondoggle" and "a circus."[31]

The world had changed so much, the political landscape had changed so much, that it was possible to rethink the extraordinary effort and expense that the Moon race was costing.

We have a secret tape that Kennedy made that shows how he was thinking about space in the fall of 1963, with a thaw in U.S.-Soviet relations well under way, at the start of a significant gap in American space launches, and facing a Congress with its own ideas about spending, and also facing his own reelection campaign starting in late 1963 and early 1964.

The second recorded meeting about space took place on September 18, 1963, in the Oval Office. The records indicate that only President Kennedy and his NASA chief, Jim Webb, were present (and only their voices are heard). On August 5, the U.S., U.S.S.R., and Great Britain had signed a partial nuclear test-ban treaty, the first limits on nuclear weapons. Two days after his meeting with Webb, on September 20, Kennedy was scheduled to give a major speech to the United Nations General Assembly.

The meeting with Webb was long—46 minutes—and Webb did much of the talking, analyzing the politics of congressional support for the NASA budget. The question the conversation wrapped itself around: how to sustain Apollo during what were clearly going to be years of spending without years of excitement.[32]

Right at the start, Kennedy said, "It's been a couple years, and . . . right now, I don't think the space program has much political excitement."

"I agree," said Webb. "I think this is a real problem."

"I mean, if the Russians do some tremendous feat, then it would stimulate interest again," continued Kennedy. "But right now, space has lost a lot of its glamour."[33]

The immediate cuts congressional committees had proposed to the NASA budget would slow America's leap to the Moon. Kennedy asked, "If we're cut that amount . . . we slip a year?"

"We'll slip at least a year," replied Webb.

"If I get re-elected, we're not going to the Moon in our period, are we?" said Kennedy.

"No. No. You're not going," said Webb.

"We're not going. . . ."

"You'll fly by it," said Webb. That is, astronauts will fly by the Moon, as Apollo 8 did, in fact, in December 1968, which would have been the end of the last year of Kennedy's second term. "It's just going to take longer than that. This is a tough job. A real tough job."[34]

President Kennedy watching a TV broadcast of the first U.S. manned spaceflight, of Alan Shepard, on May 5, 1961. With Kennedy, from left: Vice President Johnson; Arthur Schlesinger, Jr.; Admiral Arleigh Burke; Kennedy; First Lady Jacqueline Kennedy. They are in the office of Kennedy's secretary, Evelyn Lincoln.

2

President Kennedy speaks to NASA staff in Houston, in front of an early mock-up of the lunar module, during a tour of NASA facilities on September 11 and 12, 1962. Kennedy is holding a small model of the Apollo capsule that NASA staff gave him. Just to his left, behind him, is NASA administrator James Webb.

3

Charles Stark Draper, a legendary MIT professor, researcher, and pilot, led the invention and refinement of the navigational equipment that allowed Apollo to fly to the Moon.

4

Doc Draper with Ralph Ragan, one of his senior managers for MIT's Apollo program. MIT had the contract to design the computers and navigation equipment for Apollo, supervise their construction, and write the computer programs that flew the spacecraft to the Moon. Draper and Ragan are in front of Wolfie's in Cocoa Beach, Florida, a well-known 1960s deli, attached to the Ramada Inn, near NASA's Cape Kennedy launch facilities.

5

Under the shroud is a secret inertial navigation unit being installed on a B-29 bomber, part of a fleet used by MIT. In February 1953, the unit guided the plane coast-to-coast without any outside information, and without the pilot touching the controls, proving the practicality of inertial navigation.

6

A factory worker at Raytheon in Waltham, Massachusetts, hand-weaving a "rope core" memory unit for the Apollo computer. Using this painstaking technique, it took six weeks to manufacture the software for a single Apollo computer.

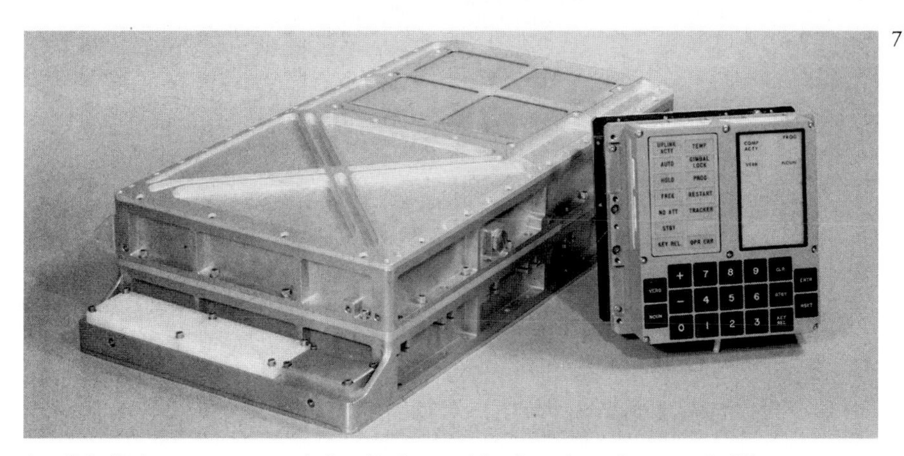

7

Apollo's flight computer and the display and keyboard used to run it. The computer weighed seventy pounds with the metal case, and was two feet long and one foot wide. At the time, it was one of the smallest, fastest, most nimble computers ever.

8

NASA engineer John Houbolt explaining an innovative technique to fly astronauts to the Moon and back. Houbolt fought a years-long battle to get NASA to seriously consider this idea for flying to the Moon. The method Houbolt advocated, called "lunar-orbit rendezvous," was ultimately the one NASA used.

9

10

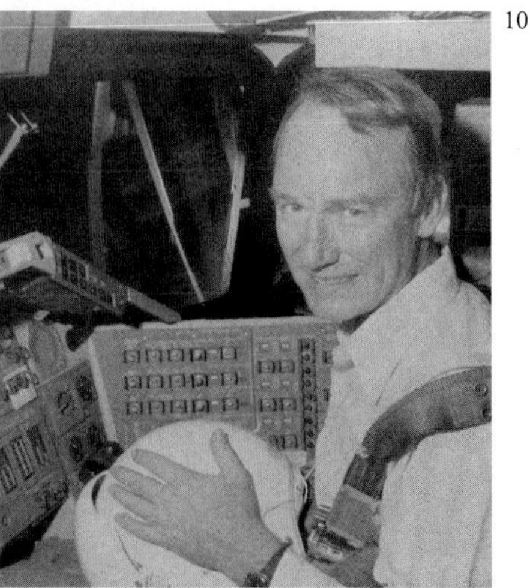

NASA engineer Bill Tindall, left, in Mission Control during an Apollo mission, with flight director Gene Kranz. Tindall—almost never photographed or written about—was critical to mastering the orbital mechanics necessary to fly to the Moon, and to helping MIT get the flight software written. Right, Tindall in a flight simulator for the Space Shuttle, which he worked on after Apollo.

11

The lunar module under construction in clean-room conditions at the Grumman factory on Long Island. Top, an LM upper stage, outer skin in place. It contained the crew compartment, and had its own engine and fuel tank (the big sphere visible uncovered in the lower photo), for blasting off from the Moon.

Below, the entire LM, with upper stage mated to the boxy descent stage. Visible are struts for mounting the LM's legs, not yet installed.

12

13

The lunar module more fully assembled, with some areas covered in the characteristic shiny Mylar insulation, suspended from a crane in the assembly building. The LM was often referred to as "spindly" or "gawky," but astronauts said that in space it handled like a sports car.

14

A rare picture of Thomas J. Kelly (white shirt, in front), chief engineer for the LM at Grumman, whom NASA called "the father of the lunar module."

In this photo from Apollo 11, Kelly is staffing a support room in Mission Control. To his left is Owen Maynard, his counterpart as LM project manager for NASA. Said Kelly of the LM, whose design and construction he oversaw: "How I wished to be a stowaway in that tiny cabin."

15

16

APPLICABLE TO: IN DESCENT, AVERAGE-G ON

| ALARM CODE | TYPE | PRE-MANUAL CAPABILITY | MANUAL CAPABILITY |
|---|---|---|---|
| 0105 MK ROUT. BUSY<br>00430 CAN'T INTG. SV.<br>01103 CCSHOLE-PROG. BUG<br>01204 NEG. WAITLIST<br>01206 DSKY, TWO USERS<br>01302 NEG. SQ. ROOT<br>01501 DSKY, PROG. BAD<br>01502 DSKY, PROG. BUG<br>00607 LAND. NO SOLN | POODOO<br>"<br>"<br>"<br>"<br>"<br>"<br>"<br>" | PGNCS GUID. LOST;<br>PGNCS/AGS ABRT/ABRT STG<br>(decision how on current rules)<br>(NO LR DATA) | PGNCS GUIDANCE NO/GO<br>(PGNCS GO for TAPE METERS, CROSS-POINTERS, CONTROL, ABORTING)<br>(NO LR DATA) |
| "O.F." = OverFlow, too many. CONTINUING OCCURRENCE OF:<br>01104 DELAY [illegible] O.V.<br>01201 EXECT. O.F. (VAC)<br>01202 EXECT. O.F. (CORES)<br>01203 EXECT. O.F. ([illegible])<br>01207 EXECT. O.F. ([illegible])<br>01210 TWO USERS<br>01211 MRK ROUT. INTRPT<br>02000 DAP O.F. | BAILOUT<br>"<br>"<br>"<br>"<br>"<br>"<br>" | DUTY CYCLE MAY DESTROY PGNCS (AGS CONTROL MAY HELP - SEE BELOW)<br>[WATCH FOR OTHER CUES, PGNCS CONDITION UNKNOWN, DSKY MAY BE LOCKED UP, DUTY CYCLE MAY BE UP TO POINT OF MISSING SOME FUNCTIONS (NAV. 1ST TO DIE). SWITCH TO AGS (FOLLOW [illegible]) MAY HELP (REDUCES PGNCS DUTY CYCLE SOME.)] | SAME AS LEFT<br>(except "other cues" which would otherwise be cause for ABORT PROBABLY AREN'T; INSTEAD IT WOULD BE PGNCS GUIDANCE NO/GO - COMPLETE MANUAL LANDING IN AGS.) |
| ISS WARNING WITH:<br>00777 PIPA FAIL<br>03777 CDU FAIL<br>04777 PIPA, CDU FAIL<br>07777 IMU FAIL<br>10777 PIPA, IMU FAIL<br>13777 CDU, IMU FAIL<br>14777 PIPA, CDU, IMU FL | LIGHT ONLY<br>"<br>"<br>"<br>"<br>"<br>" | PIPA/CDU/IMU FAIL DISCRETES PRESENT<br>(Other mission rules suffice; alarm may help point to what rule will be broken) | same as left |
| 00214 IMU TURNED OFF | LIGHT ONLY | AGS ABRT/ABRT STAGE | SWITCH TO AGS<br>PGNS NO/GO on G&C<br>(poss. NO/GO on NAV.) |
| 01107 E-Mem. Destroyed | FRESH STRT | AGS ABRT/ABRT STAGE | SWITCH TO AGS<br>PGNCS NO/GO?<br>(IMU as ref. okay) |
| CONTINUING<br>00402 BAD GUID. CMDS | LIGHT ONLY | IF ALARM DOESN'T STOP: SAME AS "POODO's" (ABRT?) | IF ALARM DOESN'T STOP: Same as "POODOs" |
| CONTINUING<br>01406 GUID. NO SOLN<br>01410 GUID O.V. | LIGHT ONLY | PGNCS GUID. NO/GO AS LONG AS ALARM OCCURRING<br>(ATT. HOLD, CONST. GTC, CONT. OK)<br>(ABRT WILL PROB. COME FROM CURRENT RULES e.g. GTC vs. V)<br>WATCH GTC ← | same as left<br>(except prob. no abort.) |

When the Apollo 11 LM computer started sounding alarms, Jack Garman had twenty seconds to assess how serious they were. Garman, at right in the top photo, had created a handwritten list of computer alarm codes (above right) so he would know how serious any alarm was. He gave the okay, helping save the Apollo 11 Moon landing seconds before touchdown.

Top, Garman receiving an award from Chris Kraft, head of Mission Control. Below, the Mission Control support room where Garman worked during flights—he is in the front row of consoles, second from left, in a sport coat.

17

An astronaut at the controls of his spaceship: Edwin "Buzz" Aldrin, Apollo 11 LM pilot and the second man on the Moon, in the cockpit of *Eagle*. Clearly visible are the LM's distinctive triangular windows, and between them the spaceship's main control panel. Above Aldrin's head is a rolled-up sunshade for the left-side window. Note the paper floating near the top of the cockpit. Apollo astronauts carried paper checklists, mission plans, and star charts to the Moon. At bottom center are two white bags, to the left and right—spacesuit helmets, stowed in their protective covers.

Aldrin is floating in the ship during an initial inspection in lunar orbit. This is a mosaic of seven photos, shot by Neil Armstrong on July 19, 1969, and assembled by Jon Hancock. (The ghostly image of Aldrin's "extra arm" is an artifact from one of those photos, in which Aldrin was facing the other way.)

That must have been a sharply felt exchange for the president. It's hard to listen to the conversation while setting aside everything we know that would come in the next 10 weeks, and the next 6 years, and just imagine it from Kennedy's point of view. This huge project he had set in motion. He wasn't even done with his first term. Congressional critics weren't just talking down the Moon landing; they were cutting the budget for the Moon landing. And Kennedy wouldn't just have to muster the political support for Apollo through the election in a year; he was imagining having to sustain support for it through his entire next *term*, to which he hadn't even been elected yet.

And even if he could do it, he wouldn't get the joy of accomplishment during his own presidency. A private moment between the president who launched America to the Moon and the man responsible for making it happen, in which Kennedy was told bluntly, *I know we talked about 1967 or 1968, but at this point you will have to pay all the cost—literally and politically—for taking us to the Moon, but you will not be president when those American astronauts set foot on the Moon.* It would have been a keen moment of disappointment, and you can hear it in Kennedy's voice. It would also have been a moment of political calculation. How do you possibly hang on to a discretionary program of such enormous scale, already under fire, through four more budget cycles?

Just after that, Kennedy asked a version of the same question he had asked a year earlier: "Do you think the manned landing on the Moon's a good idea?"

"Yes sir," replied Webb. "I think it is." But then Webb moved on to his favorite, but somewhat rambling justification, which started with building bigger booster rockets and eventually came back around to the Moon like a spacecraft on a wide looping rhetorical trajectory.[35]

To Kennedy, the broader politics were simple and discouraging: "We don't have anything coming up for the next 14 months. So I'm going into the campaign to defend this program, and we won't have had anything for a year and a half." He actually sounded disappointed, almost irritated by the timing of this flight gap. How could he talk with enthusiasm about space, when there were no spaceflights for anyone to be enthusiastic about?[36]

In fact Kennedy saw only one strategy for protecting Apollo, an extension of the very first reasoning behind the Moon race. "I want to get the military shield over this thing," he said, meaning, he wanted to be able to argue that manned spaceflight had explicit national security and defense value.[37] Kennedy had been reading and hearing the criticisms of Apollo: the spending, the relevance, the challenge of the scientists.

He said again (he and Webb track over the same frustrated terrain several times in 46 minutes): "Unless the Russians did something dramatic, and we don't have anything dramatic coming up for the next 12 months, so it's going to be the attack on the budget." Then Kennedy appeared to rehearse for Webb the arguments they'd be hearing: "That this looks like a helluva lot of dough to go to the Moon. When you can learn most of what you want scientifically, through instruments, and putting a man on the Moon really is a stunt and isn't worth that many billions.

"Therefore, the heat's going to go on unless we can say, that this has got some military justification, and not just prestige. Otherwise Eisenhower has been kicking us around, and we're going to look like he's probably right. . . .

"Why *should* we spend that kind of dough to put a man on the Moon?"[38]

Maybe the whole thing would end up looking nuts; just having serious people talk about the Moon race as a "moondoggle" and "a circus" shifted the debate. Could Apollo become Kennedy's Folly?[39]

Webb wasn't that good at sharp political messaging for Apollo. What the meeting needed—what the question needed—was Ted Sorensen, the president's phrase-maker. Webb talked about the value of Kennedy's "technological courage" in pushing the Moon race forward. He suggested including the military in a kind of exploration role, although worrying out loud that the military would simply take over the space program, something the services had in fact been angling to do since Sputnik.

Kennedy came right back to the same point, just three minutes later in the conversation. "We've got to wrap around . . . a military use for what we're doing and spending in space. Because if we don't, it *does* look like a stunt, and too much money. . . . Christ, we can't get money for slum clearance and all the rest, and people saying we're putting billions into going to the Moon. If we can show that that's true, but there's also a very significant military use.

"Now how are we going to do that?" Kennedy's tone here is sharp. He's asking Webb for his help, but he's also saying out loud, *If we can't find that kind of justification, we're going to be in trouble, and long before we're sending spaceships to the Moon.*

"How are we going to get the military to support this program on the grounds that it can be militarily useful, without them dominating it?"[40]

All in all, it was quite an hour between Webb and Kennedy.

The president twice said that "all the glamour" had gone out of space. He twice said people would come to regard the Moon landing as "a stunt," and in at least one of those moments, he appeared to agree. Without a good military justification, with most of the science achievable with robotic probes, Kennedy himself said, "it does look like a stunt." Over and over—five separate times—he said wistfully that it seemed unlikely the Soviets would have any space spectaculars during the next year either (although how Kennedy thought he knew this is uncertain). Neither NASA nor the Russians would be doing the kind of space trips that would help him keep Americans motivated.

Kennedy constantly circled back to the question of what connections could be made between the manned Moon mission and the U.S. military, without coming up with anything either pointed or useful. So much so, in fact, that Webb came up with a bolder idea: he offered to resign. "Would you be better off, thinking about '64 and a political year, if you just took a military man and put him in charge of this program? Someone you trusted completely and might ultimately want to become, say, chairman of the Joint Chiefs of Staff? This is a big, spectacular program."

Want to put a military shield over the whole thing? Just put a general in charge of it.

"I don't think that's what we ought to do now," responded Kennedy. Later he mused, "Maybe second. Who's your second man over there?"[41]

Webb went deep into the budget negotiations with Kennedy, talking about congressmen by name, but he also pulled back to remind the president of the incredible power of this kind of exploration and science for the life of Americans, for understanding how the world works, but also for the practical value of technology development, and for inspiring American students to pursue science and engineering. "The younger folks see this much better than my generation," Webb said, having visited high schools and colleges around the country. He was talking about all the things that made Americans nervous after Sputnik, all the things Kennedy himself so forcefully argued in his Rice University speech. The lunar landing, said Webb, is "one of the most important things that's been done in this nation." What will come from going to the Moon will be "staggering things in terms of the development of the human intellect."

The NASA chief concluded, "I predict you are not going to be sorry—ever—that you did this."[42]

In the whole 46 minutes, Kennedy had only one real moment of reflection and support for space: "I think this can be an asset, this program. I think in time, it's like a lot of things, it's sort of mid-journey, and therefore everybody says, 'What the hell are we making this trip for?' But at the end of the thing they may be glad we made it. But I think we've got to defend ourselves now."[43]

---

Two days later Kennedy surprised everyone—in Congress, in the world of space, in NASA itself—with a stunning proposal. He was giving his speech to the General Assembly of the United Nations, a moment of satisfaction and even triumph in a year in which he had moved from confrontation with the Soviets over nuclear weapons to the first ever treaty with the Soviets that limited those weapons.[44]

Just two years earlier, Kennedy said, when he first addressed the UN as president, "The shadow of fear lay darkly across the world."

"Today the clouds have lifted a little so that new rays of hope can break through." More than that, "we may have reached a pause in the Cold War—but that is not a lasting peace. A test ban treaty is a milestone, but it is not the millennium." He urged the nations of the world to "stretch this pause into a period of cooperation."

The UN speech was long, 30 minutes, half again as long as the Rice University address, longer even than the American University speech outlining a whole new vision for the Soviet-American relationship. And it was a sweeping address, ranging from the hope for further reductions in nuclear danger and a fresh focus on improving human rights and human freedom, to proposals for making sure every child in the world had the food he or she needed, to a worldwide program of conservation of wild lands, forests, endangered species, and marine life.

About halfway through is a single paragraph, just 90 seconds, 180 words out of 3,500, and it is about flying to the Moon:

> Why . . . should man's first flight to the Moon be a matter of national competition? Why should the United States and the Soviet Union, in preparing for such expeditions, become involved in immense duplications of research, construction, and expenditure?
>
> Surely we should explore whether the scientists and astronauts of our two countries—indeed of all the world—cannot work together in the conquest of space, sending some day in this decade to the Moon, not the representatives of a single nation, but the representatives of all our countries.

The final line about the Moon echoes Kennedy's original speech before Congress, and the echo is even more vivid in this delivery. He delivers it like the stanza of a poem:

> *Sending some day in this decade to the Moon*
> *Not the representatives of a single nation*
> *But the representatives of all of our countries.*

And there is extra emphasis on *some day in this decade*—a reminder of his vow 28 months earlier to go to the Moon *before this decade is out.*

The idea, the proposal to go to the Moon jointly was what made headlines from Kennedy's UN speech. The president who had first told Congress and the nation that going to the Moon might "hold the key to our future on Earth," the president who a year earlier had said at Rice University that the only way to secure space for freedom was to be first "and therefore we intend to be first"—that president was now suggesting that the United States go to the Moon jointly with the very adversary who had inspired the Moon race in the first place.

"Kennedy Asks Joint Moon Flight by U.S. and Soviet as Peace Step" was the headline over the main story in the *New York Times*. A story just beneath it was headlined, with admirable directness, "Washington Is Surprised by President's Proposal."

"President Urges Joint U.S.-Soviet Moon Trip" was the *Washington Post* headline, with a picture of Soviet foreign minister Andrei Gromyko applauding Kennedy.[45]

In fact, in their conversation just two days earlier, Kennedy had asked Webb if he had seen the proposal he intended to make for greater space cooperation with the Soviets in the text of the speech that had been sent over to NASA for review. Webb said he had been away but that his deputy Dryden had reviewed it, and "he feels you're taking the right line." Cooperation with the Russians, Webb said in passing, "I think that's good. I think that's good."[46]

It's not clear if Dryden reviewed the paragraph that Kennedy actually read—the *New York Times* suggests the joint-flight idea came very late to the speech itself—but regardless, the proposal stunned both space officials and members of Congress, and, despite Webb's reassurance to the president (without knowing what he would say), neither group thought it was a good idea.

Robert Gilruth, a senior NASA engineer who was then in charge of the Manned Spacecraft Center in Houston, had said just the previous week about any joint mission, "We already have trouble mating two modules by American manufacturers" and that he "trembled at the thought of the integration problems" involving space equipment produced separately by the Russians and the Americans.

The *New York Times* story on Washington's surprise—without quoting a single person by name—reported, "There was speculation in the executive branch and in Congress that the proposal represented the first step toward pulling out of the costly 'moon race' and backing away from the Presidential commitment that the United States would land a man on the Moon before the end of the decade," speculation "strengthened by the fact that the proposal came when it was becoming increasingly doubtful that it would be possible to achieve the lunar landing by 1970, and when the Administration was under strong political pressures to hold down government expenditures."[47]

Kennedy's suggestion—in the widest possible forum—to go to the

Moon with the Russians was a reversal of two years of insistence that going to the Moon was not just a uniquely American effort, but that the capabilities the U.S. would develop to do it were essential to the country's national security.

In fact just two months earlier, at his press conference on July 17, 1963, a reporter had asked Kennedy, "There have been published reports that the Russians are having second thoughts about landing a man on the Moon. If they should drop out of the race to the Moon, would we still continue with our Moon program? Or, secondly, if they should wish to cooperate with us in a joint mission to the Moon, would we consider agreeing to that?"

The answers to both left not a sliver of doubt or wiggle room.

"I think we ought to go right ahead with our own program," said Kennedy, "and go to the Moon before the end of this decade." The point of going to the Moon, he continued, is America's "capacity to dominate space," which "is essential to the United States as a leading free world power. . . . I would not be diverted by a newspaper story."

As to cooperation, Kennedy said, "We have said before to the Soviet Union that we would be very interested in cooperation." But going to the Moon jointly "would require a breaking down of a good many barriers of suspicion and distrust and hostility which exist. . . . There is no evidence as yet that those barriers will come down."[48]

It's hard to reconcile the Kennedy of that press conference with the Kennedy of the United Nations speech. No one anywhere, including Khrushchev, had done a better job of explaining why going to the Moon *should* be a matter of national competition than John Kennedy himself.

Yet Kennedy had toyed with cooperation from the very beginning. Right after the May 1961 "go to the Moon" speech, he met one-on-one with Khrushchev in Vienna over two days, and he proposed the idea of a joint effort then, with the simple line, "Let's go to the Moon together." But Khrushchev rejected the suggestion.[49]

Whatever Kennedy's political strategy with the joint-mission proposal at the UN, it was a short-term failure. If anything, it did harm to all the areas that he and Webb had so painstakingly gamed out. Newspaper stories in the days after the UN speech pointed out the utter impracticality of two nations sharing engineering responsibilities for a task so demanding and so complicated—that it might be reasonable to do a joint mission only if one nation's astronauts rode in the spaceships of the other nation.[50]

In Congress the joint-mission proposal backfired. At just the moment NASA and Webb needed maximum enthusiasm from Congress, the proposal undermined the arguments not to cut Apollo funding, and it undercut the sense of urgency that was the reason for the extraordinary funding.

In the weeks after the UN speech, a House subcommittee cut NASA's proposed budget by $250 million, on top of the $350 million it had already cut. A second House subcommittee came within a single vote of cutting the budget by $1.25 *billion* more.

And Congress was in the hands of Democrats.

The president's proposal for the joint mission, reported the *New York Times*, "is having a boomerang effect on the domestic front by encouraging Congressional efforts to cut the space budget."

Representative Albert Thomas, the Texas Democrat in whose district Mission Control was, right then, being constructed, voted to cut the original NASA proposed budget from $5.7 billion to $5.1 billion, to help make possible a tax cut and spending reductions. "Of all the spots," Thomas said, "this was a good one to cut." To be clear, the $5.1 billion was an increase over the previous year; the cut was to the increase Kennedy and NASA wanted.

That said, it was a tough year. In the budget approved by the House, 25 of 26 major federal agencies were held to their previous year's budget, or less. Only NASA's went up at all.[51]

But Webb said consistently—in private to Kennedy, in public to Congress—that a budget of $5.1 billion wasn't enough to keep Apollo on track for a Moon landing by 1970, and that cutting $250 million now would add $2 billion or $3 billion to the ultimate cost because of delays.

Webb was no political naif. On the day of the House vote on the NASA budget—Thursday, October 10—he had four astronauts sit in the House gallery for the debate and the vote: the national heroes John Glenn and Alan Shepard, along with Gus Grissom and Scott Carpenter. They were, observed the UPI account, "accompanied by their wives."

As they watched, the House passed the slimmed down $5.1 billion

NASA budget—$600 million less than Kennedy requested, at least $200 million less than Webb had said was necessary to stay on track for a Moon landing within the decade. And the House did one other thing. Before passing the budget, members passed a resolution barring use of NASA funds for any joint Moon mission with any "Communist, Communist-dominated or Communist-controlled country." Representative Olin Teague, a Texas Democrat and a senior member of the House Space Science Committee, said of the idea of going to the Moon with the Russians, "I'd just as soon cooperate with any rattlesnake in Texas."[52]

---

If John Kennedy had not been assassinated, would Neil Armstrong and Buzz Aldrin have stepped off the ladder of the lunar module *Eagle* onto the Moon on July 20, 1969?

It seems unlikely.

It's possible to argue that the Moon landing, as it did happen, would not have if Kennedy had won and served a second term. Asking where Kennedy was headed with regard to the Moon landing is important, because if he might not have pushed the program forward with urgency, then that helps us understand how the United States did in fact make it to the Moon after Kennedy's death.

What was Kennedy up to with the race to the Moon in the fall of 1963?

Inside the administration, the president kept pushing his idea of cooperating in space with the Russians, including on November 12 issuing a National Security Action Memorandum directing Webb to take responsibility for "development of a program of substantive cooperation with the Soviet Union in the field of outer space . . . including cooperation on lunar landing programs." (A poll of Americans four weeks after Kennedy was killed showed that 54 percent opposed going to the Moon with the Russians, more than had ever supported America's landing on the Moon alone.)[53]

But after the UN speech, NASA officials had quietly but swiftly reoriented Kennedy's sweeping proposal to something more like "finding areas to cooperate," including, for instance, sharing research on Moon landing sites so that routine work on a Moon mission wasn't done by both nations, as opposed to the more dramatic vision Kennedy seemed to have been suggesting, of putting a U.S. capsule atop a Russian booster.[54]

President Kennedy visited Cape Canaveral for the third time, on November 16, flying up from where he was spending the weekend in Palm Beach, for two hours of briefings and visits, including two dramatic moments. He got to see the Saturn I rocket on its launchpad, the rocket that would, a month later, finally put into orbit a payload larger than anything the Russians could launch. "It will give the United States the largest booster in the world and show significant progress in space," the president said. The Saturn I was scheduled to launch in December; it ended up being launched successfully on January 29, 1964, sending 10 tons into Earth orbit in a milestone considered so significant that the midday event was carried live by the TV networks, even though most of the cargo put in orbit on the test flight was sand.[55]

Kennedy was in good spirits during that November 16 Cape visit; he also helicoptered 30 miles out to sea to a navy observation ship to watch the launch of a Polaris missile from a submarine 50 feet below the surface. The navy had Kennedy give the firing order. When the missile burst from beneath the ocean, hung in the air, then lit its rocket engines and soared off into the sky, a reporter for the *Orlando Sentinel* wrote, "a grin, not unlike that of a youngster viewing a Christmas toy in action for the first time, spread across [Kennedy's] face." Whether on purpose or by happenstance, Kennedy's visit came on a Saturday, so the pictures of the commander-in-chief gazing up at the Saturn I rocket, and then aboard the deck of the navy observation ship, binoculars to his face as he watched the Polaris launch, ran on the front pages of Sunday newspapers across the country; the *Orlando Sentinel* devoted a full page of photographs to the visit.

But despite being trailed by a huge press corps, walking around and under the Saturn I rocket on its launchpad, and sitting for briefings on the state of Gemini and Apollo with astronauts and with Webb and von Braun, the president said nothing, unlike during his visit to the space centers the previous year. Asked how the visit had been as it was wrapping up, he answered simply, "Excellent!"[56]

After a brief return to Washington, Kennedy headed to Texas the following Thursday, to San Antonio, then Houston, and finishing in Fort Worth and Dallas. In San Antonio he dedicated a new air force research center devoted to aerospace medicine. He commented on how valuable space medical research would prove: "Medicine in space is going to make our lives healthier and happier here on Earth." He told the audience how impressed he was with the Saturn rocket he had seen the previous Saturday. "While I do not regard our mastery of space as anywhere near complete, while I recognize that there are still areas where we are behind, at least in one area, the size of the booster, this year I hope the United States will be ahead. And I am for it." There will be "pressures in this country to do less in this area as in so many others, and temptations to do something else that is perhaps easier. But . . . the conquest of space must and will go ahead." He didn't mention landing on the Moon.[57]

In the speech that had been written for him to give in Dallas at the Dallas Trade Mart—the speech he was on the way to deliver when he was shot—Kennedy would have talked with pride about reinvigorating the U.S. space program. Under his administration, the U.S. was spending more money on space each year than the entire space budget for the decade of the 1950s; 130 U.S. spacecraft had been put in orbit, including invaluable and innovative weather and communication satellites, "making it clear to all that the United States of America has no intention of finishing second in space. The effort is expensive"—Kennedy once again acknowledging the cost of the space race—"but it pays its own way, for freedom and for America." The U.S. had vanquished any fear of "a Communist lead in space." Indeed "there is no longer any doubt about the strength and skill of American science, American industry, American education, and the American free enterprise system." Space was a source of "national strength."[58]

The passages from the undelivered speech seem like reasonable campaign rhetoric, but they are in fact revealing. Kennedy was reminding the audience—and the press who would cover the speech—that every problem Sputnik had revealed, every problem that Yuri Gagarin's first human flight had revealed, had been corrected under his administration.

Those who doubted the quality of America's schools, its scientists, its ability to make and launch spaceships and astronauts, those who thought dictatorship was doing something democracy could not, all those doubters had been silenced. America had shown that it could do whatever it wanted in space, on a schedule of its own choosing.

Kennedy was in Texas, home of the Manned Spacecraft Center that would direct the Moon missions. The audience at the Trade Mart would be the Democratic elite of Dallas. The trip was the informal opening of the 1964 presidential campaign. It was, in fact, a controversial trip because of the complicated politics of Texas, so it was also a closely watched trip. But Kennedy hadn't planned to say a word about going to the Moon; the speech offered none of the thrilling and high-flown rhetoric about how Houston (or simply Texas) would be the headquarters for the explorers of the New Frontier. He would say instead that America would not finish second. But being first if the other person in the race slows down, or drops out altogether, is a much different undertaking than if you're both racing headlong for the finish line.

If you listen to the secret tapes of Kennedy's two conversations about space, if you layer on the UN speech and his effort to push space collaboration forward, if you look at what was happening to space funding and space enthusiasm in Congress, and if you look at everything else going on in the country and the world, the logical conclusion is not that Kennedy would have raced for the Moon. Just the opposite. From the evidence at hand, it's actually hard to imagine Kennedy making the Moon landing a cornerstone of his second term.

In 1962 the Kennedy of the Rice University speech and the Kennedy of the tape-recorded Cabinet Room meeting seemed at odds: a public stance on space and a private stance on space. By November 1963 the Kennedy of the Oval Office meeting with Webb and the Kennedy of the UN General Assembly, the Kennedy of the final Cape Canaveral visit and the Texas speeches, seemed to be slowly converging.

In private, in fact, Kennedy was remarkably consistent: he wasn't that interested in space—going all the way back to that dinner at the Boston restaurant Locke-Ober with his brother Bobby and MIT's Doc Draper.

His original determination to go to the Moon was to overtake the Russians. He was told, during those dramatic days in April and May 1961, after the Gagarin flight and the Bay of Pigs failure, that going to the Moon was the only way to beat the Russians in space.

In that meeting in the Cabinet Room in 1962, the frustration on both sides came in some measure from exactly that mismatch: Kennedy saw going to the Moon as the No. 1 priority, with all haste, in order to beat the Russians. Not because going to the Moon was smart or scientifically valuable or would create a wide halo of benefits back on Earth. That's the meeting where he said, "I'm not that interested in space." It's the meeting where he said directly to James Webb, "Everything you do ought to be tied to getting onto the Moon ahead of the Russians." It's the meeting where he said, "The Soviets have made it a test of the system." The Soviets had bragged from their very first space launch about how their performance in space proved the superiority of communism over democratic capitalism.

In 1962 and again in 1963, Kennedy worried about what space was costing, not as a matter of fiscal prudence or balanced priorities but as a program that was suffocating his ability to do other things he thought were urgent. The only justification for spending that kind of money, Kennedy said explicitly, was the central geopolitical rivalry with the Soviet Union. The only possible justification for the cost of going to the Moon—the actual cost, but also the very high opportunity cost—was to beat the Russians. In the president's mind, NASA and Apollo were a civilian project doing a wartime mission.

In the second meeting we also see the corrosive impact of the growing political sniping at Apollo, in which the Moon landing was starting to be portrayed as a stunt by its opponents, and also the impact of the budget cuts pushing the predicted Moon landing off beyond the end of a Kennedy presidency. That delay could well have created a self-reinforcing cycle of space frugality for NASA. Kennedy could sustain Apollo at a more modest funding level, aimed at an "early 1970s" Moon landing; once he wasn't going to get the joy and satisfaction, and political benefit, of being president during the landing, why pay the political cost himself? He could even be a little passive: speak on behalf of the Moon landing, but let Congress set the funding, and thus the pace, and take the blame for not making it to the Moon before the end of the decade. Why not fund the space program at whatever level it needed to maintain American excellence, without having to cripple everything else on its behalf? Put another way, in blunt political terms, why would Kennedy expend enormous political capital with the American public, and in Congress, to sustain enormous funding for a Moon landing that would happen in the presidency of his successor?

Even the push for cooperation with the Soviets was another way of neutering the "crash program" approach. Quite simply, if it's not a race, you don't need to move with urgency, and you don't need to worry about being beaten. In economic terms, Russia was less able than America to sustain a human lunar landing program. If both sides could assure themselves of not being beaten, by in some ways working together, neither side needed to invest the resources in a tight timeframe in order to beat the other. Cooperation may or may not have had any real value to NASA for actually going to the Moon, and some kind of joint effort might dim the triumph of a solo landing. But cooperation could do one thing for sure: it could guarantee you wouldn't lose.

Had Kennedy, or his advisors, gamed all this out in this kind of tactical detail? Maybe. Maybe not. But the outlines of a different stance on space were clear if you look at just what was happening before the visit to Dallas on November 22, 1963. In the UN speech, in the visible but quiet visit to the Cape, in the speech he was to give at the Dallas Trade Mart, Kennedy was testing out a reframing of the space race. He was de-emphasizing the Moon part of it.

More than that, during the Texas visit he was declaring victory in space. We've caught up, he said. Our system has proved not just its resilience but its brilliance. In just a few weeks we'll launch the biggest rocket any nation has ever launched, with the biggest payload anyone has ever orbited. We've got sophisticated weather satellites and sophisticated communications satellites. Not only won't we be second in space; we aren't second right now, we aren't second anymore. We can now use our position of strength to transform space from an arena of costly competition to an arena where we learn to cooperate—if the Russians want to—and can perhaps bring the lessons of that cooperation back to Earth. If they don't want to cooperate, we've proven our ability to beat them, starting from a huge disadvantage, and to benefit all mankind while we do it.

At Rice University in September 1962, Kennedy said, "We intend to win." It's possible that after seeing Saturn I on the launchpad, the president suddenly realized something: that the one thing that had been wrong all along was the one thing he had been assured was absolutely true, in the nervous few days after Gagarin's orbital leap—that the only way to win was to go to the Moon. Seeing that the U.S. had created the largest rocket ever—gleaming white, as tall as a 16-story building, just a precursor to the Saturn V Moon rocket that would be twice as tall—and knowing that the Saturn would soon be launched in the open for the whole world to witness, seeing the mock-up of the Gemini capsule and the plans for Apollo, Kennedy may have reached a very different conclusion than the NASA staff so eagerly showing him their work and their plans. He might well have concluded that the U.S. had won the space race in just over two years, that beating the Russians didn't require going all the way to the Moon after all, and that everything to come was worthwhile but could move forward at a much more reasonable, much less hectic, and much less expensive pace. Because one thing is clear from his own words: if John Kennedy didn't have to go to the Moon, he wouldn't go. He had lots of other things he wanted to do.

But none of that happened, because Kennedy was killed on Friday, November 22, 1963.

Six days later President Lyndon B. Johnson announced, in his somber Thanksgiving Day address to the nation, that he was renaming the space center in Florida the John F. Kennedy Space Center and renaming the piece of land it sat on Cape Kennedy. In a brief meeting the day before, Jacqueline Kennedy had asked Johnson to do that, and he had agreed, and immediately called the governor of Florida, Farris Bryant, to win his help and support. When Americans rocketed off to the Moon landing, they would do so from a spaceport at Cape Kennedy.

Before noon on the Friday after Thanksgiving, not even 18 hours after Johnson's announcement, painters hung a sign with the new name on it over the southern security gate for Kennedy Space Center.[59]

Eight weeks after Kennedy's death, on January 21, 1964, President Johnson submitted to Congress his budget for the next year, proposing to cut overall federal spending from Kennedy's previous budget by $500 million, including cuts to defense, agriculture, veterans affairs, and the post office. But Johnson raised spending for NASA to $5.3 billion, along with a request to immediately add back $141 million for the year already under way. Whatever Kennedy's long-term space strategy had been, his death changed the political calculation, in space as in so many other arenas. Johnson, unlike Kennedy, was an authentic believer in the space program. In announcing the NASA budget, he reaffirmed his determination to get the nation to the Moon by 1970. "No matter how brilliant our scientists and engineers, how farsighted our planners and managers, or how frugal our administrators and contracting personnel, we cannot reach this goal without adequate funds," Johnson said. "There is no second-class ticket to space."[60]

# 7

# How Do You Fly to the Moon?

I could hardly believe that this agile machine, dancing so gracefully through space, was the same crotchety beast with the broken wires and structural cracks that had given us fits.

**Thomas J. Kelly**
*Grumman chief engineer and "father of the lunar module," watching the first lunar module fly in space*[1]

By March 1964 the most sophisticated spaceship ever conceived was well along in its design. The Apollo lunar module would carry two astronauts from lunar orbit to the Moon's surface, be their base of operations on the Moon, then rocket them back to orbit and rendezvous with the command module. The lunar module—known as the "lem," which ended up abbreviated LM—was being designed and built on Long Island, at the same factory where, 20 years earlier, Grumman Corporation had produced 12,275 Hellcat fighters for World War II, averaging 14 fully assembled new warplanes a day.[2]

As Grumman conceived the lunar module, it was a two-stage spacecraft; the full ship would land on the Moon, but only the small upper stage and crew compartment would blast off from the Moon and return the astronauts to the command module, in orbit. So the lunar module had two rocket engines, a big one to land the ship, and a smaller one to blast the crew compartment back into orbit. Each of those rocket engines weighed less than the engine in a typical midsize car—and each was a marvel. The descent engine could be throttled: powerful thrust to bring the lunar module down to the Moon from orbit, and lower thrust to allow the LM to hover near the surface of the Moon while the astronauts picked a final landing spot. No rocket engine before had ever had variable power. The smaller engine, which would send the astronauts back to space after their Moon visit, absolutely had to work when the launch command was given. If it failed, the astronauts were trapped on the Moon. So the ascent engine was a study in simplicity to reduce the number of ways it could fail, and reduce any possibility it would not work.[3]

The lunar module would have sophisticated navigation, electronics, and life-support systems, and it would also have storage space for bringing home Moon rocks. The design was evolving. The cabin had already been refined to accommodate bulky spacesuits; the seats had been eliminated, and the windows made smaller, to reduce weight; the LM had gone from having five legs, which would have provided maximum stability, to having four legs, which allowed room for bigger fuel tanks.

On March 24, 1964, eighty NASA managers, engineers, and astronauts gathered with Grumman's staff in Bethpage, Long Island, for a two-day review of the lunar module as it was then designed. Grumman had built a sleek, full-size model with as much engineering and design detail, inside and out, as possible: interior lighting, environmental control equipment, radar and radio antennas, flight controls in the cockpit.

One of the big issues to be reviewed, demonstrated, and discussed was how astronauts would get from the main hatch of the lunar module, which was in the upper stage, down to the ground, a distance of 10.5 feet. "We had given considerable thought to the problem of egress to the lunar surface," said Thomas Kelly, the chief engineer and chief designer of the lunar module. Grumman's own staff had been donning spacesuits and, using a special harness that simulated the Moon's lighter gravity (one-sixth of Earth's), trying out various ideas.

Here's what they had settled on to demonstrate at that March event: a rope. A knotted rope. The Grumman engineers' idea was that the astronauts could use the rope to climb down from their spaceship hatch to the lunar surface, and then climb back up. The knotted rope was slung along the side of the lunar module, as if the astronauts were shimmying over the wall of a prison. To make the process of entering and leaving the lunar module hatch more practical, a platform had been installed just outside the hatch and immediately been dubbed "the front porch." And because climbing back into their spaceship would require both hands on the rope, Grumman had included a block-and-tackle rig so the astronauts would have a way to lower scientific equipment to the surface and haul up rock samples when they were ready to leave.

Gene Harms, a Grumman engineer with responsibility for the design of the interior of the crew compartment, and Jack Stephenson, a Grumman test pilot working on the lunar module, had both donned spacesuits in the days before the design review and, suspended in the one-sixth gravity rig, tested out the knotted rope. "They found it slow and strenuous but feasible," said Kelly. He thought use of the rope would get easier with practice.

In the whole vast engineering enterprise that was NASA's leap to the Moon, there are almost no ridiculous ideas. There are poorly thought-out ideas, and things that turned out to be impractical or simply wrong because people didn't understand the science or the challenge of flying to the Moon. Unintended consequences sometimes rendered a good idea unusable. But there are not many ridiculous ideas.

And then there was what we might call the lunar module's "egress rope."

Fly all the way to the Moon in the most advanced craft ever created by human beings, don spacesuits that were themselves highly advanced, miniature, wearable spacecraft—only to shimmy down the side of the spaceship, hand over hand, using a hefty marine mooring line, hanging on with thick-fingered spacesuit gloves as you dropped from one handy knot to the next. Then, after hours tromping around working on the surface of the Moon, doing science experiments, gathering soil and rock samples, planting the American flag—time to leave. Grab that same line and haul yourself back up to the front porch, maybe bracing your Moon boots against the flank of your spaceship to make the climb a little easier.

The rope did fulfill some of the important requirements for being a key piece of lunar module landing technology. It was simple. It was robust. It weighed almost nothing. It was reliable. It could be tested to be sure it wouldn't fail. And in ordinary use, it wouldn't fail; the astronauts weren't going to wear out a coil of robust rope with a few of Kelly's "egresses."

But the rope was also ridiculous, a completely wacky way to come and go from the cabin of a spacecraft while wearing a spacesuit and exploring a planetary body for the first time in human history. It was inconvenient, insecure, uncomfortable, strenuous, and also far from failsafe. What if one or both astronauts wore themselves out to the degree they couldn't muster the strength to climb back up? What if a spacesuit started to fail and the astronauts needed to get back to the safety of the cabin quickly? What if one were injured and couldn't climb the rope? It's not imaginable that you could haul your fellow astronaut back into the spaceship, perhaps over your shoulder, while simultaneously trying to execute the rope climb.

And you don't even have to imagine an extreme circumstance: an astronaut could easily lose his grip and slip, either going up or coming down. Yes, the gravity on the Moon is low, but no one could have imagined an astronaut falling 9 or 10 feet to the rock surface of the Moon would be a good idea.

It's not quite clear why this method so captivated the folks at Grumman. There was apparently some concern that if a ladder was used, and it got damaged during landing, the astronauts might be able to land their lunar module on the Moon but wouldn't be able to actually get to the surface, a problem the rope would likely never have. And it is, in fact, a credit to the way Kelly and the Grumman team approached the lunar module that every possible problem was imagined: it certainly would have been the height of embarrassing failure to land a ship with astronauts on the Moon, and then not have any way to safely get out and walk around.

The rope idea might seem like a joke, except that Tom Kelly, who drove the creation of the lunar module, was smart, gracious, and relentlessly determined—but he wasn't a prankster. And there's a picture of a model of Grumman's March 1964 mock-up, with the rope slung along the side, dangling right next to the front porch, with enough line so the extra coils lie neatly at the bottom, on the pretend surface of the Moon.

Perched right at the top, just stepping off the front porch, is an astronaut, with the rope in his gloved hands.

Even more amazing, there's video of a few moments of the two-day design review, and in that video, someone in a spacesuit is using the rope and the block-and-tackle rig to try to get back up to the LM hatch from the surface. It does not go well. He looks like a spacesuited longshoreman who doesn't quite know how to load cargo.

"After a full day of evaluation by Ed White," said Kelly, "Grumman's proposed use of rope, block and tackle for lunar surface egress was declared unacceptable. White found it too difficult and unnecessarily hazardous for what should be a routine activity."

In fact, said Harms, the Grumman engineer, Mercury astronaut Ed White not only found the whole thing unappealing and impractical, but he managed to injure the ligaments in one of his feet while hoisting himself with the rope.[4]

The lunar module got a ladder, attached to its front leg, with nine rungs, starting just below the hatch and the porch and ending well short of the LM's footpad, to prevent any possibility of damage to it. That bottom rung was high enough off the ground that Neil Armstrong, among others, practiced jumping back up to it, to see how hard that would be, before heading off to explore the Moon. The porch itself got handrails. And the astronauts got a practical way of coming and going to their spacecraft. Outfitted in their blazing white spacesuits and bulky life-support backpacks, they dwarf the ladder itself. But it was the right solution. And it also provided a reasonable spot from which to step off onto the footpad at the bottom of the lunar module leg, and from there onto the Moon's surface. Indeed the phrase "That's one giant swing for mankind" would not have been the words to echo down through history the way "one giant leap for mankind" has.

---

Flying to the Moon was so hard that when President Kennedy announced we were going, the people responsible for making it happen didn't know how to do it.

They didn't know what kind of spaceship to take to the Moon, what course to fly it through space to get there, how to land that ship on the

Moon, or how to take off again and head safely for home. They didn't know the answers to those big questions, and as the deliberation over the lunar module's "egress rope" shows, they also didn't always know, often for years, the right answers to the smaller, no less important, questions.

That biggest of questions—how to go to the Moon—became a major argument. The discussion and debate about whether President Kennedy should commit the United States to go to the Moon, that conversation and analysis lasted just six weeks, from the April 12, 1961, orbital flight of Yuri Gagarin to the speech by Kennedy to Congress on May 25, 1961. The analysis and debate about how to get to the Moon lasted for the next 14 months, from Kennedy's speech to NASA's formal announcement of the flight method Apollo intended to use in July 1962. Even then, the issue was so contentious, the positions so passionately held, that the White House stepped in and attempted to reopen the debate after the public decision was announced. NASA had chosen a method that science advisor Jerome Wiesner thought, quite simply, would fail; he called it a "technological travesty" and "the worst mistake in the world." Because of his objections, the debate simmered for four more months.[5]

In August 1961 NASA picked MIT to design the computers to fly to the Moon. It wasn't until November 1962 that NASA was finally able to pick Grumman to design and build the spacecraft that one of those computers would actually guide to the lunar surface. Because before you could design the spaceship to fly to the Moon, you needed to know how you were going to get there.

At the heart of the argument was something central to space travel: the technique known as rendezvous, the ability to bring two spacecraft together while they are in orbit, whether around the Earth, or around the Moon or any other planetary body.

When the Space Shuttle flew to the International Space Station, and then docked with it, that's rendezvous. The Space Shuttle did that over and over, of course, dozens of times.

When Neil Armstrong and Buzz Aldrin were finished with their visit to the Moon—which lasted just 22 hours—their job was to blast off in the upper stage of the lunar module and fly back up to meet the command module, which had stayed in orbit around the Moon, piloted by Michael Collins. The lunar module had to lift off from the Moon, fly up to the right altitude and the right orbit, and then zero in on Collins and dock with the command module, at which point Armstrong, Aldrin, and their Moon rocks would transfer over to the capsule for the ride back to Earth.

In simpler terms, rendezvous is no different from a tugboat leaving the pier and heading out to a freighter it needs to guide into port. The tugboat has to know the location of the freighter, it has to know its own location, it has to know in what direction the freighter is moving and how fast it's going. With all that information, the tugboat can find the freighter in the bay, catch up to it, pull alongside, tie up, and do its job. Even if you don't pilot ships on the ocean, you can imagine how that would work—not so different from riding a bike or driving a car.

But although we live in a three-dimensional world, that rendezvous between a freighter and a tugboat is really happening in two dimensions: on the surface of the water. You can sense what direction the freighter is moving, or see what direction it is moving in by its bow wave and the things it passes. You can sense its speed. You can sense your own speed, and you have decades of experience judging the distance between you and other objects in the world. To catch the freighter, you throttle up and make sure you're going faster than it. As you approach, you throttle back so you don't overtake the ship, and ease in, matching its speed to pull alongside.

How to take your tugboat out to rendezvous with that freighter is not a mystery. But rendezvous in space is completely different.

Here's how different it is, in one disorienting example. Say you're flying the Space Shuttle in orbit, and you want to dock with the International Space Station. Well, if you simply zoom toward it—giving the Shuttle some extra rocket thrust to swoop in alongside the docking port—you'll instantly discover two things: the more power you give the Shuttle, the faster you will fly but the farther behind the Space Station you will fall and the higher you will get in orbit. That is, if you rocket *toward* the thing you want to catch in orbit, you will end up farther from it: above its orbit and farther behind it. The way to catch up to something in orbit is, typically, to slow down. As if the way to catch the freighter was to slow down your tugboat.

And that is just the start of the weirdness of navigating in space. Your human intuition—based on 30 or 40 years of living on Earth and rendezvousing with things all the time: a doorway, the curb in front of Starbucks, entrance ramps on freeways—is not only useless in space; it tells you to do the wrong thing. Piloting a spaceship requires putting aside those instincts, which aren't just in the muscles of your feet, controlling the speed of the car, and in the muscles of your hands on the wheel—they are in the way you see the world, instincts imprinted on your brain.

The people in charge of figuring out how to fly to the Moon knew all that. They knew that rendezvous in space was going to be something humans had to figure out how to do, on paper, with equations—first. And then would have to struggle to learn in space itself.

Dave Scott, the astronaut who commanded Apollo 15 and who became proficient at the Apollo computer, spent four years flying jet fighters for the air force. Then from 1960 to 1962 he went to MIT and got two graduate degrees, in aeronautics and astronautics. He took Dick Battin's legendary astronautics course. He knows flying. He knew the theory and practice of orbital mechanics before becoming an astronaut. Rendezvous was so daunting, Scott remembers, that "when I was studying at MIT, the ability to rendezvous in space was an issue for debate. It wasn't clear whether it was possible to develop the mathematics and speed of computation necessary to bring two vehicles together at a precise point in space and time."[6]

In fact rendezvous in space is a kind of puzzle, where you have to start at the ending and work backward. If there's a space station in orbit, and I want to get my spaceship to it, where do I have to put my spaceship so that as I either slow down or speed up, the space station gets closer and closer to me, and we are able to meet at a point in space?

Rendezvous requires a lot of math. It requires thinking ahead. You can't simply launch spaceships into orbit and get them to rendezvous. You have to have a plan in advance, and the plan involves where you are launching from, where you're going to, what orbit the spaceship you're chasing is in, what time of day it is, how fast everything is going. You have to work out the plan in advance, and then you have to follow it—or you stand no chance of connecting with that spaceship.[7]

In the early days of spaceflight, rendezvous wasn't just hard and counterintuitive; it was scary. As Apollo got under way, rendezvous was the problem that confronted NASA, and bitterly divided it.

---

Among the first people to start thinking about how to rendezvous in space was a group at NASA's Langley Research Center in Hampton, Virginia. Langley is where NASA's predecessor, NACA, had been headquartered, and before NASA created the Manned Spacecraft Center in Houston, Langley was where the early manned spaceflights were developed and managed.

Langley was also an aeronautics research center, but it had been mostly focused on aircraft rather than spacecraft before it became part of NASA. In the weeks after Sputnik, one Langley scientist went looking for books on orbital mechanics—on how to fly in space—and in the Langley technical library he found exactly one: Forest R. Moulton's classic, *An Introduction to Celestial Mechanics.* In 1958 Langley was in possession of the most recent version of Moulton: the 1914 update of the 1902 edition. The newest text Langley had about movement through space dated to before World War I.[8]

But it was clear at Langley and elsewhere that any serious effort in space would require figuring out rendezvous. If you wanted a space station, you'd have to be able to send spacecraft up to dock with it. If you wanted to service or retrieve satellites, you'd have to be able to chase them down and capture them in space. If you wanted to be able to launch pieces of a bigger project—a spaceship or a space station—and assemble them in space, you'd have to be able to rendezvous.

By the summer of 1959 Langley had formed two different committees to study how to do rendezvous, and both were chaired by a man named John C. Houbolt, a scientist, engineer, and analyst regarded at Langley as brilliant. Houbolt was chosen to, among other things, teach the Mercury astronauts about space navigation.[9]

Houbolt became fascinated by rendezvous, by the mechanics of it, the math, by things like the timing rendezvous required: when did you have to launch your spacecraft from Earth to be able to rendezvous with a particular spacecraft already in orbit? Although he had colleagues at Langley also working on rendezvous, Houbolt became known as "the rendezvous man."[10]

In particular, it looked like rendezvous could really help with the logistics of a flight to the Moon. By stunning coincidence, on the very same day in January 1960, one of Houbolt's colleagues at Langley and a group of visiting engineers and scientists from the aeronautics firm Chance Vought presented variations on the same idea of how to use rendezvous to make a Moon trip, suddenly, more practical.

One of the hard things about flying to the Moon is balancing weight and fuel. If you fly a single rocket from the Earth to the Moon and back, it has to be awfully big, because it has to launch from Earth, have enough fuel to get to the Moon, then it has to land on the Moon, have enough fuel to launch itself back off the Moon, and enough fuel to fly back home, where it has to have the heavy heat shield necessary to protect it on reentry. A single rocket performing all those jobs would, frankly, have to be gargantuan—it would be carrying fuel all the way from Florida to the surface of the Moon, which is to say, it would be carrying fuel in order to carry fuel. It would also be carrying all the way to the Moon and back a heavy protective heat shield that would be used in the last 30 minutes of the mission, and it would have to carry the fuel to launch that heat shield off the Earth, and then off the Moon, and fly back to Earth.

But what if you flew to the Moon, and then left most of the heavy equipment in lunar orbit and used a little shuttle craft to fly down to the Moon and back? That was the idea that William Michael of Langley came up with—he called it a "parking orbit"—and it was also the idea that the visitors from Chance Vought presented. Only take to the surface of the Moon what you needed on the Moon; leave everything else—the fuel to fly home, the big spaceship with its heat shield—in lunar orbit, so that at least you didn't have to fly it down to the Moon and then launch it back up.

This version of rendezvous—this use of rendezvous—absolutely captivated John Houbolt. It became known as "lunar-orbit rendezvous," because once you left part of your spaceship in lunar orbit, you had to rendezvous back with it when you lifted off the Moon. "It became clear that lunar-orbit rendezvous offered a chain-reaction simplification," said Houbolt. Everything involved in spaceflight—"development, testing, manufacturing, erection, countdown, flight operations," as Houbolt ticked them off—would be easier if you didn't have to figure out how to make a single ship do every job and also carry all the fuel necessary to do every job. "The thought struck my mind, 'This is fantastic!' I vowed to dedicate myself to the task."[11]

Houbolt sketched out the steps of a basic flight to the Moon before John Kennedy had even become president, while NASA was already doing some long-term Moon mission planning. And Houbolt's calculations were arresting. He figured you could save almost half the weight of a Moon rocket by using lunar-orbit rendezvous, compared to hauling everything all the way to the Moon and back. In spaceflight terms, being able to cut the total weight of your launch rocket and its spaceships in half was huge. That, in fact, is part of why Houbolt found lunar-orbit rendezvous so compelling. It really did make a Moon flight seem possible.[12]

Starting in the second half of 1960 and well into the middle of 1961 and beyond, Houbolt became a missionary on behalf of lunar-orbit rendezvous, "LOR," as it became known inside NASA. He talked to any group that would listen, and he sometimes talked to groups that weren't that interested. One document shows that starting in September 1960, with a presentation to Robert Seamans, the second-in-command at NASA, and running through September of 1961, Houbolt gave 15 presentations about LOR.[13] And he was consistently surprised that NASA's engineers and senior staff didn't immediately see what he saw: if you were going to the Moon, lunar-orbit rendezvous was the way to go. To Houbolt, it was head-smackingly obvious.

In fact Houbolt didn't just encounter indifference to LOR; sometimes he was met with outright hostility. In December 1960, just before Kennedy took office, Houbolt gave a presentation to the senior NASA headquarters staff; NASA's first administrator, T. Keith Glennan, was there, as was Seamans, Wernher von Braun, and the brilliant and volatile spacecraft designer Max Faget. When Houbolt was done, Faget "jumped up" and said to the group, "His figures lie! He doesn't know what he's talking about!" Faget's outburst was intemperate and uncollegial, not to say disrespectful. But he was channeling a broad current of opposition to Houbolt's LOR proposal. Lunar-orbit rendezvous suffered from just one problem: rendezvous. And not just any rendezvous; LOR required two spacecraft to rendezvous in orbit around the Moon, 240,000 miles from Earth.

Perhaps because Houbolt had spent so much time working on rendezvous itself—on the techniques, the timing, the equations that would make it possible—he was completely comfortable with the idea of rendezvous. More than comfortable, he was confident. Rendezvous, the technique, wouldn't be a problem.

But in that December 1960 meeting with the most senior staff at NASA, he was speaking five months before Americans would launch a single man into space, and then for just 15 minutes. Houbolt was speaking 14 months before America would manage to put an astronaut into Earth orbit and bring him home safely. He was advocating the most sophisticated spaceflight technique to a room full of engineers, scientists, and managers who were still struggling to get the basics of the American space program into operation.

Rendezvous then remained almost unimaginably challenging, even to sophisticated space managers. Today it has become routine; figuring out rendezvous trajectories and launch windows is just part of space operations. We have the math, we have the computers, we have the rockets, we know how to do it. We also launch single jet planes, each carrying 300 or more passengers, every hour that cross the Atlantic and the Pacific without a thought and with almost never a problem. We send mail that goes from Chicago to Shanghai in a second or two. Both of those things too would once have seemed unimaginable.

The space historian James Hansen published an extensive study of Houbolt's impact on Apollo in 1995, pointing out that it can be hard, now that rendezvous is routine, to appreciate how uncertain and risky it seemed in the early 1960s. The fear was simple: if one computation was done incorrectly, if one engine was fired too long or too briefly, if anything went wrong, rendezvous was unforgiving. The little shuttle ship leaving the Moon—already scientists at Langley had nicknamed it a "lunar bug" or a "lunar schooner"—would never be reunited with the mother ship orbiting above. "The fear that American astronauts might be left in an orbiting coffin some 240,000 miles from home was quite real," Hansen wrote. "The morbid specter of dead astronauts sailing around the Moon haunted the dreams of those responsible for the Apollo program. It was a nightmare that made objective evaluation of the LOR concept by NASA unusually difficult."[14]

The Moon mission planning that had been going on in 1960 and early 1961 got turbocharged, of course, once Kennedy gave NASA the mission to get to the Moon by the end of the decade. NASA officials at that moment, from James Webb well down into the ranks, took both parts of the charge with equal seriousness: to get astronauts to the Moon and back safely, and to do it before the end of the decade. Do it, do it well, do it with speed.

There were, in fact, two other ways NASA was studying to go to the Moon. One was that classic method: one big rocket and spaceship all the way and back, like a schoolboy might have drawn. That method went by the name "direct ascent," and it had enough backers inside NASA that the rocket engineers in Huntsville, Alabama, were working on the design of a massive rocket called Nova that might have been capable of pulling it off. It would have had 50 percent more lifting thrust than the giant Saturn V did.

Another method went by the name Earth-orbit rendezvous (EOR). This was the favored technique of von Braun, among others. EOR too involved using a big spaceship to go to the Moon, but instead of the big spaceship being launched on a single big rocket—which might take years to develop and test—it would be launched on two or more smaller rockets and assembled in orbit. In one scheme, two Saturn V rockets would launch fully fueled spacecraft modules—too big for a single rocket—that would then be assembled in orbit and head for the Moon. In another version, one Saturn V would launch the fully assembled Moon ship, the other would launch its fuel; once in Earth orbit they would rendezvous, and the fuel would be transferred to the Moon ship so it could head out on its mission. Earth-orbit rendezvous at least kept the astronauts close to home, so if something went wrong with the rendezvous, they could simply fire their retrorockets and return to Earth.

In hindsight, both these ideas have a completely impractical science-fiction air about them, even 50 years later. How in the world would a crew of three astronauts land a massive spaceship, backward, safely onto the Moon? As one engineer pointed out, NASA was having enough trouble trying to get much smaller rockets, from modern, fully staffed launch facilities, to fly successfully in the other direction.[15]

Put aside the landing of such a big rocket on the Moon—or assume it was through some miracle successful. Skeptics quietly noted that the simple question of how to get the astronauts down to the Moon's surface from the cabin atop an 80-foot rocket, and then back up to the cabin—that problem alone was unsolvable. (Rope would not have worked there either.)

As for Earth-orbit rendezvous, by what method would astronauts be able to assemble a massive spaceship from components in Earth orbit? How would they transfer highly volatile fuels, which needed to be kept super-chilled, from one orbiting spaceship to another?

And yet rendezvousing in lunar orbit seemed so daunting that through much of 1961, direct ascent and Earth-orbit rendezvous were treated as the serious contenders for a way to go to the Moon, and LOR was an afterthought at best.

In mid-July 1961 Houbolt was to give a major presentation at an Apollo planning conference. As was typical for NASA, the presentations were practiced beforehand. Houbolt had given a lot of presentations by this point and had developed his own approach to being persuasive. For this event he focused much of his talk on rendezvous as a technique and came to the power of lunar-orbit rendezvous at the end. A senior official from Langley, attending the rehearsal, told Houbolt, "That's a damn good paper, John. But throw out all that nonsense on lunar-orbit rendezvous."[16]

Houbolt had joined Langley in 1942, and in 1961 he was 42 years old. He was not exactly mild-mannered, but neither was he excitable. He was a 1950s rocket scientist. He was careful, he was precise, he was matter-of-fact. But by the summer of 1961, he was fed up. So he did something he thought might get him fired—or might get someone to make the right decision.

---

On November 15, 1961, Robert Seamans, the second-in-command at NASA, received a letter. "Somewhat as a voice in the wilderness," John Houbolt opened to Seamans, "I would like to pass on a few thoughts on matters that have been of deep concern to me over the recent months."

The two had met not long after Seamans had taken the job as associate administrator of NASA, during a tour of Langley more than two years previously, and Houbolt had given Seamans an early version of his LOR run-through. Houbolt had also written Seamans a brief letter six months earlier. By way of opening in this letter, Houbolt wrote that he needed to pass on "ideas and suggestions which are so fundamentally sound and important that we cannot afford to overlook them."

> I fully realize that contacting you in this manner is somewhat unorthodox; but the issues at stake are crucial enough to us all that an unusual course is warranted. Because you . . . do not know me very well, it is conceivable that after reading this you may feel that you are dealing with a crank. Do not be afraid of this. The thoughts expressed here may not be stated in as diplomatic a fashion as they might be. . . . The important point is that you hear the ideas directly, not after they have filtered through a score or more of other people.

The letter is nine single-space pages. It reviews the three Moon-flight plans under consideration, and also the bureaucratic politics of committees and planning groups that refused to take LOR seriously. The letter is often impolitic. "I have been appalled at the thinking of individuals and committees on these matters," Houbolt wrote of study groups that Seamans himself had set up. "For some inexplicable reason, everyone seems to want to avoid simple schemes. The majority always seems to be thinking in terms of grandiose plans."

At one point, without naming the person involved, he recounts having lunar rendezvous dismissed contemptuously, without serious consideration, by a senior NASA official, someone it's possible Seamans could have easily identified. "I am bothered by stupidity of this type being displayed by individuals who are in a position to make decisions which affect not only NASA, but the fate of the nation as well."

Houbolt made the case for LOR: "The lunar rendezvous approach is easier, quicker, less costly, requires less development, less new sites and facilities." The question is "Do we want to get to the Moon or not?" He underlined it for emphasis.[17]

Houbolt was about a dozen levels down the organizational chart from Seamans. A serious PhD scientist, 19 years at NASA and its predecessor, but the letter itself, and especially its tone, bordered on the insubordinate. "Unorthodox" at the least.

Seamans was a calm boss, a careful and thoughtful manager. He found the letter irritating. "My first reaction was, I'm sick of getting mail from this guy! I thought of picking up the phone and calling Tommy Thompson, Houbolt's superior at Langley, and telling Tommy to turn him off. Then I thought, 'But he may be right.' . . . [NASA] may not be very keen about [the idea], but it makes a lot of sense to me."[18]

Seamans did something smart: he took the letter down the hall to Brainerd Holmes, one of his deputies who was then in charge of all manned spaceflight, including Apollo. (Holmes himself would be fired the following summer because of his somewhat prickly personality.) "I got another one of these zingers from John Houbolt," Seamans told Holmes. "I'd like to have you read it while I'm here."[19]

Seamans insisted that Holmes give LOR a serious, thoughtful assessment. At the beginning of January 1962, Holmes had hired a new deputy named Joe Shea, and his first major responsibility was to sort out this "mode question": how NASA was going to get to the Moon. Time was pressing. You couldn't start designing and constructing your Moon spacecraft—you couldn't train astronauts, ground crews, flight controllers—if you didn't know how you were going to the Moon.

Shea recalled that soon after he was hired he ended up in a meeting with Holmes and Seamans where the topic of Houbolt's letter and the right way to go to the Moon came up. "You know," said Seamans, "I don't think we really yet know how we're going to the Moon." Shea, the newcomer, replied, "I was beginning to get that same suspicion."[20]

The disagreement didn't just revolve around Houbolt, of course. The partisans of direct ascent and of Earth-orbit rendezvous had also been in fierce dispute. It was time to end the brawl. Shea made the rounds of the centers: he went to Langley; he went to the Marshall Space Flight Center, the base of von Braun in Huntsville, Alabama; he went to Houston, where the vast Manned Spacecraft Center was just starting to be assembled and constructed. In fact, as 1962 got under way, the direct ascent option faded because of the logistics on the Moon—how exactly would three astronauts, alone, launch a rocket that at Cape Canaveral required 1,000 people to launch?—but also because it just looked like creating the Nova rocket would take too long.

The team in Houston in charge of manned spaceflight had relocated down from Langley in the middle of the debate, and they had gradually come around to the value and elegance of Houbolt's idea. At one point in early 1962, Houbolt made a trip to Houston and, under the wing of a colleague, visited with "almost everyone with some interest in the mission mode issue," explaining lunar-orbit rendezvous in detail.[21]

It really did make all the engineering and design easier if one spaceship could be dedicated to the trip to the Moon, and a different ship could be used solely for the trip from lunar orbit to the Moon's surface and back to lunar orbit. The main ship—what became Apollo's capsule-shaped command module—would be aerodynamic and robust and have the heavy, high-tech heat shield necessary to come home at 25,000 mph. The "lunar schooner" could be lightweight, carrying only enough fuel, equipment, and supplies for a couple of days' stay on the Moon, and it could be designed specifically with the idea of landing on the Moon and taking off again, without help from a support staff or a launchpad. The command module would need controls only for flying to the Moon and back to Earth. The lunar schooner would need controls only for flying to a Moon landing and back to orbit.[22]

Sometimes you need a freighter, sometimes you need a ski boat, and the things those two boats do well cannot in fact be very elegantly combined. The separation of functions, which did of course require making two completely different spaceships, nonetheless dramatically simplified the jobs of creating each of those ships and of making it possible to build them and fly them.

Von Braun's group at NASA's rocket center in Huntsville was devoted to defending Earth-orbit rendezvous, in part because they were in the rocket business, and with EOR each Moon mission would require at least two Saturn V launches. EOR might also require the construction of a basic orbiting work platform—some kind of early space station—as a place to assemble the parts of the Moon rocket or fuel it. Von Braun had a long-standing passion for creating an orbiting space station.[23]

Shea spent much of the first half of 1962 working through the "mode question." Seamans, Holmes, Shea—none of them wanted to impose a decision. The challenge was too great to have one of NASA's huge regional space centers spend years resenting that its choice of how to get to the Moon had "lost." Shea hired an outside firm to assess the options—he hired, in fact, Chance Vought, the aeronautics company that two years previously had developed its own early concept for lunar-orbit rendezvous. One of Chance Vought's conclusions was that a complete Moon mission could be done using LOR with one Saturn V launch, but that EOR would require at least two.

Shea also assigned fresh analyses of both the LOR and EOR methods inside NASA, but he made Houston do the assessment of EOR, the choice they didn't favor, and he made Huntsville do the assessment of LOR, the choice they resisted. He wanted each side to get to know the option they didn't favor, not just as a cartoon opponent but in engineering and scientific detail.

There were lots of meetings where Shea pulled together staff from both Houston and Huntsville and, in what would become standard for Apollo, forced them to stop arguing and work through specific technical issues with each method. Shea himself, at his first briefing from Houbolt at Langley in January, had quickly developed an appreciation for the power of LOR.

The breakthrough came at a surprising moment, at a meeting in von Braun's headquarters in Huntsville on June 7, 1962. It was a daylong session devoted to having von Braun's staff present the virtues of Earth-orbit rendezvous to Shea.

At the end of the day, von Braun himself stood up to speak. The text of his remarks runs 11 single-space pages. But the stunning news was in the opening moments: he was putting his considerable weight and authority behind lunar-orbit rendezvous. "Why?" von Braun asked. "[Because] we believe this program offers the highest confidence of successful accomplishment within this decade." By all accounts even von Braun's own scientists and engineers were as surprised as anyone else by the rocket pioneer switching sides.

Von Braun was persuaded by the practicality of the lunar-orbit method, by the fact that it could be done more quickly than Earth-orbit rendezvous, and that it was cheaper. He even nodded to John Houbolt, pointing out that now both his space center and Houston "have actually embraced a scheme suggested by a third source," and so the partisanship could be put aside. Moving forward with clarity and urgency was vital, von Braun insisted. "We are already losing time."

In the end, the specialization of vehicles was refined one step further after LOR was chosen: not only would the lunar module be specialized to do just the Moon landing, but most of the lunar module would be left on the Moon. It would be separated into two vehicles itself, so that just the crew cabin would blast off back to orbit, leaving behind as much equipment as possible: the entire lower stage, including legs, ladder, fuel tanks, plumbing, and the big descent engine. In fact before each of the six lunar modules lifted off from the Moon to head back to orbit, the astronauts depressurized the cabin, opened the hatch, and tossed out onto the surface the backpack life-support units from their spacesuits that they had needed to walk around. No point blasting off with big pieces of equipment you were done using.

Here's how powerful the idea of lunar-orbit rendezvous would prove to be in the end. The heat shield that covered the bottom of the command module to protect the astronauts as they came blazing back through the atmosphere—the high-tech resin, applied meticulously by hand, and the framework in which it was mounted—weighed 3,000 pounds. The entire ascent module of the lunar module—the crew cabin with its ascent engine—weighed 4,700 pounds. If you had taken a single spaceship all the way to the Moon, carrying just the heat shield would have been the equivalent of taking almost an entire extra spacecraft to the Moon's surface just to blast off again.

Houbolt was not in Huntsville that day von Braun made his announcement, and he hadn't been much a part of Shea's effort to pull the senior officials and their staff into consensus. Not long after the Huntsville meeting, though, he was at NASA headquarters on a day when the Houston staff had gathered to practice their presentation of lunar-orbit rendezvous for Jim Webb; they were rehearsing to persuade Webb that LOR was the way to go to the Moon, and the rehearsal was being reviewed by Seamans. Houbolt asked if he could attend and was invited in.

When it was finished, Seamans turned to Houbolt and said, "Well, John, how does that answer your letter?"[24]

The decision to use lunar-orbit rendezvous and a "lunar ferry" (as the *New York Times* called the lunar module) was announced at a news conference in Washington on July 11, 1962. The urgent Moon mission had been under way for 14 months, and it could at last commission a Moon-landing spaceship, even as NASA chief Webb would spend months fending off the complaints of White House science advisor Wiesner.

On the Wednesday of the announcement, Houbolt was in Paris giving a paper at a scientific conference. His Langley boss, Ed Garrick, was with him and happened to notice news of the LOR press conference in the Paris edition of the *New York Herald Tribune*. Garrick showed Houbolt the newspaper story, then shook his hand. "Congratulations, John," he said. "They've adopted your scheme. I can safely say I'm shaking hands with the man who single-handedly saved the government $20 billion."[25]

The one thing Houbolt's critics would turn out to be right about was that rendezvous would prove to be just as devilishly counterintuitive as everyone guessed. It was so exotic that when NASA had announced publicly in May 1961 that it was going to *attempt* to rendezvous vehicles in space sometime between 1965 and 1970, both the *New York Times* and the *Washington Post* wrote news stories about it.[26]

And NASA's first effort to use astronauts to rendezvous failed, in just the way one might have imagined for a species of gravity-bound, novice space travelers.

Gemini 4 was a four-day mission launched June 3, 1965, with astronauts Ed White and Jim McDivitt. It was only the second Gemini mission with astronauts, and it was the first multiday U.S. space mission; NASA wanted to learn how to operate effectively in space, and manage space missions on the ground, for longer durations. It featured the first U.S. spacewalk, with White floating outside the capsule for 20 minutes and enjoying it so much that he had to be personally ordered back into the spaceship by flight director Chris Kraft taking to the radio and hollering, "The flight director says, Get back in!"

McDivitt and White were the first astronauts to attempt a rendezvous in space. Their goal was to rendezvous their capsule with the used second stage of the Titan rocket that had helped launch them. Just after separating from the booster, as they entered orbit, McDivitt made a first stab at rendezvous. The official NASA history of the Gemini program, *On the Shoulders of Titans*, tells what happened: "McDivitt braked the spacecraft, aimed it, and thrusted toward the target. After two bursts from his thrusters, the booster seemed to move away and downward. A few minutes later, McDivitt pitched the spacecraft nose down and the crew again saw the rocket, which seemed to be traveling on a different track. He thrusted toward it—no success—and stopped. McDivitt repeated this sequence several times with the same luck."[27]

To Mission Control, McDivitt reported, "The booster fell away quite rapidly and got below us like there was a considerable difference in our velocity." The booster wasn't moving away, of course; it was McDivitt's spaceship that was moving away, following the laws of orbital mechanics.

McDivitt and White tried again after passing through darkness. No luck. The booster stage had only two lights on it, and McDivitt had trouble assessing how far away it was. Their Gemini capsule had no radar to help them judge the distance.

McDivitt estimated that he started out a few hundred feet from the booster. After 45 minutes of trying to get to it, he radioed CapCom Gus Grissom at Mission Control: "I think we ought to knock it off, Gus. It keeps falling. It's probably three or four miles away, and we just can't close on it."

"Right, knock it off," Grissom replied. "No more rendezvous with the booster." In the first 90 minutes of a four-day flight, McDivitt had used up half the capsule's maneuvering fuel futilely "chasing" the booster in defiance of the laws of physics and motion.[28] The failure of the rendezvous was mostly overlooked, however, by the triumph of White's spacewalk, which came just after.

McDivitt was a deeply experienced air force pilot—10 years, including 145 combat missions in the Korean War. In the crew debriefing after the flight, two things were clear: he had learned the theory and math of orbital mechanics, and he had been taught what the correct maneuvers were to be able to rendezvous. But when it came down to it, when he was at the controls of his spaceship looking at that booster out the capsule window, he just didn't buy those instructions. He tried to fly in space using instincts honed on Earth.

"I thrusted right at the booster again," he said in the debriefing, "did it three or four times. . . . I thrusted some more right at the booster, trying to just overcome orbital mechanics with brute force. . . . It was too late to start playing fancy games with the orbital mechanics."

Eventually, McDivitt said, "it looked like a hopeless task and that we had better stop this stuff or we were going to lose all the fuel for the whole mission."[29]

NASA would perform a successful rendezvous between two Gemini spacecraft just six months later. And Jim McDivitt would go on to command Apollo 9, the first test flight of the lunar module in orbit with astronauts at the controls, a huge success in which McDivitt flew the lunar module 111 miles away from the command module in Earth orbit, and then returned for a pinpoint docking. McDivitt had learned to fly using the rules of orbital mechanics rather than trying to beat them.

---

The creation of the lunar module would prove to be as challenging, as awkward, even as harrowing for Grumman and for NASA as any element of Apollo's creation.

The LM was, in fact, perhaps the strangest flying craft ever created by human beings. It was the first spacecraft designed solely for use off Earth, and is still the only such spaceship. It would never have to fly through an atmosphere, so it didn't need the structural robustness that would require. It also didn't need to be aerodynamic. It would only ever fly in space, and then part of it would be left in space, and part of it on the surface of the Moon.

The lunar module is often described as "spindly" or "spider-like." Before the design was close to finalized, people inside NASA had taken to calling it a "bug." On the day of the lunar module's first test flight—conducted without astronauts—the *New York Times* called it "the ugly duckling of the Apollo project vehicles." A month before it landed on the Moon, *Life* magazine called it "utterly and lovably different" and "its own grotesque self."[30]

Its wild shapes came from the fact that it didn't need to be sleek or symmetrical, and it didn't need exterior panels protecting it. The lunar module was what happens when form follows function: Where there was a bulge, there was a fuel tank. The legs only needed to do their job—cushion the LM as it settled onto the Moon; they would never encounter air resistance, so they didn't need to be folded or stowed or hidden. The antennas poked out all over the top of the spaceship, ready to be positioned where they needed to be.

The lunar module had two other strange, even unsettling qualities. First, it could never be test-flown. There's no place on Earth to take a spaceship designed for flight in a zero-gravity vacuum and fly it around. The first time a lunar module was flown, it was in the middle of a Moon mission. And that was also the last time that that particular lunar module would be flown. And second, that meant the people who would pilot the lunar modules to the Moon never got to learn to fly them, or practice flying them, except in simulators, which were themselves designed and built by people who had never flown a lunar module.

And yet, as with every other element of Apollo, the lunar module had to be absolutely reliable. How could you guarantee the reliability of a system you couldn't actually use in the way it was intended?

First, you took incredible care with how you assembled the spaceship. That was something Grumman had to come to grips with, after decades of assembling warplanes in long ranks in open factory spaces. The lunar modules ended up assembled in a hangar-scale factory building that had been turned into a giant clean room; it was 200 feet long, 80 feet wide, with a 35-foot-high ceiling, the whole interior painted white and the room sealed and supplied with highly filtered air and maintained at positive pressure, so nothing could slip inside. Everyone entering had to put on coveralls, booties, a cap, gloves, and goggles. Those working directly on the lunar modules always wore gloves. Anyone working inside the crew compartment had to log and then account for every item they took into the cabin, going in and coming out, including whatever might have been in their pockets. Eventually tool kits inside the assembly area had Styrofoam cutouts so Grumman could be sure that no tools, or even something small like a wrench socket or a drill bit, accidentally got left inside the cabin. That kind of floating debris could cause havoc in space.[31]

After the lunar modules were assembled, each stage was mounted in a huge contraption known as "the Tumbler." It turned each half of the lunar module upside down and rotated it in midair, shaking it out. The goal was to rattle loose anything left inside—even something as small as the shavings from a rivet. Grumman workers spread a canvas sheet beneath the Tumbler to collect whatever fell out.[32]

The other way to assure reliability was testing. Grumman developed a near obsession with testing that ended up baked into the culture of the 6,000 people designing and building the lunar modules in Bethpage, Long Island. "It didn't take that long to assemble the vehicle," said chief engineer Tom Kelly, "but it took a long time to test it. . . . That's what we were doing most of the time in Bethpage." Each lunar module "was tested for two years for a three-day mission. We practically wore them out."[33]

Individual parts were tested, components were tested, systems were tested. Astronaut Fred Haise—he would be the lunar module pilot on the ill-fated Apollo 13 mission—was assigned to help. Although Haise lived in Houston, like all the astronauts, during the development of the lunar module he spent 14 months out of 17 at the Bethpage factory, "most of the time in a trailer that was attached to the clean-room area in Plant 5."

Grumman had started out a year and a half behind other parts of the Apollo system, just because it took so long to decide on how to fly to the Moon. The company not only never caught up, it continually fell further behind its own production deadlines, and NASA's. At one point in 1964, astonishingly, NASA calculated that for every five weeks of work Grumman was doing on the lunar modules, it was falling four weeks further behind schedule. The work was growing faster than Grumman could do it.[34] The pace was unrelenting, even for an astronaut assigned to the effort, like Haise. "There were times I'd be here a whole week," he said, "and never get any further than Vito's Deli, which is just across the street, to get a hero sandwich or something. . . .

"Tests ran around the clock. I've run tests consecutively, as long as 23 hours. When we ran the first radar test, that ran for 27 hours straight." During those stretches, Haise said, "rather than leave the vehicle, I'd just lay down on the floor, and say, Call me when we're ready to go again." And then he would sleep on the floor of the cockpit of a not-quite-finished lunar module.[35]

The LM was plagued with fuel leaks, so every joint in the fuel system was x-rayed. The testing resulted in a credo at Grumman: There are no small anomalies, there are no random failures. Anything that happens that isn't what it should be needs to be tracked down and understood. Because if it happens in space, everyone could be in trouble. Kelly's staff logged the "anomalies"; over the life of the lunar module project there were 14,000. "At the end of the program," he said, "we only had 22 of those that were unexplained failures."[36]

In December 1967, for instance, a lunar module that had been finished was being given a pressure test; the cabin was sealed and pressurized, as it would be in spaceflight. At Grumman this vehicle was designated LM-5; it was, in fact, *Eagle*, the lunar module that would be the first to land on the Moon during Apollo 11. As the pressure was brought up, one of the two distinctive triangular windows in the front shattered.

It was a stunning moment, and not a good one. That window was just a single part in a spaceship with one million parts, but if a lunar module window shattered in flight, the two astronauts would die instantly.

"The window shattered without anybody touching it," said Joe Gavin, who was the senior executive in Bethpage in charge of the lunar module, and Tom Kelly's boss. The windows were specially made, of course, by Corning, regarded as the best glass company in the world (and the company that went on to make the windows for every NASA spacecraft). The glass was single-pane, three-eighths-inch thick.

Was there a flaw in the window when the glass was cast and finished? Had it somehow been installed incorrectly? Had it been accidentally damaged or banged after being installed?

"We went back to Corning," said Gavin, "and reviewed the whole process." In fact, as a result of the shattered window, NASA developed and performed a whole set of tests on lunar-module window glass—11 distinctive tests. A new pre-acceptance testing procedure was developed as well, to make sure the windows Grumman was getting from Corning really were robust and correctly made. A protective cover was developed for the windows once they had been installed in the lunar module that showed a mark if anyone so much as touched them. "It was one of the last things done before launch at the Cape," Gavin explained, "to tear off those protective covers." And, he added, the staff at Grumman ended up "finding out a lot about glass that I don't think any of us realized."

The newly rigorous window screening tests proved invaluable: eight lunar module windows failed during acceptance testing over the life of Apollo. In flight, though, said Gavin, "we never had a . . . problem."

But despite that intensive investigation, they also never figured out why that window had shattered in the first place: it ended up as one of the 22 mysteries. "The solution was never one of those that gave you a completely warm feeling," said Gavin.[37]

The "anomalies" often validated the incredible demands being made on the space equipment itself, the care NASA and its contractors took to unravel the failures, and also to track how the spacecraft and its parts were assembled, which could seem ridiculous, except when it turned out to be critical.

The command module, the service module, and the lunar module were all filled with high-pressure tanks of all kinds—for holding fuel, helium (used to pressurize other tanks), oxygen. On the three spacecraft modules, there were 71 tanks total, each a potential disaster.

Not long before the window failure, earlier in 1967, a propulsion tank for the lunar module's descent engine, designed to hold helium at extremely cold temperature, failed while it was being tested at the company that made it, AiResearch in California. That was the point of testing, of course, to catch the problems. The failure happened at a weld between two hemispheres of the tank.

The welding rod used to seal the tank's parts was found and examined; it was the right material. But a microscopic examination of the place where the failure originated in the tank showed tiny cracks. Tom Kelly tells the story of what happened next in *Moon Lander*, his memoir of building the lunar module. Henry Graf, the manager of supercritical helium systems at AiResearch, "became obsessed with finding the cause of this failure."

> He led his engineering and quality staff through a minute examination of every step in the manufacturing process, starting with the receipt of the titanium forgings and the quality pedigree that accompanied them. At each step of the process, they looked at what had been done on the failed tank, and asked whether anything in this step was different from their process on the previous tanks. Graf's careful detective work paid off, discovering a cause so trivial that a less observant investigator would surely have overlooked it. Graf noticed one minor difference in the process for this tank and those that had preceded it: instead of using new cloth pads to wipe the tank surfaces prior to welding, washed, re-used cloths were employed. Examination of the washed cloths showed traces of detergent, and test samples [of titanium] that were wiped with them failed under combined stress and humidity testing. The trace detergent attacked titanium! There could be no more gripping example of the extreme sensitivity of highly stressed tank material and welds to contamination.[38]

A change in procedure that might well have been thoughtful and well-motivated—and that, as Kelly pointed out, could easily have been overlooked—put at risk a future Moon landing in a way no one could have predicted, even if they had known about it in advance. Which is why the "There are no trivial failures" culture turned out to be so critical.

Some parts of the lunar module couldn't be tested at all. The batteries that supplied power throughout the lunar landing and the surface stay were specially made for Grumman; they were high capacity, but they could not be recharged. So if you "tested" them, you also used them up. In cases like that, Grumman resorted to a "test and sample" technique: it had more batteries manufactured than it needed, so it tested a sample from each batch, and if they were the right quality and worked correctly, the engineers randomly made a selection of flight batteries from the rest of the batch.[39]

In the end, Grumman manufactured 14 flight-ready lunar modules. The company that during World War II had been able to produce 14 Hellcat fighter planes a day needed a decade to produce 14 spaceships. That's a measure of the learning curve, to be sure, but it's also a measure of the difference in complexity between a high-performance warplane and a high-performance spacecraft.

Of the 14 flight-ready lunar modules Grumman built, ten flew in space, and six landed on the Moon. The total cost of the lunar modules was $1.6 billion ($11 billion in 2018 dollars); each one cost $100 million, although by the time lunar modules were flying to the Moon, Grumman said it could produce a new one for just $40 million, if anyone wanted one.[40]

For all the gawkiness of the lunar modules when you see them here on Earth, what they really look like is a working spaceship. The space capsules of Apollo, but also of Mercury and Gemini, are aerodynamic and bear the scars of spaceflight. But there is something inscrutable about a capsule. It's hard to know what you might do with it. The lunar modules don't look gawky so much as they look useful, like they were designed with work in mind, with a specific mission. They are no more awkward than a submarine or a crane or an aircraft carrier.

Kelly, for one, never lost affection for the flying machine he was so instrumental in creating, and never lost sight of its extraordinary mission. "Sometimes," Kelly wrote in his memoir, "if it was not crowded I would go inside the LM cabin."

> Crawling on all fours through the hatch, I stood up inside the crew compartment at the flight station, about the size of a modest walk-in closet. There were usually one or two technicians inside. . . . But sometimes I was there alone and could let my imagination fly ahead to the very day when that very LM, that very square foot of cabin flooring where I was standing, would descend to the Moon's alien surface in the final test of all our efforts and dreams. How I wished to be a stowaway in that tiny cabin![41]

The lunar modules were specialty flying machines in a way that is rarely noticed. They did their mission perfectly, in the end, but they didn't actually have to do much flying—they just had to do that flying well. Put aside for the moment the lunar module from Apollo 13, which saved the astronauts but never did fly independently. The other eight lunar modules that flew in space with astronauts together totaled 54.5 hours of flight time. The typical space-flight time was a little more than six hours, from lunar orbit down to the Moon, then back up to rendezvous with the command module. After landing, of course, the bottom half of the spaceship was left on the Moon, where they all remain today. (In fact the bottom half actually flew for only three hours.) The crew compartment, the ascent stage, took the Moon walkers back to orbit, where they transferred themselves and their Moon rocks to the command module, sealed the docking tunnel, then jettisoned the upper stage to either orbit the Moon or be crashed into it by remote control from Mission Control, as part of one final scientific test. The lunar modules got neither a ceremonial nor a particularly dignified ending. They got, in fact, exactly the kind of ending that matched their mission and personality: utilitarian.[42]

Given how novel the machine was, and how novel its flight profile was, one thing that's surprising is how little the astronauts talked about what the experience of actually flying it was like. When you read the mission transcripts during the time astronauts were in the lunar module and flying it, the experience itself is so demanding and so absorbing that there's almost no idle time and no idle exchanges with Mission Control. When the upper stage is jettisoned off into space, not one astronaut expresses the slightest sentiment.

Neil Armstrong, the first man on the Moon, just after blasting off from the Moon in Apollo 11's lunar module to head home, said, "The *Eagle* has wings." And as Pete Conrad and Alan Bean rocketed back off the Moon in Apollo 12, in the lunar module they had given the call sign *Intrepid,* Conrad radioed perhaps the only line that the folks at Grumman, or the lunar modules themselves, needed: "I tell you, Houston, I sure do enjoy flying this thing."[43]

---

*Life* magazine did a cover story on the lunar module in May 1969, about four months before Apollo 11 would head for the Moon. The headline was "How an Idea No One Wanted Grew Up to Be the LM," and the story was a profile of John Houbolt and what was then the seven-year-old story of his quest to get NASA to take the idea of the lunar module seriously.

The *Life* story was part of the media hubbub over the first Moon landing that—even with the perspective of the media-saturated world of 50 years later—was astonishing in its range and imagination. One of the best examples was a cover story that *Esquire* magazine did for the month of the first Moon landing, July 1969: "What words should the first man on the Moon utter that will ring through the ages?"

"At the great moments of discovery and invention in the past," William Honan wrote for *Esquire,* "men have risen, or stumbled, to the occasion with everything from instant eloquence to stupefied silence." *Esquire* asked for suggestions for the first words to say on the Moon from 50 luminaries of all stripes, among them Vladimir Nabokov, Hubert Humphrey, Leonard Nimoy, Anne Sexton, Truman Capote, Isaac Asimov, Timothy Leary, and Ayn Rand. It was an inspired and playful idea; in the opening spread, the headshots of each of those invited for comment are in little space helmets. But what is so striking is that, at least from a half-century's remove, almost none of the 50 came up with anything particularly witty or moving or even intriguing. Muhammad Ali cracked wise: "Bring me back a challenger, 'cause I've defeated everyone here on Earth." But he didn't seem to quite understand the assignment. George McGovern was not atypical in his pedestrian suggestion: "I raise the flag of the United Nations to claim this planet for all mankind." Isaac Asimov came closest. His suggestion, "Goddard, we are here!" was a tribute to the American engineer Robert Goddard, who fired the first rocket into the air in 1926. The only problem: unlike Asimov, the typical Earthling was unlikely to know who Goddard was. In the whole roster of *Esquire*'s famous and important, not one came close to what Armstrong came up with: "That's one small step for man, one giant leap for mankind." *Esquire*'s snarkiness notwithstanding, it does indeed echo down the decades of history.[44]

The *Life* profile of Houbolt, so many years and so much work after the fact, occasioned some grumbling inside NASA. The lunar module, lunar-orbit rendezvous, wasn't of course "Houbolt's idea," nor had he ever claimed it was. But it was unequivocal that, for more than a year, he made himself the main advocate of it, and not just a lonely voice but one sometimes treated dismissively by his own colleagues. Would NASA have come around to lunar-orbit rendezvous, and the lunar module it created, without Houbolt? It's impossible to say. But what is absolutely true is that many people inside NASA—senior officials and also frontline staff engineers and scientists—first learned about what looked like an unconventional path to the Moon from John Houbolt. Houbolt's unorthodox letter to Seamans may have had an aggrieved tone, but it also unquestionably changed the conversation about the technique Houbolt had been trying to get people to take seriously.

George Low, who was second-in-command in Houston and then head of the Apollo spacecraft office in the wake of the Apollo 1 fire, thinks that lunar-orbit rendezvous was the only way NASA would have gotten to the Moon, and thinks Houbolt was the reason NASA managed to pick LOR. "Had Lunar Orbit Rendezvous Mode not been chosen," Low would write a decade after the Moon landings, "Apollo would not have succeeded; and without Houbolt's letter to Seamans (and the work that backed up that letter), we might not have chosen the Lunar Orbit Rendezvous Mode."

Houbolt left NASA in 1963, not long after the contract for the lunar module with Grumman was finalized, to go to work for a private aerospace consulting firm. But he was invited to witness that first Moon landing, on July 20, 1969, in a place of honor: inside Mission Control, as a guest in the VIP section behind huge glass windows overlooking the flight control positions.

Also in the VIP viewing room that day was Wernher von Braun, still a power inside NASA, whose Saturn V rocket had gotten the astronauts to the Moon. When Armstrong radioed back, "Tranquility Base, here—the *Eagle* has landed," joy unleashed itself inside Mission Control and in the VIP section as well. "Turning from his seat," wrote James Hansen, the historian who studied Houbolt's impact, "NASA's master rocketeer, Wernher Von Braun, found Houbolt's eye among all the others, gave him the okay sign, and said to him simply, 'John, it worked beautifully.' "[45]

# 8

# NASA Almost Forgets the Flag

There's no atmosphere on the lunar surface and unless they starch the flag, it is going to hang from its staff like a limp dishrag.

> **United Press International story**
> *describing congressional legislation forbidding astronauts from planting any flag other than the American flag on the Moon, June 11, 1969*[1]

Neil Armstrong had been walking around on the surface of the Moon for 45 minutes when he and Buzz Aldrin teamed up to erect the American flag they had brought with them.[2]

The flag was a simple but clever contraption designed to display an off-the-shelf Stars and Stripes, fully extended, as if waving in a breeze that doesn't exist on the Moon. The flag was three feet tall and five feet wide, and it was attached to a vertical pole for planting in the Moon's rough surface. A seam had been sewn along the top edge of the flag, and a second pole ran through it, like a curtain rod. The two poles—the one for planting the flag, and the one for extending it like a curtain—were hinged at the upper corner of the flag.

Aldrin stepped back to watch Armstrong set up the flag—he hadn't actually planted it yet—and Aldrin gave a quick salute, which was almost lost in shadow.

"See if you can pull that end off a little bit," Armstrong asked him.

Aldrin stepped back in and took the far corners of the flag in his gloved hands and tugged to see if he could give it a fuller display.

"It won't go out," he said.

The flag ended up about 85 percent extended on its curtain rod, with a charmingly rumpled quality, as if it really were rippling in a breeze.

Satisfied they'd unfurled it the best they could, Armstrong raised the pole and gave it a two-fisted jam into the ground. But the Moon's gritty surface proved resistant to the spiked end of the pole. In the TV scene beamed back to Earth, Armstrong was clearly dissatisfied with how the flag was standing. He didn't quite let go of the pole, sensing that the whole thing was wobbly. It would not be good to make it all the way to the Moon, then have the American flag topple over into the dirt.

Armstrong lifted the pole up and jammed it in again. He was standing with his back to the TV camera, but you could see the top of the flag and the pole rise above his helmet over and over. He removed and replanted the flag five times before getting the pole deep enough that the flag was in no danger of falling. In the end, Armstrong said, he angled the pole slightly backward to get it to stay up.

As obvious and iconic as the flag-raising moment has become—particularly in the color photos of the Moon walks, the American flag is an arresting burst of color and emotion against the grayscale Moon—it was an afterthought. In the last push to get American astronauts to the Moon before the end of 1969, no one had thought to pack a flag.

The final lap really was a race: the first five manned Apollo missions were flown in just nine months. Apollo 7, the first flight after the Apollo 1 fire, was launched October 11, 1968. By the time Collins, Aldrin, and Armstrong blasted into orbit the following July 16, five missions had been assembled and launched, including two trips all the way to the Moon and back.

The pacing was furious: NASA compressed five Apollo flights into 40 weeks, and every one of those missions was really a test flight, each breaking new ground, each refining new navigation and flight techniques, pushing the spacecraft, the astronauts, and the ground crews through one demanding shakedown flight after another. By comparison, the Space

Shuttle flew for 31 years, and in 21 of those years it launched five times or fewer, the pace Apollo managed to hit during that leap to the first Moon landing.

One thing NASA had become good at during the eight years since May 1961 was planning: gaming out the future, thinking not three or five or seven steps ahead, but 17 steps ahead, to imagine what would happen, what could happen, what might happen, and developing contingency plans and training to support those contingency plans. Except for one thing.

As 1969 got under way, no one at NASA had given any thought to what kind of ceremony should accompany the first Moon landing—right there on the Moon. Armstrong and Aldrin would arrive, the first people to ever visit the Moon. More than that, they would be the first living creatures to land safely on any other planetary body in the solar system. It was going to be momentous.

What would Armstrong and Aldrin do to commemorate it? There were experiments and tools assembled, detailed timelines for collecting rocks and soil samples, taking photos and setting up instruments to be left behind. But what about the moment of sheer human achievement? Something for history, and for the cameras? And what about the triumph of the United States being first to make it to the Moon?

As late as January 1969, there was no plan. Which was not just un-NASA-like; it was a problem.

---

Richard Nixon, of all people, nudged NASA into thinking about what the astronauts might do to celebrate arriving at the Moon. In his first inaugural address, on January 20, 1969, Nixon drew inspiration from the Apollo 8 trip around the Moon on Christmas Eve, just four weeks earlier. "We shared the glory of man's first sight of the world as God sees it," he said, "as a single sphere reflecting light in the darkness."

Without mentioning the Soviet Union by name, Nixon continued, "Those who would be our adversaries, we invite to a peaceful competition—not in conquering territory or extending dominion, but in enriching the life of man. As we explore the reaches of space, let us go to the new worlds together—not as new worlds to be conquered, but as a new adventure to be shared."[3]

The rivalry with the Russians had fired NASA's determination to get to the Moon. The Apollo 8 mission itself was an effort to make sure that Americans were, unequivocally, the first *to* the Moon, before any cosmonauts could loop it.

With it rapidly becoming clear that Apollo would put astronauts on the lunar surface in 1969, Nixon sounded a generous internationalist tone, a reminder that the United States wasn't going to the Moon like Columbus went to the New World. There would be no claiming of territory or sovereignty or control.

Nixon's brief allusion—worlds not to be conquered but to be shared—apparently jarred officials at NASA headquarters. Three days after the inauguration, George M. Low, the legendary Apollo spacecraft manager, wrote a single-paragraph memo to Robert Gilruth, director of the Manned Spacecraft Center in Houston.

Low had just gotten a call, he wrote, "indicating that, in light of Nixon's inaugural address many questions are being raised in Headquarters as to how we might emphasize the international flavor of the Apollo lunar landing. Specifically, it was suggested that we might paint a United Nations flag on the LM descent stage instead of the United States flag. My response cannot be repeated here.

"I feel very strongly that planting the United States flag on the Moon represents a most important aspect of all of our efforts." Low agreed there'd be nothing wrong with carrying some small UN flags to the Moon and back, "provided, of course, they don't weigh too much."[4]

There was already an American flag painted on the descent stage of the lunar module, much the way there's an American flag painted on most U.S. aircraft and spacecraft, but it was in the manner of ships at sea flying their national flag, as a national designator, not a symbol of exploration and arrival. (On the Moon, as it turned out, that flag on the lunar module's descent stage was almost never visible on TV or in pictures.)

Any possible Moon landing commemoration slipped through NASA's meticulous planning culture for a simple reason: it wasn't anyone's responsibility. Planting a flag wasn't, strictly speaking, part of the mission. On February 25, 1969, NASA administrator Thomas Paine fixed that by appointing what came to be called NASA's Committee on Symbolic Activities for the First Lunar Landing, with instructions to review suggestions from inside NASA as well as from places like the White House and the State Department for what to take to the Moon in celebration.

At some point in the next several weeks, the group turned to Jack Kinzler of the Manned Spacecraft Center in Houston, who had the mundane job title of head of technical services, a group of 185 craftsmen and technicians. In fact Kinzler, a self-taught engineer who never went to college, had started at NASA's predecessor, NACA, in Langley, Virginia, straight out of high school, building airplane and spacecraft models for wind-tunnel testing. He so impressed Bob Gilruth that when Gilruth moved from Langley to Houston to set up the Manned Spacecraft Center, he asked Kinzler to come with him and build the technical services group. Over an almost 40-year career at NASA, Kinzler earned the nickname "Mr. Fixit" for his ability to solve sometimes urgent problems of spaceflight on the fly. After watching a Russian cosmonaut flailing around during the first-ever spacewalk in 1965, for instance, Kinzler's group in Houston created a handheld unit that used small cylinders of nitrogen to give Ed White the ability to use bursts of gas to maneuver in space during the first U.S. spacewalk, three months later during Gemini 4. It was an inspired and nifty maneuvering tool that solved a problem before it happened.[5]

In the spring of 1969 Kinzler got a call from Gilruth inviting him to the first meeting of the Committee on Symbolic Activities; Kinzler called it "the committee on how to celebrate the first Moon landing." Gilruth told him, "I'd like you to come over with some suggestions of what we might do."[6]

Kinzler and his deputy, David McCraw, did some brainstorming. "Well, they need a plaque," Kinzler said, and they agreed. "Why don't we use some stainless steel? It'll be . . . long-lived. And certainly it has to have a message on it; it has to have crew names on it; and it might have to have the landing site and that sort of thing."

Kinzler headed off to that first meeting in Gilruth's office on April 1 with a prototype of his plaque idea.[7] Among those in the meeting was Neil Armstrong, although he wasn't an official member of the committee. Kinzler and McCraw had created a curved piece of metal that could be mounted around one of the legs of the lunar module, right behind the rungs of the ladder. The idea, Kinzler explained, was that "it would be there forever more, because it was mounted on the descent stage, which stays on the moon."

"They were thrilled with that idea of having a plaque," Kinzler recalled. He had engraved an American flag in the middle of the prototype, with red, white, and blue enamel paint baked into grooves for the flag's colors, along with space for a message and the signatures of the astronauts.[8]

Gilruth wanted to know what else Kinzler had in mind. Kinzler thought the flag painted on the spacecraft was totally inadequate, "a terrible way to celebrate a major event that the crew would be achieving." So he pitched the group his other idea: "What you need is a freestanding American flag."[9]

Real explorers plant flags. During the first landfall of his first voyage, Columbus recorded carrying three flags ashore in the Bahamas: the royal standard of Spain in his own hand, and with two of his captains each carrying the green cross, the banner of the king and queen of Spain. The great explorers Lewis and Clark, whose U.S. government expedition through the Louisiana Purchase had daunting ambition and logistical challenges, carried dozens of 17-star American flags with them, including small flags they flew on their boats, a large flag that flew over their camp, and a stock of flags they brought to distribute to Native American chiefs as gifts and as symbols of American territorial claim.

Roald Amundsen planted a Norwegian flag at the South Pole, and Robert Peary an American flag (sewn by his wife) at the North Pole, not because they were claiming the poles but as a symbol of national pride and accomplishment at having been the first to reach each one. The pictures of Amundsen with the Norwegian flag, its staff coming out of a tent, and of Peary's crew with the American flag, are iconic. They mark the moment in a way that a picture of a bunch of explorers standing around simply cannot. The flag is a visual punctuation mark, to be sure, but it also connects the achievement to all the effort it required. The flag provides the emotional heft, especially in photos where the explorers' faces are obscured by protective clothing.[10]

The American astronauts weren't claiming territory, but they were explorers as ambitious as any that had come before, representing a nation that had invested as much effort as any nation ever had in an expedition. How could they not raise a flag? As Kinzler recalls, his idea for a freestanding flag gave Gilruth another idea. Gilruth looked at the plaque prototype and said, "Suppose we took the flag . . . off your plaque and put two hemispheres on the top: one the Eastern Hemisphere and one the Western Hemisphere. [Then] if any creatures from outer space should land on the moon at a later time and look over back toward Earth, they would see this kind of an outline and they would know where this craft came from."

Kinzler told exactly the same story of the plaque and the flag on at least two occasions, always with a gee-whiz enthusiasm. He reported his own reaction to replacing the flag with outlines of the continents identically each time: "Boy, that's a great idea!" And he also reported his boss suggesting that future alien visitors might pit-stop at the gray, barren Moon and check out discarded lunar module parts before heading on over to Earth itself. [11]

Kinzler's enthusiasm for being able to give the astronauts a flag they could erect on the Moon was undeniable. In the files of NASA's Manned Spacecraft Center is a sketch Kinzler made back at his office on a sheet from a lined legal pad: one pole, planted 18 inches deep in the lunar surface, a flag roughly three feet by five feet, attached along its top edge to a second pole. Where the two poles join, in the upper corner of the field of blue, is handwritten the word "pivot."[12]

The idea was to create a slim, light, easily packed and toted piece of equipment, almost like a pair of tent poles with a tent fly attached to them, that could be deployed by the astronauts without a lot of effort but would result in a flag that was "flying" on the Moon. To do that, said Kinzler, "we used the concept of a telescoping tube that's inside of a drape in a window." He was inspired by his own mother making drapes when he was growing up. For that second tube, "the idea was to hinge it up . . . and have a stop, so it would catch and would just stand horizontal." Fabricating a couple of collapsible tubes, hinged to each other, wasn't a challenge for Kinzler's group. He quickly hit on the idea of attaching the flag kit to the outside of the lunar module; there was no room inside for yet another piece of equipment, nor was there time to test it in the way that would be required for something riding inside with the astronauts. Mounting it with clips to one of the vertical struts of the ladder was the perfect solution.[13]

There were two problems: how to protect the flag, and whether, even just clipped to the ladder, the flag could somehow pose a hazard to the lunar module.

Inside its tube on the ladder, the flag would be exposed to the full range of space temperatures as the lunar module made its way from Earth orbit to the surface of the Moon—up to 250ºF. And then, as the lunar module settled to land on the surface, riding the cushion of thrust from its descent engine, rocket exhaust up to 3,000ºF might deflect up along the legs and ladder. Kinzler wasn't worried about the 250º portions of the flight to the Moon, but the heat from the engine exhaust was another matter. It would last only 10 or 20 seconds, but it would vaporize the nylon-cloth flag.

At the moment Kinzler was sketching the flag contraption for the Moon landing, Thomas Moser was working as a design engineer at the Manned Spacecraft Center, assigned to the command and service modules. Moser went on to big jobs at NASA in the 1970s and 1980s. He was a senior manager on the development of the Space Shuttle orbiter; he was the first program director for the International Space Station; and he was deputy associate administrator of NASA, based in Washington, D.C. But in the spring of 1969 he was six years out of graduate school, working on the Apollo capsule from a desk in Houston.

"I was sitting in my office late one afternoon," he remembered, "and my boss came in and said, We're going to put a flag on the Moon. And you can't talk to anyone about it." In April and May the Committee on Symbolic Activities was actually hoping that the plaque and the flag would be surprises during the Moon walk, to be unveiled for the worldwide audience.[14]

Moser's job: help make sure the flag survived. And make sure the flag kit itself didn't break anything—or, as Moser put it, "Make sure we don't 'fail' something by this last-minute change. It had to be done in a hurry. I dropped everything and just did that."

Moser's quick analysis concluded that the area of the ladder where the flag would be mounted would be subject to 2,000ºF for 13 seconds.[15] Protecting the flag for that 13 seconds had a relatively simple solution: the flag assembly was covered with a long metal half-tube, a shroud. After being folded up, the flag assembly itself was wrapped in a piece of Mylar insulation, no thicker than a sheet of household tin foil. The Mylar was exactly the same insulation that the exterior of the LM itself was wrapped in. If it could protect the vital components of the LM, it could protect the flag.

The flag—an ordinary, off-the-shelf version—weighed just 9.7 ounces, a little more than half a pound, and the entire kit—flag, poles, insulation, shroud, mounting clips—weighed about 9.5 pounds. It seems improbable that something so modest could somehow break the LM's metal descent ladder or interfere with the flight. But not quite as improbable as it sounds. And without a functioning ladder, getting from the LM's hatch to the Moon's surface would have been harrowing. Getting back up inside the cabin might have been impossible.[16]

The lunar module's ladder—like everything on the Moon lander—had been slimmed and trimmed and thinned to have minimum weight while still doing its job. At some critical points in its design and assembly, the lunar module was running thousands of pounds overweight, so much so that it could not have been launched, let alone made it to the Moon.

As part of the weight scrubs Grumman did, the strength and structure of the ladder were designed only for use on the surface of the Moon, in one-sixth lunar gravity. The Apollo spacesuit, including the life-support backpack, weighed 180 pounds, so on Earth a fully outfitted Apollo astronaut weighed about 350 pounds.[17] On the Moon he weighed 60 pounds. The ladder might have had a safety margin of 100 percent or more; it might have supported 120 pounds, or 150. But no ordinary adult, even just wearing street clothes, could climb on it on Earth without damaging the integrity of the ladder itself.[18]

It was Moser's task "to make sure the flag was not going to cause the ladder to fail during all of the loads it experience[d], all the way to landing. Whether or not the shake, rattling and rolling—plus the 8Gs going uphill—could damage the ladder."

On launch from Cape Kennedy, in fact, the spacecraft (and the astronauts) went through the opposite of the Moon's gravity experience. The acceleration of launch put everything through high-gravity pressure, although not quite as high as Moser remembered. The maximum g-force on the Saturn V rocket and everything it was carrying was 4Gs, meaning the flag assembly added about 40 pounds of force to the ladder. "You hang something on there that was not very heavy, but with vibration, with the static load during lift off—we just needed to make sure it did not fail that ladder," explained Moser.

He did a lot of engineering and stress analysis. He also took a mock-up of the ladder, with the flag kit attached to it, and mounted it in a "shake apparatus" at the space center that mimicked the wild shaking and vibration of launch to orbit—all to test the safety of a four-foot-long tube that a fifth grader could easily have shouldered. That kind of care was routine at NASA by 1969; it was a reflex, an instinct. To guarantee the safety of the flight and the success of the flight, you had to imagine the ways that something inconsequential might end up having devastating consequences. It would have been perhaps the most embarrassing moment in the whole history of human exploration to land successfully on the Moon, only to find that you couldn't make it the final 10 feet to the surface because the ladder had failed because you decided to take an American flag with you.[19]

Moser's calculations and experiments concluded that Kinzler's flag kit wasn't a problem. The thing that surprised him: "Nobody checked what I did. Nobody checked my calculations. That scared the bejesus out of me."[20]

---

As Kinzler and Moser were working on adding the flag to the lunar module, the politics of putting a flag on the Moon were ricocheting from Washington to many corners of the country.

At a closed budget hearing in April, NASA administrator Paine told House members, "As far as the question of planting a flag and this sort of thing, we have had a group for some time quietly considering what would be appropriate for the United States to do." NASA was working with the State Department "on the international implications," Paine said, and State had suggested perhaps erecting a United Nations flag. "We haven't made any decisions at all on this," Paine told the congressmen. "We welcome any suggestions, incidentally."

The NASA boss got an earful of guidance, right then. "I hope nobody has given any serious consideration to putting any flag on the Moon when we get there other than the U.S. flag," responded Representative

Charles R. Jonas (R-NC). "If you think you will attract some sympathy from the rest of the world by putting some world flag up there, I will remind you that that was never done at the North Pole or any other place that I have ever heard of being discovered on the face of the Earth."

Representative Burt L. Talcott (R-CA) warned that NASA could see a serious loss of support in Congress, including budget cuts, if astronauts planted any flag except Old Glory. "You might have some nice international implications by using somebody else's flag, but I think you would have some very bad internal political reactions and a great reduction in funds for NASA if anything like that happened."

Paine's testimony wasn't made public for two months, but when it was, the exchange made news. "Congress Furor: Will Moon Crew Plant U.N. Flag?" was the front-page headline in the *Detroit Free Press.* "Lawmakers Warn Space Chief against U.N. Flag on the Moon," said the *Tampa Tribune.*[21]

Apollo—the mission, the scientists, the engineers, the astronauts—tended to operate in what was almost a separate universe from the cultural and political tumult of the late 1960s in America. In the spring of 1969, as NASA's Committee on Symbolic Activities was meeting, opposition to the Vietnam War was cresting. It had been barely a year since the assassinations of Martin Luther King and Robert Kennedy, the riots at the Democratic Convention in Chicago, followed by the thorough defeat of Democratic candidate Hubert Humphrey by Richard Nixon. In April 1969 students took over the administration building at Harvard and ejected the university's administrators. The previous spring, New York City police broke up protests at Columbia University using tear gas. (Woodstock happened just a month after the Apollo 11 Moon landing.)

Richard Nixon had campaigned on restoring law and order to the United States—campus protests were often called "disorders"—and the American flag was a vivid and emotional symbol, burned so often in antiwar and student protests that just the previous summer Congress had passed the first law making it a federal crime to desecrate, mutilate, or burn the American flag.

Senator Wallace F. Bennett (R-UT) wrote Paine a letter after his testimony that had an almost beseeching tone: "Why not, just for a change, as Americans and a government, display our pride in our own accomplishment." Dismissing the idea of erecting a UN flag, Bennett wrote, "Let's look out for ourselves for once. It is getting a bit tiresome to be kicked around for failure so let's broadcast success."[22]

Members of Congress decided not to leave the question of exactly whose flag was raised on the Moon to the judgment of the people responsible for flying there. Representative Richard L. Roudebush (R-IN) introduced an amendment to NASA's 1970 funding authorization forbidding the agency from erecting the flag of any other nation or international organization on space missions in which the United States had provided all the funding. He was setting the standard for flag-raisings on the Moon, and perhaps on to Mars and beyond. Roudebush also proposed an amendment to the NASA funding bill that forbade the space agency from providing research grants to college or university students participating "in campus disorders."

"In all due fairness to the American taxpayer," Roudebush stated during the June 1969 debate over the flag amendment, "it does not seem too much to ask that our flag—Old Glory—be left on the lunar surface as a symbol of U.S. pre-eminence in space to which the citizens of this nation can refer with just pride."

Both amendments were approved, as was the funding bill containing them—but the bill hadn't been to the Senate yet, so Roudebush's amendments had no actual legal force.

What's more, Paine didn't need to know the results of any congressional votes to get the message clearly from the closed hearing in April. On the very day of the debate and approval of NASA's budget in the House—June 10, 1969—NASA notified congressional leaders that astronauts would erect only an American flag.[23]

Roudebush's flag restriction, despite being little more than a bit of congressional flag waving, made a splash. The *Muncie (Indiana) Press* headlined its story, "Roudebush Helps Put Old Glory on the Moon." The *Orlando Sentinel* put the story across the entire top of its front page, with the headline, "Apollo 11 Crew Can Plant U.S. Flag Only on Moon." A few days later, the *Philadelphia Inquirer* called the debate "the great flag foofaraw of 1969."

NASA did manage to hang on to one element of surprise. The UPI story on the passage of the NASA funding bill and Roudebush's flag rules ended with this prediction: "There's no atmosphere on the lunar surface and unless they starch the flag, it is going to hang from its staff like a limp dishrag."[24]

---

If there was ever any consideration of Armstrong and Aldrin erecting a United Nations flag, it was fleeting. In the scant documents about the deliberations of the Committee on Symbolic Activities for the First Lunar Landing, the final report to NASA administrator Paine says simply that the UN flag would not be planted on the Moon.

And the whole debate overlooked two small details. The Soviets had landed their banner on the Moon a full decade earlier, albeit robotically. And in an almost completely overlooked moment, an actual, intact American flag was already sitting on the Moon before Armstrong and Aldrin landed.

On September 13, 1959, the Russians put a probe onto the Moon. Luna 2 was the first human-made object of any kind to hit the surface of another planetary body in the solar system. At the moment it happened, it was another first for the Soviets, another humbling for the United States. (The *New York Times* included in its front-page coverage of Luna 2's landing a separate four-paragraph story headlined "U.S. Has Tried 5 Times to Send Rocket to Moon.")

At that moment, just two years into the Space Age, the largest successful U.S. launch was Explorer 6, a satellite that went into Earth orbit just a month earlier and weighed 142 pounds. Luna 2 (known in the press coverage of the time as Lunik 2) weighed 858 pounds and traveled 240,000 miles to the Moon. It crashed into the Moon going an estimated 7,500 miles an hour—2 miles a second—and so even at the time the scientific view was that it was vaporized as it impacted.

But Luna 2's symbolic cargo had been given a lot of care. The spaceship contained two metal spheres—one 3 inches in diameter, one 5 inches—that looked like small soccer balls. The surface of each was composed of small pentagonal panels. One panel had the U.S.S.R.'s hammer-and-sickle symbol, surrounded on each side by sheaves of wheat, with the Cyrillic initials C.C.C.P. below. The other panel had a single star at the top, above the initials, and the date, September 1959. The spheres were designed to blow apart and scatter the individual "pennants," as the Russians called the metal facets, onto the surface of the Moon.

The hammer and sickle made the front pages in the United States. "860-Pound Red Missile Hits Moon, Plants Soviet Union's Coat of Arms" was the two-deck headline in the *Washington Post*, spread across all eight columns. "Flags in Vehicle," the *New York Times* said in a sub-headline, and the second sentence of the account read, "The first object sent by man from one cosmic body to another bore pennants and the hammer-and-sickle emblem of the Soviet Government."

The day of Luna 2's successful impact on the Moon, the U.S. State Department announced that simply because the Soviet Union had "planted" its flag on the Moon didn't mean that the U.S.S.R. had any legal claim to "rule" the Moon itself.

Two days after Luna 2 scattered its hammer and sickles on the Moon's surface, Premier Nikita Khrushchev landed in the United States for an official state visit, the first visit ever to the U.S. by a Russian leader. As a gift, Khrushchev gave President Dwight Eisenhower an identical version of one of the metal balls, covered with Soviet symbols, that had just hit the Moon. In his arrival remarks, alongside Eisenhower, Khrushchev could not resist a humorous taunt. "We have no doubt," he said, "that the excellent engineers and workers of the United States of America who are engaged in the field of conquering the cosmos will also carry their pennant over to the Moon. The Soviet pennant, as an old resident, will then welcome your pennant and they will live there together in peace and friendship."[25]

At a press conference two days after Khrushchev's arrival, amid laughter, a reporter asked Eisenhower what he was thinking when Khrushchev handed him the little metal sphere. "I found it very interesting," Eisenhower replied, pointing out to reporters that it was composed of dozens of little pentagons. "I suspect, in view of the speed with which it was running and hit, that the whole thing was probably vaporized, but nevertheless it was there."

The *New York Times* couldn't resist a little muttering of its own, noting in a news story that if the metal balls and their pentagons with the hammer and sickle had, in fact, been vaporized, "the pennants . . . may not be waiting on the moon's surface to 'welcome' similar pennants from the United States, as proclaimed by Soviet Premier Khrushchev on his arrival here Tuesday."[26]

The Soviet flag didn't come up in the congressional debate a decade later. And neither did the story of an actual American flag that had already made it safely to the Moon, in 1966, through a mix of pride, secrecy, and sleight of hand.

On June 2, 1966, Surveyor 1 settled into a crater in an area of the Moon called the Ocean of Storms. Surveyor 1 was the first U.S. spacecraft to make a soft, powered landing on the Moon, swooping to within 40 miles of the Moon's surface at 5,800 mph, before firing its retrorocket, then using landing thrusters to slow further, until the 640-pound craft coasted the last few feet to drop safely onto the surface.

Surveyor 1 was a triumph. The spacecraft sent back TV images and 11,237 photographs of the Moon's surface. It came on the heels of the U.S. Ranger missions, a set of spacecraft that were designed to photograph the Moon as they approached it, and send those photos back to Earth, right up to the moment each Ranger spacecraft crashed into the surface, destroying itself.

Each of the first five Ranger probes failed—some on launch, some en route to the Moon. Ranger 3 famously missed the Moon completely, by 22,860 miles. And although the last three Ranger missions were successful, even Surveyor's scientists had given Surveyor 1 only a 10 to 15 percent chance of a successful soft landing.[27]

Surveyor 1 started a process of examining directly the surface of the Moon, to see what kinds of challenges it would pose to the lunar module and to astronauts walking around. It took pictures of its own landing pads, which showed they had sunk just an inch or so into the surface. (One theory of geologists was that after eons of meteorite strikes, the surface of the Moon might consist of many inches, or even feet, of powdery material into which spacecraft would simply sink, like dry quicksand.)

Surveyor was a mission run by NASA's Jet Propulsion Lab (JPL) in Pasadena, a series of seven identical robotic spacecraft headed to different places on the Moon. The distinctive, three-legged lunar landers were built by Hughes Aircraft. The day Surveyor 1 landed on the Moon and started sending back pictures, Sheldon Shallon held a press conference and revealed that the ship carried a small, secret cargo that neither NASA nor JPL knew about: an American flag.

Shallon was the chief scientist at Hughes for Surveyor. On a Saturday afternoon 10 days before Surveyor 1's launch, he went to a Sav-On discount drugstore on Sepulveda Avenue in Los Angeles and bought an American flag on a wooden stick, the kind you would get for an eight-year-old to wave at a parade. The flag cost 23 cents.

In a memo written months later, during an internal investigation into "the flag incident," Shallon details with the austere precision of a Cold War spy-mission after-action report the subterfuge necessary to get the five-and-dime flag onto the Moon:

> The staples that held the flag on a wood staff were removed with a pocket knife.
>
> In the afternoon of 21 May 1966, the flag was placed between the pages of a report that was to be carried [to the spacecraft preparation area the next day].
>
> In the evening of Sunday, 22 May 1966, the flag was removed from between the pages of the report and given to [Surveyor] personnel.

A second, attached memo details the care with which the wispy flag was then prepared for flight. A Hughes staff member named D. C. Smith took responsibility for cleaning the flag, a two-hour process he described in a one-paragraph memo acknowledging his participation:

> The flag was delivered to me at approximately 20:00 hours [8 p.m.] on Tuesday 24 May 1966. I removed all loose fibers and threads from the seamed edge, then washed the flag in clean solvent [Freon TF], followed by thorough rinsing in a flowing stream of solvent filtered through a membrane filter rated at 0.45 microns. The flag was dried by blowing with clean gaseous nitrogen, then immediately sealed into a nylon bag. I delivered the flag to the vehicle test crew . . . at approximately 22:00 hours [10 p.m.] on Tuesday, 24 May 1966.

The Hughes flag plotters were nothing if not meticulous in their effort not to let the flag unintentionally hurt the mission.

The next day, Wednesday, May 25, 1966, Shallon's account continues, "the flag was removed from the plastic bag and inserted with the aid of a thin screwdriver through a small hole into the lower spaceframe member below the Surveyor TV Camera."

The flag—at four by six inches, just a little bigger than an index card—had been rolled up and slipped into one of the hollow tubes of Surveyor 1's structural frame. Surveyor was launched five days later, the following Monday, and landed on Wednesday.

The press loved the story. The tiny rolled-up flag made the front page of Washington's *Evening Star*, the inside of the *New York Times*, and the front pages of dozens of newspapers from one side of the country to the other. The *Tallahassee (Florida) Democrat*: "A 23-Cent U.S. Flag Is on the Moon." The *Denton (Texas) Record Chronicle*: "'Old Glory' Flying High."[28]

NASA and JPL were not pleased, but we know this only secondhand, from the Hughes "investigative" memos. The flag is not mentioned in any NASA documents or oral histories and is absent from the official JPL report of the mission, which runs 626 pages across three volumes and says Surveyor 1 "achieved a perfect soft-landing on the moon."[29]

In a final internal Hughes Aircraft memo on the topic, written just two weeks before the launch of Surveyor 2 and distributed widely to Surveyor spacecraft staff, R. R. Gunter, a manager, wrote, "Considerable criticism and discussion has risen from the SC-1 American Flag incident. NASA and JPL have taken a very strong position with regard to such actions. As a result, you are explicitly directed to see to it that no similar actions are taken on SC-2, 3, 4, or other spacecraft. It would be extremely embarrassing to the Company if such actions were taken."[30]

NASA and JPL were so irritated by Shallon's stowaway that they required Hughes to buy two more flags from Sav-On, clean them exactly the same way, then test them for contamination. Shallon allowed as how the flag was likely the cleanest item on the spacecraft. The flag slipped into the framework of Surveyor 1 would not be the last, or even the most interesting item snuck onto spacecraft headed to the Moon.

---

Flags have emotional heft, dramatically magnified by the moment they are hoisted or struck, and they carry more than symbolic meaning.

Columbus didn't just step ashore in the "New World"; he brought the flags of Spain. The U.S. Marines took Iwo Jima during their drive across the Pacific in the last months of World War II and sent six Marines to the top of the island's highest point to raise the American flag; the flag-raising, captured in a Pulitzer Prize–winning photograph, was so instantly iconic that the U.S. Postal Service issued a commemorative stamp with the image just five months later, before the Japanese had even surrendered.[31]

So the committee on how to celebrate the Moon landings, to use Jack Kinzler's name, wanted to be careful about the message it sent to the world from the Moon.

In 1967, as the Americans and the Russians raced to land probes on new worlds, the UN Outer Space Treaty was ratified; both the U.S. and Russia signed on the day the treaty was first opened to signature. The UN treaty laid a foundation of basic rules for the exploration of space, and its second section, Article II, forbids any Christopher Columbus–style maneuvers. "Outer space, including the Moon and other celestial bodies, is not subject to national appropriation by claim of sovereignty, by means of use or occupation, or by any other means," the treaty states.[32] You can't stake your claim to any celestial body, or any part of any celestial body, simply by showing up there.

The memo to Thomas Paine from the committee summarizing its recommendations suggested that any symbolic activities "should be simple, in good taste from a world-wide standpoint, and have no commercial implications or overtones. The overall impression to be conveyed by these symbolic activities and by the manner in which they are presented to the world should be to signalize the first lunar landing as an historic forward step of all mankind that has been accomplished by the United States of America."

The plaque would do the first job, with "a short and simple inscription commemorating the first landing of men from the Earth as a human achievement of all mankind." The flag would do the second, with care taken so the flag-raising would "symbolize the American effort that made the landing possible but should avoid the implication that the U.S. is or intends to exercise sovereignty on the Moon," which "would be contrary to our national intent" and violate the new UN treaty.[33]

So the plaque would state, with great purposefulness, "We came in peace for all mankind." Not only does the Apollo 11 plaque not have an American flag on it, but the only mention of the United States is in passing, underneath Richard Nixon's signature, where he is identified as "President, United States of America."

The memo does say, "The U.N. flag or other international or religious symbolism will not be used." Indeed the only thing that jumps out from the committee's report is phrasing, repeated twice, to describe the dramatic nature of the landing itself—first, "an historic forward step of all mankind," then, a paragraph later, underlined and in quotes, "the 'forward step of all mankind' aspect of the landing." That phrase would be echoed a few months later in Neil Armstrong's legendary words as he stepped onto the Moon: "That's one small step for [a] man, one giant leap for mankind." But Armstrong was always a little vague about where and when he came up with that phrase; in his authorized biography, he said he didn't think hard about those words until after he had safely piloted the lunar module to the surface of the Moon. When he was asked specifically whether it was suggested by the language of the committee report, he couldn't recall ever reading the memo, or even seeing it.[34]

Kinzler and his team had designed a cover for the plaque, to hide its words until it was unveiled on the Moon. But Kinzler was given word a few days before the launch that Houston would tell reporters about both the plaque and the flag before the Moon landing and that he could dispense with the plaque's cover altogether.[35]

Still, that first flag for Apollo had a kind of improvisational feel right up to the end, which is remarkable given how significant it ended up being. Kinzler himself came up with the procedures for folding the flag (12 steps) and packing it in its container (4 steps), which was done quietly by him and a handful of staffers in his office days before Apollo 11's July 16, 1969, launch. "Since it was under wraps," he recalled, "I used my division office for the final assembly and folding. You wouldn't ordinarily do that—we have clean rooms and all sorts of things."[36]

In fact Kinzler himself took the plaque and the tightly packed flag, in its protective shroud, from Houston to the Cape aboard a Gulfstream jet with George Low, senior manager of the Apollo spacecraft program, and Low's secretary. On the morning of July 9, 1969, at 4 a.m., Kinzler supervised installation of both items on the lunar module. That was a delicate period for Apollo 11, which had long since been moved from the protective cover of the Vehicle Assembly Building to Launchpad 39-A. It was launched a week later. Kinzler had, of course, written the installation instructions for mounting the flag package on the lunar module ladder (11 steps).[37]

---

On the Moon both Armstrong and Aldrin had a printed checklist sewn onto the broad cuff of their left spacesuit glove, so they could glance down and see where they were in their tasks during their two-hour Moon walk. The lists of tasks were just cryptic lines of type, in two columns.

Setting up the flag wasn't listed. It had come too late to be included.

Armstrong and Aldrin didn't make a big show of the flag. Indeed over the Moon-to-ground link, they didn't talk about it at all, nor did they turn it into much of a ceremony, even as 600 million people watched.

The process didn't go as well as they would have liked. "It took both of us to set it up and it was nearly a disaster," Aldrin said. "To our dismay the staff of the pole wouldn't go far enough into the lunar surface to support itself in an upright position." That's what can clearly be seen on the TV broadcast, Armstrong trying over and over to get the flag planted securely in the rugged ground. "After much struggling," continued Aldrin, "we finally coaxed it to remain upright, but in a most precarious position. I dreaded the possibility of the American flag collapsing into the lunar dust in front of the television camera."[38]

It didn't, at least not then, and not on camera. In fact the reason Armstrong and Aldrin had so much trouble with the flag was that they somehow abandoned Kinzler's instructions and training. The flag was specifically designed so the bottom section of the vertical pole could be hammered into the ground separately, using a geology tool. The top edge had been hardened so it could be hammered on. And near the bottom two red lines had been painted, to give the astronauts a sense of how deep they needed to pound it in so the flag would stand up securely. Armstrong and Aldrin skipped all that; they assembled the flag pole, including extending the rod that allowed the flag to "fly," before trying to get the pole itself into the ground.

Walter Cronkite, narrating the first Moon walk for CBS News, explained to TV viewers how the flag worked. "That flag is on a frame, there is no wind to hold it out like that, of course. It's a three-by-five flag. It's got a frame of its own to hold it out like that."

Cronkite paused. "Nothing more really is needed here, but it does seem like there ought to be some music," he said, chuckling. He explained that the astronauts weren't "claiming" the Moon with the flag, and couldn't anyway, because of the UN treaty. "So this planting of a flag is not the old 15th, 16th, 17th century business of planting a flag and claiming territory. It's to put the United States flag there to let the world know that we are there. To sense the pride the American people feel in this tremendous accomplishment and the contribution they have made to it."[39]

Right after they got the flag upright, Armstrong stepped out of TV camera range to take Aldrin's picture alongside the flag. Aldrin stepped back and saluted, although the salute is hard to see on camera, because as he faces the flag, his right hand is on the far side. Unknown to the TV audience, at the moment the astronauts got their flag planted, a second flag was hoisted in Mission Control, where it was on display until the last flight, Apollo 17.[40]

The flag-raising, unceremonious as it was, was almost immediately overtaken by a phone call from Earth: President Richard Nixon, calling from the Oval Office, at what on Earth was 11:49 p.m. ET, July 20, 1969. Nixon's congratulatory phone call was an odd mix of ceremony and public relations, secrecy and politics. The astronauts had not been told of it in advance, although Armstrong had been told there might be a surprise caller during the Moon walk. It was a remarkable feat: Nixon, in suit and tie, talking on a classic green push-button AT&T phone, with a row of lights along the bottom, straight into the helmet headsets of the astronauts standing on the Moon.[41]

But while the astronauts were en route to the Moon—without any access to the media—the *New York Times* learned of the planned phone call and the day before the Moon landing published an editorial titled "Nixoning the Moon," a contemptuous dismissal of Nixon's calling the astronauts. The *Times* editorial writers were clearly angry that Nixon—the nemesis and rival of John F. Kennedy—had the honor of presiding in the White House as Americans landed on the Moon, the mission Kennedy had launched with such eloquence and determination: "By accident of the calendar, President Nixon is now the nation's chief executive as the moment approaches for realization of the dream for which his two predecessors worked so effectively."

A phone call to the Moon was "unseemly," said the editorial, "[an] attempt to share the stage with the three brave men on Apollo 11," really just a self-indulgent waste of time, squeezed into a Moon walk schedule "so crowded with assigned tasks that the full schedule of scientific activities may not be realizable. . . . Such an intrusion looks suspiciously like a publicity stunt of the type Khrushchev used to indulge in. It strikes us as unworthy of the President of the United States."[42]

With 50 years of hindsight, the editorial is ridiculous, a sign of the politics of the late sixties. Why wouldn't the president of the United States call the astronauts and salute their courage—and their colleagues—while they were on the Moon itself? If the person in office had been Kennedy himself, would the *Times* have been fretting about slipping the call into the busy Moon walk schedule?

Four minutes after the flag had been erected, Mission Control called Armstrong and Aldrin back from various tasks to stand in front of the TV camera. Bruce McCandless, the astronaut who was CapCom for the first Moon walk, said, "Neil and Buzz, the President of the United States is in his office now and would like to say a few words to you. Over."

"That would be an honor," said Armstrong.

"Go ahead, Mr. President. This is Houston. Out."

As Nixon started to speak, the TV transmission was divided to show the president, live in the Oval Office, on the left, and the astronauts, live on the Moon, on the right. Armstrong and Aldrin had positioned themselves facing "the audience," that is, facing the camera, with the American flag between them.

Nixon called it "the most historic telephone call ever made" from the Oval Office. "I just can't tell you how proud we all are of what you have done. For every American, this has to be the proudest day of our lives. . . . For one priceless moment, in the whole history of man, all the people on this earth are truly one. One in their pride of what you have done. And one in our prayers that you will return safely to Earth."

Although caught by surprise, Armstrong replied with eloquence. "Thank you, Mr. President. It's a great honor and privilege for us to be here representing not only the United States but men of peace of all nations, and with interest and a curiosity and a vision for the future." Nixon signed off by saying he looked forward to meeting the astronauts on the aircraft carrier *Hornet* in a few days. Aldrin replied, "I look forward to that very much, sir."

The *New York Times'* toe tapping notwithstanding, Nixon kept the call to two minutes—actually, five seconds less than two minutes. As the call ended, both Armstrong and Aldrin paused on the surface and saluted the flag and their commander in chief.

And then without so much as an acknowledgment of what had just passed, McCandless mics back in with a stream of technical data for Mike Collins, orbiting the Moon in command module *Columbia*: "*Columbia, Columbia*, this is Houston. Over. . . . I got a P22 AUTO optics—AUTO optics PAD for you."[43]

And both the presidential phone call and the flag were left to history.

---

Some at NASA thought the Apollo 11 flag deployment had served its purpose. George Low, the Apollo project manager in Houston, wrote a memo to Bob Gilruth, the head of the Manned Spacecraft Center, saying he and his colleagues had "tentatively decided not to emplace any more flags on the lunar surface. There will, of course, always be a flag painted on the descent stage, but my view was that it would not be necessary to go through a flag-raising ceremony each time we go to the moon."[44]

But the power of the flag deployment overruled Low. Kinzler's records show that in October 1969, NASA managers had decided to send a flag with each mission, and Kinzler's team assembled the kits.

Training to get the flag planted securely was included in subsequent mission preparation, and four of the other five missions that made it to the Moon did what Kinzler had designed: they pounded the bottom section of the flagstaff into the Moon dirt before setting up the flag itself. John Young, commander of Apollo 16, couldn't get the poles to do what he wanted. He fell to his knees at one point, in his spacesuit, in order to prevent the flag from hitting the ground. At his technical debriefing, Young said, "I can honestly say I had as much trouble putting the flag together in one-sixth gravity as I did in one gravity. My main concern was with the TV sitting there watching us: that we'd end up with the flag in the dirt and us standing on it."[45]

None of the other crews made much of a ceremony of the flag on TV, any more than Armstrong and Aldrin had, but there were some moments. On Apollo 15, Jim Irwin started out pushing the bottom pole section in by hand, then used the geologic hammer to finish the job. "I'll hit it a few times so it will stay up a few million years," he said.

Apollo 17 took to the Moon the flag that had been hanging in Mission Control during the previous Apollo flights, mounted like the others on Kinzler's flag-flying contraption. Gene Cernan, Apollo 17's commander, did the work to get the pole into the Moon's surface, pushing the staff into the dirt, then taking 16 whacks with the geologic hammer. Cernan asked his crewmate Harrison Schmitt to help him adjust the flag itself. "Hate to touch it," Schmitt replied, "my hands are so dirty."

Cernan stood and saluted the planted flag. "This has got to be one of the proudest moments of my life," he said.

The only flag deployment that got even a hint of narration was also the one no one on Earth could watch. Apollo 12, the second Moon landing, sent Pete Conrad and Alan Bean to the Moon's surface. Not long into their first Moon walk, while positioning their color TV camera, Bean accidentally pointed it at the Sun for just a few seconds too long, and the intense, unfiltered sunlight burned out the camera's sensor, leaving the mission with no TV coverage back to Earth.

Conrad and Bean went on with their work, and almost immediately after giving up on getting the camera working again, they planted the flag. The event was captured on the audio recording of their call to Mission Control.

***Conrad:*** Okay, the flag is up.

***CapCom:*** Roger, copy. The flag is up.

***Conrad:*** We hope everyone down there is as proud of it as we are . . . as proud of it as we are to put it up.

***CapCom:*** Affirmative, Pete. And we're proud of what you're doing.[46]

For Apollo 12, which had two Moon walks, each of which was almost four hours long—double the length of Apollo 11's single Moon walk—the checklists of what to do on the Moon's surface were much more elaborate than they had been for Armstrong and Aldrin. The Apollo 11 checklists fit on a single panel, sewed onto the flaring cuff of each astronaut's spacesuit glove. For Conrad and Bean, with eight hours of Moon walking, the list of tasks was much longer: Conrad's checklist was 34 pages long, Bean's was 30 pages, each page a 3.5-inch square. Each checklist was assembled into a small spiral-bound book, each page printed on fireproof paper and laminated, the little flip-book itself secured around the cuff of the glove with a strap. At the one-hour mark in each is the command "DEPLOY FLAG."[47]

The checklists were typical NASA technical documents: cryptic acronyms, brief instructions, the occasional diagram. But on about half the pages someone has sketched lighthearted cartoons of astronauts going about their tasks on the Moon—setting up a radio antenna, using a hammer to pound the flagpole into the lunar surface, photographing each other.

For Conrad and Bean, the cuff checklists contained a cheekier surprise. Each one has pictures of two Playboy playmates, the photos taken directly from the magazine, photocopied down in size, put on the special paper, laminated, and secretly bound into the checklists after they had been reviewed by the astronauts.

Bean was apparently the first to find one of the playmates in his checklist; she's about nine pages in, smiling and wearing a Santa hat and nothing else, the caption supplied by Bean's NASA colleagues: "Don't forget—describe the protuberances."

"It was about two and a half hours into the extravehicular activity," said Bean. "I flipped the page over and there she was. I hopped over to where Pete was and showed him mine, and he showed me his."

There's not a whisper of the discovery on the radio exchanges between Apollo 12 and Mission Control. "We didn't say anything on the air," Bean said. "We thought some people back on Earth might become upset if they found out we had Playboy playmates in our checklists. They would have said, 'This is where our tax money is going?' "

But the astronauts appreciated the prank. "We giggled and laughed so much," said Conrad, "that people accused us of being drunk or having 'space rapture.' "[48]

Word about the Playboy playmates reaching the Moon apparently never reached reporters, during Apollo 12's flight or after. The first story about it appeared in *Playboy* itself, in December 1994, on the 25th anniversary of the flight. But there is inadvertent photographic evidence of the prank right from the surface of the Moon. One of the classic photos from Apollo 12 is a close-up portrait of Commander Pete Conrad, in his spacesuit, facing the camera and holding another camera. The lunar module is visible in the distance over his left shoulder. Alan Bean, taking the picture, is reflected in the visor of Conrad's spacesuit helmet. And on Conrad's left arm his cuff checklist is open, and it just happens to be open to the page where a playmate is reclining on a hay bale (caption: "preferred tether partner"). In the photo, the image is too grainy to make out if you don't know it's there, but if you zoom in on the photo, you can just make her out. Conrad had that photo framed in his home but didn't notice for years that on his wrist was Reagan Wilson, October 1967 Playmate of the Month. A Playboy playmate not only flew to the Moon; she was photographed there.[49]

As brisk as the flag-raisings were on the Moon, and as improvisational and hurried as getting that first one arranged for Apollo 11 had turned out to be, they made a dramatic impression back on Earth.

The raising of the flag by Armstrong and Aldrin is the Moon landing moment marked by history. Dozens of papers picked for their front pages the image of Armstrong holding the flagpole and Aldrin tugging the flag out from its far corner—two otherworldly Earthlings, wearing bulky white spacesuits, standing on the barren surface of the Moon, their spaceship in the background, and between them the American flag. That picture made the front pages of the big papers—the *New York Times*, the *Washington Post*, the *Philadelphia Inquirer*, the *New York Daily News*, the *Boston Globe*, the *Detroit Free Press*, the *Denver Post*, the *San Francisco Chronicle*, the *Los Angeles Times*. And the flag made the front page of the little ones too—the *Fairbanks (Alaska) Daily News-Miner*, the *Great Falls (Montana) Tribune*, the *Tallahassee (Florida) Democrat*, the *Nashville Tennessean*, the *Huntsville (Alabama) News*, and *Cocoa Today*, the hometown paper of Cape Kennedy.

The caption for the picture in the *Washington Post* explained how the astronauts had made the flag "fly" on the airless Moon, using the rod across the top. The *New York Times* put the Stars and Stripes in the headline: "Men Walk on the Moon: Astronauts Land on Plain; Collect Rocks, Plant Flag."[50]

The flag was the exclamation point to the adventure itself. It was a moment that perfectly combined achievement and pride. It was also a striking composition; the only really familiar thing in that picture from the Moon's surface is the flag itself, whether or not you are an American. In a sense, the flag put all the rest of us on the Moon.

President Kennedy had foreshadowed the moment. In September 1962 when he made his sweeping tour of space facilities to understand the progress of Apollo. Over two days he visited the launch facilities at Cape Canaveral, the rocket facilities in Huntsville, Mission Control in Houston, and the factory where the Mercury space capsules were being manufactured in St. Louis.

In Houston, where Kennedy gave his historic speech in the football stadium at Rice University, laying out his most powerful argument for a manned space program, he used the flag to represent human inspiration and human aspiration. "No nation which expects to be the leader of other nations can expect to stay behind in this race for space," said Kennedy. "We have vowed that we shall not see [space exploration] governed by a hostile flag of conquest, but by a banner of freedom and peace."

That first flag planting left a powerful impression on American culture. When the pioneering—even, for its time, radical—music video channel MTV debuted on August 1, 1981, its signature video logo featured none other than Buzz Aldrin, standing in salute alongside the American flag on the Moon. The Stars and Stripes had been swapped out for the animated MTV logo, which constantly changed design. That snippet—Aldrin saluting MTV on the Moon—played at the top of the hour on MTV, through the whole programming day, every day for five years, tens of thousands of times.

MTV was aiming to recapture and reenergize the revolutionary spirit of rock 'n' roll from the 1960s. And to MTV's founders, nothing said revolution like walking on the Moon. The network identified with that moment of the Moon landing, and 1960s iconography, so closely that when it debuted its MTV Video Music Awards in 1984, the award statuette it designed was an astronaut, holding a flag with the MTV logo, called the "Moon man." Thirty-five years later, the "Moon man" is still the statuette of the VMAs.[51]

# 9

# How Apollo Really Did Change the World

No one can predict with certainty what the ultimate meaning will be of mastery of space.

> **President John F. Kennedy**
> *announcing the Moon mission to Congress, May 25, 1961*[1]

One day in early 1969 two engineers from General Motors were standing in the corridor just outside the office of NASA rocket maestro Wernher von Braun, holding what looked like a toy car. Von Braun was the former Nazi who had run Adolf Hitler's V-2 rocket program during World War II. As Germany collapsed in defeat, von Braun had contrived to surrender, along with many of his pioneering rocket group, to the advancing U.S. Army instead of falling into the hands of the Russians.

Von Braun was a charismatic force for space travel in the United States. Just 33 at the end of World War II, he and many of his German colleagues ended up in Huntsville, Alabama, at what by the mid-1960s was the Marshall Space Flight Center, headquarters for NASA's effort to design and test the biggest rockets the world had ever seen, including the Saturn V, which sent Apollo to the Moon.

As it happened, in a bureaucratic quirk, Marshall was also in charge of "Moon mobility" vehicles—lunar rovers.

The two GM engineers outside von Braun's office that day were

Sam Romano and Ferenc Pavlics. They had come to Huntsville in a last-ditch effort to persuade NASA that the astronauts had to have a car on the Moon for at least some of the Apollo missions. It was late to be making that argument—the first Moon landing was just weeks off—and NASA had several years earlier shelved rover development. In the early 1960s NASA had developed elaborate lunar rovers that were like enclosed Moon minivans. But it canceled the projects because the rovers were too big, too heavy, and too costly.

Romano and Pavlics were so determined that the astronauts have a Moon vehicle that they kept working, using GM's own money, even after NASA decided not to send any kind of car to the Moon. "I decided it can be done, it should be done, and we want to do it," said Romano. "If there's going to be a vehicle on the Moon, it's going to be a General Motors vehicle, and I'm going to make sure it happens."

The men quietly talked to engineers at Grumman, where the lunar module was being designed and built, and got the dimensions of a storage compartment on the outside of the lunar module that was empty and that they could use to stow a lunar vehicle if they could design one to fit. The whole idea was silly on its face: that compartment was shaped like a tall wedge of pie: five feet wide at the wide end, five feet tall, and five feet deep, narrowing to a point. An odd shape, and they would be trying to design a Moon car that could somehow fit into a space no bigger than the trunk of a typical Earth car, while also being useful once it was on the Moon.

That day in early 1969 they had brought with them what looked like a child's toy car, with the lines of a sleek, open-topped dune buggy. It was, in fact, a scale model of the lunar rover Romano and Pavlics wanted to send to the Moon. Pavlics had designed it and built the scale model, with meticulous detail, including seats sewn by his wife, and the 18-inch car motored along using batteries, operating by radio remote control. As he was finishing the model, Pavlics noticed that his young son's latest GI Joe was a new version, "Astronaut GI Joe," wearing a shiny Mercury spacesuit. For the trip to Huntsville, Pavlics had borrowed Astronaut GI Joe and put him in the little rover's driver's seat. The men set the model down in the corridor outside von Braun's office. "I guided the little model with radio control into his office," said Pavlics, "right to his desk. He was on the telephone, looking at what's coming into his office."

The NASA rocket chief, who was also director of the Marshall center, immediately hung up. "What have we here?" he asked.

Said Romano, "That gave us the opportunity to tell him what we could do."

Half an hour later, von Braun was convinced. He slapped his hand on his desk and declared, "We must do this."

Romano and Pavlics, who had already been told no by NASA, with their determination and their captivating motorized Moon car, had just changed the history of space exploration.

Just weeks later, von Braun created a project office to run creation of a lunar rover. It was April 1969, ridiculously late to imagine adding something as complicated as a car to the Moon flights. Spaceships, spacesuits, experiments, procedures—not only were they all designed, built, tested, and flight-qualified, but the astronauts had been practicing with their Moon equipment for months or even years.

But von Braun was true to both his word and his influence. A quick competition was run to pick companies to design and build the rover, and the companies took the competition seriously, with Grumman, builder of the lunar module, entering a credible challenger to the one Romano and Pavlics had designed.

But in the end GM was picked to design the rover, working with Boeing, which built the GM design. The ramp-up to get the work done was astonishing: Romano and Pavlics's group of a half-dozen expanded to a team of 400 in just weeks, with Pavlics as the chief engineer. The key to GM's victory was an almost magical system Pavlics had designed for folding up the car like an elaborate metal origami. The seats folded down, and the front end of the rover was hinged and folded flat onto the center of the vehicle—wheels, suspension, and all. The rear end did the same, like a pool lounge that could be folded flat. Once front and rear were folded into the center, the wheels unlocked and angled in as well, to make a package in the shape of that wedge storage compartment. On the Moon, the car would unfold out of the side of the lunar module and plop onto the surface of the Moon, almost ready to drive.

The whole thing weighed 460 pounds, including the batteries that powered it, a color TV camera, seat belts for the astronauts, and four quarter-horsepower electric motors, one driving each wheel. The rover was 10 feet long and 6 feet wide, and it could carry 1,050 pounds of astronauts, gear, and rocks across the surface of the Moon at 8 mph. As with much else for Apollo, the rover was an improbable combination of high-tech and handcrafted. The wheels had an inner rim of titanium; the tread was a cleverly engineered metal mesh that rolled along the soft lunar surface, allowing the gritty lunar soil into the wheel, but then flexing open as the wheel turned so the lunar dirt fell right back out. That metal tire "tread" was woven by hand out of piano wire.[2]

The first Moon road trip had Apollo 15 lunar module pilot Jim Irwin in the observer seat and commander Dave Scott at the wheel. The rover was operated with a single joystick control that worked exactly as we've come to know them: push it forward and the rover went forward; the harder you pushed, the faster it went. You angled the stick left and right to turn the rover, which had innovative dual front and rear steering to give it maximum maneuverability on the bumpy lunar surface.

The rover brought exuberance to lunar exploration. Within minutes of heading off on their first expedition, Irwin and Scott were laughing with the sheer fun of driving on the Moon. "Man, this is really a rocking-rolling ride," Scott said to Mission Control.[3]

In 15 minutes of driving on that first trip, Scott and Irwin went farther than any of the previous three Apollo landing crews had been able (or been allowed) to walk in hours on the surface. And just on that first jaunt, one of three using the rover, Scott and Irwin stayed out for two hours, driving around, getting out, gathering specimens, filming geological features, then hopping back in the buggy and racing off to the next place. They not only covered terrain; the pair gave a nonstop narration of the geology they were seeing and that the rover's camera was transmitting in real time back to Earth. The live TV coverage had a rapt audience of, among others, geologists and scientists who felt like they were looking over the shoulders of the lunar astronauts from the back seat, as it were, seeing an astonishing display of never-before-seen alien geology.

"Keep talking, keep talking," CapCom Joe Allen said. "Beautiful description."

Heading home from that first excursion, Scott and Irwin got going so fast down a lunar hill they accidently did a sudden 180-degree spin in the rover, going in an instant from zooming downhill to being pointed back uphill.

It sent them both into gales of laughter, which Mission Control took a moment to appreciate. Said Scott to CapCom Allen, "Boy, I'll tell you, Joe, this is a super way to travel."[4]

The rover transformed the scientific value of the last three lunar landings, missions about which there was deep skepticism after the near-disaster of Apollo 13. Apollo 14 had been successful and restored America's faith that we could go to the Moon and come home safely—and it was the third landing. With the safety of the crews at such vividly demonstrated risk after Apollo 13, what exactly were we doing going back three more times after that?

Apollo 15, 16, and 17 each carried a rover, and the two-man crews ended up being able to explore wide swaths of terrain with the confidence that came from the experience of the previous four Moon missions. Each of the last three Moon missions had the astronauts on the Moon for three full days, and on each, the two astronauts did three long excursions using the rover.

The confidence and comfort was evident just in the basic statistics. Apollo 11's lunar module, *Eagle*, was on the surface of the Moon with Armstrong and Aldrin for a total of 21 hours and 36 minutes, touchdown to liftoff. They were outside on the surface for two and a half hours. By comparison, Apollo 17's two astronauts—the last people to visit the Moon, Gene Cernan and Harrison Schmitt—spent 22 hours walking and driving around the Moon, in three 7-hour stints. The astronauts of the final Moon mission were outside their spaceship exploring for more hours than the first two astronauts spent on the surface total.[5]

And so those last three missions did a good deal more science and geology than the first three. The rover extended the range the astronauts could cover by a factor of 10 and made it easier for them to work; trekking in bulky spacesuits from place to place, the astronauts found, was surprisingly wearing, despite the low gravity, and used up limited air and water in their suits just to get them to the next task. Driving from place to place allowed them to rest and conserve their resources, so they hadn't tired themselves out just reaching a spot worth exploring.

As with everything else about the Moon missions—and almost all of NASA's subsequent space missions—the rover explorations were carefully mapped and scripted. But they allowed the astronauts, for the first time, to go places they otherwise couldn't and exercise both their curiosity and their judgment.

On the second day of their motorized excursions, Scott and Irwin drove their rover up the slope of the Apennine Mountains, stopping several hundred feet up. There they almost immediately spotted a glittering white rock, positioned somewhat oddly atop a pedestal of dirt.

Irwin: "Look at the glint!"

Scott, to Mission Control: "Guess what we just found [laughter]. Guess what we just found! I think we found what we came for. . . . What a beaut!"

The rock was something they had trained to look for, a mineral called anorthosite, and they knew it was something special. The astronauts, and geologists watching on Earth, thought it might actually be a rock unchanged since the formation of the Moon's primordial crust. As they bagged the rock, Scott said to Irwin, "Make this bag, 196, a special bag." On Earth that day, with perhaps an excess of enthusiasm, Apollo 15 flight director Gerald D. Griffin declared, "We have witnessed the greatest day of scientific exploration that we've ever seen in the space program."[6]

The half-pound rock turned out to be 4.15 billion years old, and it became instantly famous—nicknamed the Genesis Rock by reporters—as one of the older and more geologically revealing samples to come back from the Moon. It would have been completely out of reach without a Moon buggy.[7]

The rovers, like so much else, stayed behind on the Moon. Before they left, the astronauts of Apollo 15, 16, and 17 would park their rover several hundred feet from their lunar module, with the color camera aimed back at the spaceship. Operated from Mission Control, the camera would then beam back video of the lunar module's upper stage, and its astronauts, rocketing off the Moon and back to lunar orbit, the ascent engine spraying debris everywhere as it sent the top half of the lunar module zooming up out of the picture.

The rover project, completed in a hectic 17 months, wasn't cheap. It cost $38 million total in the early 1970s ($236 million in 2018 dollars), and each of the four flight rovers individually cost $1.5 million ($9.2 million in 2018 dollars). Three of these went to the Moon; the fourth was reserved for spare parts. The significance of the rover for the missions themselves was instantly appreciated: it was honored with its own U.S. postage stamp, issued on Earth while Apollo 15 was on the Moon.[8]

In that first speech on May 25, 1961, that launched America to the Moon, President Kennedy had said, "No one can predict with certainty what the ultimate meaning will be of mastery of space."[9] By the end of the final Moon mission, Apollo 17, we had mastered a slice of space travel; we had mastered going to the Moon, exploring the Moon, and coming home from the Moon. We hauled a car to the Moon—an electric, all-wheel-drive car designed by General Motors—and the astronauts drove it around and had a great time doing it.

So what did it get us?

---

It's 50 years since those first Moon landings. And yet you can't go spend a year doing PhD research at the U.S. Moon base or the international Mars station, or even spend a few prosaic months on a work rotation doing zero-gravity medical research on an orbiting space station. You can't even take a tourist trip into orbit for a couple loops, just to see what the Earth looks like from on high.

The promise of space has, in fact, blossomed, just as was predicted in that first decade of space missions in the 1960s. Today space exploration is indispensable to how we live on Earth every minute of every day, in terms of weather, commerce, communications, agriculture, navigation, the safety of planes and ships and nations, not to mention for directions to the nearest Waffle House.

But what we haven't managed to do is enter what we have always imagined would be the Space Age. We rely on space; we just don't get to go there.

The Apollo Moon landings were a spectacular achievement, a demonstration of technological excellence, of design brilliance, of American ingenuity, of organizational skill and individual courage, and of national determination. The Moon landings—with astonishing fidelity—did exactly what John Kennedy said they would do, seven years before they happened: going to the Moon tested and measured the best of us. Arthur Schlesinger, the historian who predicted that the Moon landings would still be marked 500 years from now, is unquestionably right. When everything about the 20th century has faded to insignificance, the dawn of human spaceflight will still be remembered.

The Moon landings were a triumph on the Moon and a unifying and thrilling and also humbling moment on Earth. They were quite simply an unqualified success.

And yet we think of them now with a certain wistful nostalgia, as if they'd somehow become a disappointment, as if the decades since landing at Tranquility Base had not managed to fulfill their promise. The future we live in somehow doesn't live up to the future the Moon landings suggested. So maybe the Moon landings themselves were different than we thought—fruitless or overrated, a waste of money, a dead end. As Kennedy worried in private: nothing better than a stunt.

A lot of historians, space policy experts, and public officials would say that's exactly what happened. Apollo was a remarkable achievement, yes, but it distorted the entire space program, and it left U.S. space exploration adrift. What next? Going to the Moon when we did it, going to the Moon the way we did it, simply didn't have a good rationale that related to space travel, to science, to exploration. And so once we did it, we looked around and couldn't quite figure out what to do next.

John Logsdon, one of the legendary space historians, wrote, "The impact of Apollo on the evolution of the U.S. space program has on balance been negative. Apollo turned out to be a dead-end undertaking in terms of human travel beyond the immediate vicinity of this planet; no human has left Earth orbit since the last Apollo mission in December 1972." Far from becoming the foundation of the next step in space travel, the Apollo flight hardware "quickly became museum exhibits to remind us, soon after the fact, of what once had been done."[10]

We've turned Apollo into a series of historical displays. But you don't go to a Civil War museum or a World War II museum and think, *If only we made muskets like that now. If only our military could produce a plane as good as the B-29.*

I had an encounter of a similar sort in person with the dean of space historians, Roger Launius. Launius was for 12 years the official historian of NASA, and went on from that job to be a senior curator at the Smithsonian Air and Space Museum. Launius knows NASA as well as one person can; he knows the details of the race to the Moon, and also the sweep of it. He's written or edited 20 books and more than 100 journal articles on the subject. If you go looking for a topic relating to space, you might find him.

"Apollo: A Retrospective Analysis"—Launius.

"Public Opinion Polls and Perceptions of U.S. Human Spaceflight"—Launius.

"Heroes in a Vacuum: The Apollo Astronaut as Cultural Icon"—Launius.

*Spaceflight and the Myth of Presidential Leadership*—Launius (coeditor).[11]

Launius was gracious enough to meet and talk about writing about space. He's an intimidating figure, with a well-earned air of authority. Over lunch in the summer of 2016, he rocked back in his chair, looked me in the eye, and growled a question: "Was Apollo a success or a failure?"

I was 55 years old, so I knew a trick question when I saw one. I also knew I might as well answer quickly, because I was doomed to answer incorrectly. "Well," I said, "if the goal was to do as Kennedy said—to send a man to the Moon and return him safely to Earth before the end of the decade—it was a success. An unqualified success."

"That's true," said Launius. "But if the real goal was to open the solar system to human exploration, to human settlement, then it was a failure."

Quite so. Since Apollo, our planetary exploration has been spectacular and spectacularly revealing; we have ranged to the most exotic destinations in the solar system—but robotic probes have done it all. Our human spaceflight—the Space Shuttle, the International Space Station—has been complicated by politics and competing agendas, but none of it has gone beyond Earth orbit, and that's true for the United States and for the rest of the world as well. The Russians never did send cosmonauts to the Moon, and although China has an ambitious space program, its astronauts too have only gone into Earth orbit (six crewed missions, 11 people in space total over 15 years, through 2018).

No one in the world of space thought we were going to the Moon simply to go to the Moon. We were teaching ourselves to fly anywhere in the universe we wanted to go. And so in the world of space history and space policy, Apollo became regarded as a cul de sac. A brilliantly executed failure.

The point that Launius and Logsdon make can, in some ways, be inverted with even more power: If human spaceflight is so important and so valuable, as we argued in the 1960s, well, in the past 50 years, why has no nation sent astronauts to the Moon and beyond? It can't be that important if no one is doing it.

The historians bring to bear a sense of historical perspective. But in the case of the Moon landing, you didn't need to wait to hear the dissenters.

Earl Warren, the chief justice of the U.S. Supreme Court, gave a commencement address on June 1, 1969, seven weeks before the first Moon landing and a month before his own retirement. Warren was one of the most important and progressive officials in the U.S. government and in the shaping of the nation as the sixties came to a close; he wrote the Supreme Court's *Brown* decision that struck down segregated schools, and the decision that established Miranda rights for criminal defendants. And he was speaking to college graduates, young people embarking on a life of adventure and contribution. He was also a man who couldn't have been less removed from the world of space travel. Warren was 78 at the time; he'd been born in 1891. He told the graduates, "We're going to be on the Moon—perhaps by July, they tell us. But it would be better if our universities taught us how to live in our great cities."

Senator Edward M. Kennedy, the youngest brother of John Kennedy, gave a speech on May 20, 1969, while Apollo 10 was just beginning its pioneering flight to the Moon with a lunar module. Kennedy said that with the Moon in reach by his brother's deadline, "a substantial portion of the space budget can be diverted to the pressing problems here at home"—and that was before we'd actually landed. It was a bold moment to speak out; Kennedy was one of the honored speakers at the dedication of a new library named after the rocket pioneer Robert Goddard at Clark University, and in the audience were Wernher von Braun, NASA's own rocket chief, and Buzz Aldrin, who would be flying to the Moon himself in 60 days. It was 1969, and the division and social unrest across America—because of Vietnam, because of civil rights, because of poverty—could easily make flying to the Moon seem like an indulgence as opposed to an achievement.[12]

The Moon trips—which could, with a moment of reflection, seem like the realization of a dream of all humanity since we'd first looked up at the twinkling night sky and the full Moon with puzzlement and curiosity—the Moon trips consistently inspired the need to be infused with meaning.

At the press conference of the Apollo 11 astronauts after they had spent 21 days in quarantine after returning from the Moon—three weeks of isolation to prevent any possible contamination of the Earth by Moon microbes—a reporter asked, "Many of us and many other people in many places have speculated on the meaning of this first landing on another body in space. Would each of you give us your estimate of what is the meaning of this to all of us?"[13]

Three years later, on the Sunday just before the launch of Apollo 17 in December 1972, the final Moon mission, the *New York Times* printed an essay from Amitai Etzioni, the Columbia University scholar who had published the book *The Moon-Doggle* in 1964, in the middle of the Moon race. Seeing the lunar rover leaping across the dunes of the Moon had not charmed him. If anything, five actual Moon landings had left Etzioni choleric. His essay opens, "The most hopeful epitaph for Project Apollo might be: This was the last gasp of a technologically addicted, public-relations-minded society, the last escapade engineered by an industrial-military coalition seeking conquests in outer space, while avoiding swelling needs on earth."[14]

In that same issue the *Times* had asked its veteran space reporter, John Noble Wilford, the man who wrote the main page-one story of Apollo 11 and of every subsequent Moon landing, to talk to scholars, thinkers, poets, and scientists worldwide and ask them for the meaning of Apollo. The *Times* put Wilford's story with those reflections at the top of the front page. The reporter asked two dozen people "for their assessment of the probable place of space exploration in the broad sweep of history and in the evolution of man and man's perception of himself and his universe." He talked to Margaret Mead, C. P. Snow, and the British historian Arnold Toynbee. The answers ranged from the anodyne to the memorably idiosyncratic. Almost all were, in some way, focused on the payoff or the cost.

John R. Platt, a well-known biophysicist at the University of Michigan, spoke of the "Earthrise" photo, that first color image of Earth floating in dark space, taken by Bill Anders of Apollo 8: "That great picture of Earth taken from the Moon is one of the most powerful images in the minds of men today and may be worth the cost of the whole Apollo project. It is changing our relationship to the Earth and to each other."

Claude Lévi-Strauss, the influential French intellectual and anthropologist, spoke of television and boredom: "I never look at TV except when there is a Moon shot, and then I am glued to my set, even though it's boring, always the same and lasts a long time. Still, I can't turn away."[15]

The American philosopher Eric Hoffer said, "Except for unmanned vehicles, we are not ready yet to go beyond Apollo. It is time that we returned to Earth to see what we can do about strengthening the weakest link, man."

Daniel J. Boorstin, the Pulitzer Prize–winning historian, said, "The great thing about space exploration is that we don't know what its payoff will be. This symbolizes the American civilization. The people who settled America had no idea what the payoff would be. They settled it before they explored it."

Jacob Bronowski was a British scientist, mathematician, and historian of science, and the writer and narrator of the legendary BBC TV series *The Ascent of Man*. In that series, Bronowski traveled the world to trace the development of human society through its understanding of science. Of the Moon missions, Bronowski said, "I am not at all impressed with people who tell me it is useless. It is only useless if we do not know how to use the experience."[16]

---

Bronowski was right: The Moon missions are useless only if we don't know how to use them. If we don't understand them and understand their impact.

That's what we've utterly failed to do.

Their value wasn't in the future of space travel. The race to the Moon helped unleash all the forces of the technological age in which we live, the culture of technology which is the hallmark of late 20th-century and early 21st-century America. We revel in that culture and take pride in it; we identify with it, we rely on it, and we also see it as an American creation that we have shared with the world, not unlike the Moon landings themselves.

What's amazing is that we don't connect it to the decade of the sixties or credit the race to the Moon for any of it. But the race to the Moon focused the early days of that digital innovation in the labs and factories where it was just being born, and the race to the Moon showed Americans and the world what it could do.

In the world of Apollo, we could ask computers to do anything, and they would help us do it. It was Apollo that so visibly and so dramatically transformed technology from being a tool of war to a tool of amazing accomplishment, and also a tool of everyday life. The Moon race did that for the idea of technology, and also, quite literally, for the word "technology."

In his inaugural address in January 1961, John Kennedy called science a "dark power": the U.S. and the U.S.S.R. must "begin anew the quest for peace, before the dark powers of destruction unleashed by science engulf all humanity." The Americans and the Russians "need to invoke the wonders of science instead of its terrors."[17]

That's exactly what the race to the Moon did: it invoked the wonders of science, with about as much drama as could be imagined.

At the dawn of the 1960s, "there was no everyday idea of 'technology' the way we think of it today," according to Eric Schatzberg. He is a professor at the Georgia Institute of Technology, specializing in the history of technology and in tracking how the word itself has revealed attitudes about science, technology, and society.

People in the 1950s and early 1960s loved household gadgets; it was the blooming of the age of TVs, lawnmowers, automatic clothes washers and dryers, dishwashers, disposals. But Americans didn't think of those gadgets as "technology," just a part of modern life. Technology, to the degree it was even an idea and a word in people's consciousness, was military technology: the Manhattan Project, the atomic bomb, the hydrogen bomb, and the missiles created to deliver them. Technology was *Dr. Strangelove.*

Even as technological advances seemed to be improving life—from sleeker and more powerful cars, for instance, to the green revolution in food production—they were often shadowed by dreadful consequences. Rachel Carson's book *Silent Spring*, published in 1963, detailed the poisonous and long-lasting damage to health and to the environment from the new wave of pesticides. Ralph Nader's book *Unsafe at Any Speed*, published in 1965, showed how carmakers consistently put style and cost ahead of the safety of people riding in their cars. Both books had a galvanic impact in teaching Americans to balance the benefits of technological advance with testing, regulation, and safety rules. Nader's book led directly to seatbelt laws in 49 states and to the creation of the U.S. Department of Transportation. Carson's book came out as she was battling breast cancer, and she died less than two years after publication. But *Silent Spring* sparked the modern environmental movement (and modern environmental journalism), led directly to the campaign to ban the pesticide DDT, and was in part the inspiration for the creation of the Environmental Protection Agency in 1970.[18]

What became clear in the 1960s, said Schatzberg, is that when it came to technology, "the benefit and the damage were often in the same advance." Indeed the Space Age itself was born with a sense of unease: the eerie beeping of Sputnik as it raced across the sky over the United States and the rest of the world.

In the environment of the nuclear age, with atomic bombs being exploded out in the open and hailed as patriotic accomplishments, President Eisenhower's decision to remove the U.S. space program from the hands of the military and create an all-civilian agency devoted to space—to satellites, space research, and space travel—turns out to have been far-sighted and to have had a powerful cultural impact.

Throughout the 1960s, culture, technology, and space travel shaped one another. The Space Age—the charmingly technologized world of *The Jetsons*—came to seem like a version of the world mostly free of the dark shadow that "technology" had carried since Hiroshima. "Apollo had a powerful cultural impact," said Schatzberg. "It's absolutely certain that it generated enthusiasm for high technology. It resonated with people."[19]

NASA and spaceflight became the easy embodiment of the age: the astronaut in a gleaming white spacesuit, grinning behind the wide glass helmet; the tall white rocket, floodlit or sunlit on the pad; the ranks of engineers in white shirts and ties, sitting at computer consoles in

Mission Control. Or, at least, it became *one* embodiment of an age in which many facets can be instantly conjured with a single image: Martin Luther King delivering the "I Have a Dream" speech at the Lincoln Memorial; the crowd at Woodstock; Elvis or the Beatles onstage.

NASA's openness—scientifically, but also to the press—magnified the cultural power of the dawning Digital Age. The astronauts were heroic explorers. But the engineers and scientists and technicians who sent them on their way—they were smart people, using cutting-edge technology to power a great adventure.

In 1965 *Time* magazine did a cover story titled "The Computer in Society" about the rapid infusion of computers across America. The story came early enough in the dawning of the Digital Age that it included a count of the number of computers in the entire United States at that moment: 22,500. (That comes to 450 computers per state, although the federal government alone was reported to have 1,767 computers.) For comparison, during the Christmas season in 2017, Apple sold 35,800 iPhones an hour. It took Apple 38 minutes to sell as many handheld computers as the U.S. had in total 52 years earlier.

The picture spread across the opening page of that *Time* story on computers in America was a fisheye-lens photograph of NASA's Mission Control, with dozens of consoles and computer screens. The opening anecdote was about the network of NASA computers spread among 15 locations around the world that "guided, watched, advised and occasionally admonished the Gemini astronauts."

Four years later, just before the first Moon landing in July 1969, *Fortune* did a story entitled "The Unexpected Payoff of Project Apollo," and the opening picture for that story is as breathtaking as anything the Apollo 11 astronauts would beam back to Earth. It's a wide shot of the Apollo launch control room at the Kennedy Space Center, and the ranks of computers and staff appear to run to the horizon. In fact there are eight tiers of computer consoles, running left to right, that recede into the distance at the far end of the room, each row with between 12 and 20 positions. Across the back of the room, in tall dark cabinets, are the actual computers driving the consoles that the dozens of staff people sit at. Television monitors and clocks hang from the ceiling. It was the digital future, come to life, in a single image.

Scientists who grumbled about Apollo—the amount of money it

took, the number of highly skilled people it absorbed across the country—were missing the much bigger picture. The race to the Moon created an aura around science and technology for the first time. Yes, the single "Earthrise" picture was important, but so were all those images of Mission Control.

We spent a decade watching spaceflights on TV, which is to say, watching scientists and engineers and technicians sitting in Mission Control, using computers to fly spaceships. They weren't in uniform; there was no military mission; there was none of Kennedy's "dark powers of science" at work here. Just the opposite. Apollo helped us reimagine the computer. Spaceflight showed us technology at work in a completely different environment, and for a completely different purpose, than we'd seen previously. And not just any purpose: technology, science, computers were the key to the greatest adventure ever undertaken. Spaceflight was thrilling, and the thrill was in part powered by and was reliant on smart people sitting at computers. It was the dawn of the Digital Age, and it was also the dawn of the age of the nerd. "The computer," said *Time* in the opening of that 1965 story, "is, in fact, the largely unsung hero of the thrust into space."[20]

No question. But the space program would turn out to be the completely unsung hero of the computer.

Whatever the power and influence NASA was having on the popular perception and acceptance of technology in everyday life, deep inside the industry NASA was having an even greater effect, a singular impact, on the immediate future of computing.

NASA needed computer chips to fly to the Moon; it needed integrated circuits, and it needed them to work, to be absolutely dependable. Indeed, it needed every individual integrated circuit to be perfect. They were flying people to the Moon.

When Eldon Hall first bought integrated circuits from Texas Instruments to test for MIT, they cost $1,000 each. When he made the pitch to NASA that the leap to the Moon required integrated circuits, in November 1962, MIT was buying them for under $100 each. By the middle of 1963 MIT's chips were $15 each—a reduction in price of 98.5 percent in just three years. That $15-per-chip price would fall another 50 percent by 1965, to $7.28. So, as NASA ramped up purchasing from 1960 to 1965, the price of computer chips fell 99.3 percent.

Then, between 1965 and the Moon landing in 1969, exactly the same thing happened again. The 1965 price of $7.28 became $1.58 in 1969, falling another 78 percent in the second half of the decade. In the space of nine years, the cost of computer chips had fallen from $1,000 per chip to $1.58 per chip—and the $1.58 chips were many times more powerful and orders of magnitude more reliable.[21]

That was Moore's Law: that the computing power of integrated circuits would double every two years for at least the next 10 years, even as the cost came down an order of magnitude—kicking in before almost anyone had heard of it—or had heard of computer chips.[22] But there was nothing foreordained about what happened in the 1960s. In fact the more likely path was very different. In 1960 almost no one wanted to buy integrated circuits. Who could afford to put a $1,000 chip in a product? Or a $500 chip? Or a $250 chip? And then not be certain how well that chip would work? Companies just stuck with transistors, which weren't tiny but had two advantages: they worked, and they were cheap. (Those $1,000 chips that Hall bought for MIT seem expensive on their face, but $1,000 in 1960 is the equivalent of $8,500 in 2018. Imagine pitching the boss on buying just 64 of those, as MIT first did, just to start testing—$544,000 to test out a new possibility.)

We have a perfect case study of how wary the U.S. economy was of integrated circuits. As MIT was designing the Apollo computer, IBM was designing its new series of mainframes, the IBM 360. In 1961 IBM had two-thirds of the U.S. market in computers. The new 360 was designed to break open general-purpose computing for businesses, to let companies use computers in all the ways they could imagine, as opposed to only for dedicated functions. IBM's revenue at the time was $2.5 billion a year, and the 360 cost $5 billion to develop. In his study of the early days of the computer chip industry, economist Richard C. Levin wrote, "By 1962 there was still considerable debate among potential military and civilian users about whether the integrated circuit could be made sufficiently reliable to gain wide acceptance. No major commitments had yet been made by private-sector customers."[23]

Levin, a professor of economics at Yale who went on to be president of that university for 20 years, published his analysis in 1982, just a decade after the end of Apollo. For IBM's key new product line, at the moment the company recognized as the blossoming of the computer era in business, IBM looked hard at integrated circuits. "After careful study," wrote Levin, "IBM opted against the use of . . . integrated circuits in its new 360 series of computers." The IBM 360 series was a precise contemporary of the Apollo computers: it was announced in 1964, and customers started buying it in 1965. It was a huge hit; the business scholar Jim Collins ranks the IBM 360's impact with that of the Ford Model T and Boeing's first passenger jet, the 707. Among the customers for the IBM 360: MIT and NASA. The computer was used to write software for the Apollo flight computer, and the IBM 360 was the core of the computing power in Mission Control during Apollo, in the Real-Time Computer Complex. Just without integrated circuits. Integrated circuits might have been the future, but not even the biggest, most powerful computer company in the world was ready to use them.[24]

There was only one big customer for integrated circuits in the first half of the 1960s: the U.S. government. And within the U.S. government, only two groups needed the chips badly enough to shoulder their risks: the Air Force, for its Minuteman missile, and NASA, which is to say MIT on behalf of NASA. As Levin put it, "[Those] two key procurement decisions of government agencies were responsible for moving the integrated circuit into production on a significant scale."[25]

In 1962, the federal government bought 100 percent of integrated circuits produced in the world.

In 1963, the U.S. government bought 94 percent.

In 1964, 85 percent.

In 1965, 72 percent.

And even as government purchasing fell as a share, the volume soared. The federal government was buying only 72 percent of total production in 1965, but the volume in those three years had gone up by a factor of 20.[26]

Inside the government, the big customer was the Apollo guidance computer. In 1963 alone, 60 percent of all integrated circuits were purchased for NASA.[27] And MIT didn't just want those chips to be reliable; it insisted on it. Recall that every batch of chips was tested, and if even one chip failed, the whole lot was returned. MIT, on behalf of NASA, drove up reliability by a factor of 100 or more across the industry.

The idea that Apollo helped lay the foundation for integrated circuits has somehow been lost to computer history. Yet it's very clear that

NASA wasn't just a pioneer with the computer that MIT designed and built; NASA drove the rapid development of the underlying technology.

When MIT decided to use integrated circuits, only Fairchild Semiconductor could produce the right chips in the quantity, and with the quality and punctuality, MIT needed. "Large scale purchases of [Fairchild's] circuits by NASA in 1963 contributed to substantial learning economies," Levin concluded in his study. "The following year, Fairchild announced substantial price reductions in the commercial market, which led to wide use of these circuits in the computer industry."[28]

The computer chips that flew to the Moon created the market for the computer chips that did everything else.

Moore's Law, which has become the key benchmark of the pace and impact of computing power, was born in 1965. Gordon E. Moore, a pioneer and one of the giants of the computer business, the cofounder and CEO of Intel, opens his seminal 1965 paper: "The future of integrated electronics is the future of electronics itself. . . . Integrated circuits will lead to such wonders as home computers—or at least terminals connected to a central computer—automatic controls for automobiles, and personal portable communications equipment." In the paper, the only customer for integrated circuits that Moore identifies by name is "Apollo, for manned Moon flight."[29]

And what was Moore doing in 1965, as he was imagining with incredible prescience the future of computer chips and computing across American society and the rest of the world? He was the director of research and development at Fairchild Semiconductor (which he had cofounded with Robert Noyce), whose most significant customer was Apollo.[30]

By the time Armstrong and Aldrin were walking on the Moon, the market for integrated circuits was 80 times larger than it had been in 1962, but companies were buying more than 60 percent of that production. If chips were good enough to fly human beings to the Moon, they were good enough for whatever you could do with computers on Earth.

Integrated circuits, and their modern offspring, are as important to the economy of the U.S. and the world today as concrete, as electricity. You can't do anything without them except take a walk, and many people don't even take a walk without their computer chips.

Would the computer revolution have happened without NASA? Of course. Would Jeff Bezos have created Amazon without Apollo? Probably. Would we be hypnotized by our iPhones without the race to the Moon? Almost certainly. But that doesn't in any way diminish NASA's contribution to each of these.

Apollo launched rockets to the Moon. It also launched America into the Digital Age. NASA didn't invent the integrated circuit. NASA didn't invent the culture of perfection and continuous improvement—"the learning economy"—that were the key to their acceptance and use in the rest of the business world and the economy. But NASA's needs forced the semiconductor companies to create the perfect chip, and the continuously improved chip, on which the modern digital economy is built. And American semiconductor companies were the first to achieve that level of virtuosity. So it isn't just that semiconductors have become indispensable; it was American companies that drove that transformation and that dominated the industry for most of the past 50 years.

At the moment when the modern computer industry was being born, the most important customer, the most influential customer, the biggest customer—almost the only customer—was NASA's spaceship, headed to the Moon.

And so while NASA has always searched for its own impact and come up wanting; while the space historians look to the heavens and conclude that Apollo didn't have anything like the impact everyone hoped; while the critics of spaceflight bemoan the wasted billions that could have gone to alleviating hunger or poverty or improving education (but never would have)—the most obvious and most important impact has always been there, perhaps so large, so coincident in time, that it's always been overlooked. No, Apollo didn't usher in the Space Age, but it did usher in the Digital Age. It helped lay the foundation of the technology that created the digital revolution, and it helped give Americans a sense of excitement and anticipation about the Digital Age, a sense of excitement that had been completely missing when the 1960s began.

That excitement is still with us. Today, when you first unwrap a new iPhone, when you first boot up a new and newly powerful laptop, that little frisson of excitement you feel is a small echo of the thrill of spaceflight itself. Hey, those guys sitting at their computer workstations in Mission Control were using computers to fly people to the Moon. Let's see what we can make this laptop do.

---

Setting aside for a moment this remarkably powerful impact that Apollo had across the economy and culture of America, which has historically been overlooked, what of the criticisms of Apollo?

There are principally two. The first is that human spaceflight isn't worth the expense compared to the other things you can do with that money.

From start to finish, Apollo cost $19.4 billion, in real dollars, in the years they were spent, if you simply add them up. In inflation-adjusted dollars, through the end of the program, and calculated so all the costs are in 1974 dollars, Apollo cost $25.4 billion ($126.4 billion in 2018 dollars).[31]

That Apollo was too expensive, that we could have made better use of that money, is a reasonable argument, if you oppose spaceflight already. Fly to the Moon, or feed people who are hungry? Fly to the Moon, or provide better schools? It's not a contest.

The day before the launch of Apollo 11, NASA administrator Thomas Paine met with the protesters at Cape Kennedy who were led by the Reverend Ralph Abernathy. As Paine recounted the meeting, Abernathy told him that one-fifth of Americans lacked adequate food, shelter, clothing, and medical care. "The money for the space program, [Abernathy] stated, should be spent to feed the hungry, clothe the naked, tend the sick, and house the shelterless." Paine told the protesters, "If we could solve the problems of poverty in the United States by not pushing the button to launch men to the Moon tomorrow then we would not push that button."[32]

But that's not the way spending decisions in the federal government are made (except perhaps at the margins). There's never a vote balancing money for a new nuclear aircraft carrier against money to raise teacher salaries. It doesn't even work that way within categories; no one in Congress in the 1960s insisted that the right thing to do was to spend the money we were spending on the space program but divert it directly to cancer research instead. No one even suggested that.

Apollo 12 commander Charles "Pete" Conrad examines an earlier Moon lander, Surveyor 3, which landed on April 19, 1967, and sent back pictures and information that helped prepare for Apollo.

NASA originally hoped to navigate the Apollo 12 LM to a landing within a mile of Surveyor 3. Conrad and LM pilot Alan Bean landed less than 600 feet away.

On their second EVA, Conrad and Bean walked over to examine Surveyor and used tools to remove its TV camera and other equipment, so scientists on Earth could see how two and a half years on the Moon had affected them. In this picture, taken by Bean on November 20, 1969, Apollo 12 LM *Intrepid* is visible in the distance.

2

NASA engineer Tom Moser, above, during testing of the American flag that was sent to the Moon, at the Manned Spacecraft Center. NASA almost forgot to take a flag—planning for a flag that could be erected didn't start until just weeks before the first Moon landing. The flags that went to the Moon, purchased off-the-shelf, were rigged with a pole along the top so they could be extended curtain-style, even on the airless Moon.

Below, Apollo 14 commander Alan Shepard with the flag mounted and extended, on the Moon at Fra Mauro, February 5, 1971.

3

Apollo 16 commander John Young jumping while saluting the flag, taken April 21, 1972. The lunar rover, in which Young and Charlie Duke drove around the Moon, is parked in front of their lunar module, *Orion*.

At right, Apollo 12's flag hangs limp. The latch at the corner of the two poles didn't work properly. Photo taken November 19, 1969.

The Apollo 16 base camp in the Descartes Highlands. John Young is working on the lunar rover, alongside lunar module *Orion*. Apollo 16 was the fifth Apollo landing, and operations on the Moon had become more confident, more routine, and more ambitious. Young and Charlie Duke did three Moon walks, totaling twenty hours, and drove 16.6 miles to explore the geology of the region.

In the foreground, boot prints from Young and Duke, and tracks from the lunar rover. The insulation underneath *Orion* was removed so the astronauts could unload equipment. Picture taken by Duke, April 21, 1972, during the first Moon walk.

7

Apollo 16 commander John Young shows off the performance of the lunar rover, above, during a series of maneuvers designed to give engineers a sense of how the rover performed, while Charlie Duke filmed, during their first Moon walk, May 8, 1972 (still from video). The astronauts, who wore seat belts, said the rover gave a great sense of speed, in part because of the low gravity. The rover's top recorded speed on the Moon, during a downhill run, was 10.5 mph.

Below left, Apollo 17 LM pilot Harrison Schmitt examines a boulder near the North Massif, on the third EVA by him and Commander Gene Cernan, during which they drove 7.5 miles. Photo by Cernan, December 13, 1972.

Below right, the right rear fender of the Apollo 17 rover. It broke when Gene Cernan accidentally dropped a geology hammer on it, and the astronauts found driving without it sent dust everywhere. They rigged an improvised fender using four plastic-covered maps, held together with duct tape and two clips. Photo by Cernan, December 12, 1972.

8

9

Apollo 17 commander Gene Cernan, after his third Moon walk, inside LM *Challenger*. Cernan is covered in Moon dirt, as are the spacesuits, stowed on the left. This photo gives a sense of the unromantic reality of some aspects of Moon missions—the lunar module cabin was cramped, it had no sleeping bays, although the later missions stayed on the Moon three days (the astronauts just curled up and made themselves as comfortable as possible), and Moon dirt was clingy and hard to remove. Moon dirt also had a smell that some astronauts described as being like the smell of a just-fired gun or firecracker. This is a mosaic of two images, taken by Jack Schmitt, December 14, 1972.

11

12

13

After the first Moon landing, the Apollo 11 astronauts were a worldwide sensation. Top, Neil Armstrong visiting troops in Long Binh, Vietnam, with Bob Hope. Twenty-thousand GIs came to the Christmas show on December 29, 1969. Center, the astronauts, in sombreros and ponchos, swarmed by fans in Mexico City, September 29, 1969. During a worldwide tour, they visited 27 cities in 24 countries in 39 days, using Vice President Agnew's plane. Bottom, the astronauts and their wives meet Pope Paul VI, in the Papal Library at the Vatican, October 16, 1969.

The Apollo 17 astronauts blasting off from the Moon, December 14, 1972. Video of the ascent stage in flight was taken using the TV camera on the lunar rover, operated from Mission Control. Cernan and Schmitt had a brief ceremony as they left the Moon's surface, on the last Apollo mission. "We leave as we came and, God willing, as we shall return, with peace and hope for all mankind. Godspeed, the crew of Apollo 17," said Cernan.

The ascent stage used the LM's lower stage as its launch platform. No flames are visible because there is no air on the Moon, but debris is scattered widely by thrust from the ascent engine.

The rover camera was operated from Mission Control by Ed Fendell, who managed to record the entire liftoff, sending commands to angle the camera up in sync with the LM's swift ascent. Fendell started sending commands three seconds before the astronauts gave the blastoff command, to account for the time it took the radio signals to reach the Moon (still from video).

It was a common critique in the 1960s, and it still is today: that somehow, by devoting time, money, and energy to space travel, we must of necessity neglect other things. If we go to space, we can't have good schools or accessible health care or clean water or a strong spirit of community. It's like saying art museums cause poverty.

President Kennedy, for one, not only didn't try to hide the cost of Apollo, he talked about it incessantly. He almost never mentioned the race to the Moon without noting how expensive it was. In private, the cost clearly worried him. But Kennedy also made the case that America could afford to go to the Moon, and he often did so insisting that we could no longer afford to be second in space—a line perfectly designed to remind Americans of those embarrassing years of Sputnik overhead and exploding American rockets on Earth.

And President Kennedy reminded us that we really could afford what we wanted to do, and what we needed to do—we could have good schools and strong defense and well-fed children and Apollo too. As he pointed out at Rice University, space was still costing "somewhat less than we pay for cigarettes and cigars." The peak years for spending on Apollo were 1966 and 1967, when the cost each year was about $3 billion. In each of those years, American smokers spent more than $9 billion on tobacco products. Clearly we could indeed afford Moon rockets, and also our Marlboros. Going to the Moon was a decision about policy, priorities, and national goals. Space was expensive, but that was a different issue than whether the U.S. could afford it or not.[33]

That would become stunningly clear—really, devastatingly clear—just as the actual Apollo flights were getting under way in 1968 and 1969. Those were the most expensive years of the Vietnam War; in each of those years, the war alone cost more than $19 billion. That is, in each of those years, the U.S. spent more on the fighting in Vietnam than the total cost of going to the Moon over 11 years. Vietnam was astonishingly expensive, costing more than $14 billion a year for five years in a row. Apollo, specifically, never cost more than $3 billion in any give year.

Put aside for a moment the almost unmeasurable human cost of the Vietnam War: 58,200 American service members dead, 1.3 million Vietnamese soldiers dead on both sides, another 2 million Vietnamese civilians killed, and decades of pain and adaptation for those who fought and survived. And then there was the cost of the war at home, the cost to American society.[34]

But just in terms of money spent:

Apollo cost $19 billion.

Vietnam cost $111 billion.

They happened alongside each other. They were both events of such significance that they shaped the future of America, although in very different ways.

Vietnam turned out to be not just a military defeat but a geopolitical defeat as well; we set the stakes of saving Vietnam from communism, and communism won. The Vietnam War, and everything that attended it, was also corrosive to Americans' confidence in government, in the ability of the government to do what it said, and also to, quite simply, tell the truth—about American aims, American performance, and the lives of American service members.

Apollo was a success. It was a demonstration of American technological prowess, a demonstration of engineering and manufacturing excellence; it was a reminder of American economic power and also American determination. In Apollo we also set the stakes—the almost unreachable goal of landing people on the Moon and bringing them home safely. And then we did it. Many times.

And yet the way the two are often summed up:

Vietnam was a mistake.

Apollo was a waste of money.

One thing Apollo was not was a waste of money. We spent a lot of money, by any measure, but we got our money's worth. We taught ourselves to fly in space, and that turned out to be just as hard as the people who had to do it thought it would be. The science the astronauts did, the science the astronauts enabled Earthside scientists to do, completely remade our understanding of the formation of the Moon and its composition and geology, and by extension, it changed our understanding of the early years of the Earth and the relationship between the Moon and the Earth.

Apollo also accomplished that mission which John F. Kennedy first set for it: it powered America into the leading role in space. It took most of the decade, in fact, but it turned out that democratic capitalism could not be overmatched, even in space. Kennedy also said in that speech in

May 1961, "No single space project in this period will be more impressive to mankind." That too turned out to be true, at a level we couldn't even anticipate, around the world.

Imagine for a moment that the first astronauts to land on the Moon had been Russian and that they had unfurled and planted the flag of the Soviet Union, the hammer and sickle on a field of solid red. The Soviet flag was the international emblem for communism. But it was also a symbol of oppression, of tyranny, of simple lack of individual freedom, the inability to speak your mind, to pick your own destiny. The Soviet banner represented the supremacy of the state over the individual. That isn't just Western chauvinism; in 1969 the people living across Eastern Europe behind the Iron Curtain knew that more vividly than any American. The problem wouldn't have been Russian dominion over the Moon; it would have been the assertion of power required to get to the Moon, reflected back to the rivalry on Earth. It would have been a great achievement had the Soviets put people safely on the Moon, but it would have been chilling in a way we could only have imagined in the days of Sputnik's beep-beep-beeping overhead: a Red Moon. First, forever.

If the Soviets had made it first, of course, it wouldn't have changed what happened in the decades to follow; it wouldn't have signaled some alternative future, where the Berlin Wall did not fall and the Soviet Union did not unravel, any more than the triumph of Yuri Gagarin's first flight had.

But it was not, in fact, simply two nations racing for the Moon. The Soviets had made it "a test of the system," as Kennedy put it. Which system had the resources and the skill and the grit to get to the Moon—communism or democracy? The landing of the *Eagle* on the Moon, the moments when Armstrong and Aldrin stepped off the ladder onto the gray lunar ground—those represented a soaring accomplishment of human ingenuity. The moment when they unfurled the American flag, for all the complexity of America's role in the world, that underscored that it was also an achievement of human freedom. The American flag meant something very different from the Soviet flag. Instead of the triumph of tyranny, it was just the opposite: going to the Moon is forever the symbol of what freedom can accomplish, of how far human aspiration can take you.

As the astrophysicist Neil deGrasse Tyson said of Apollo 11, "No other act of human exploration ever laid a plaque saying 'WE CAME IN PEACE FOR ALL MANKIND.' "[35]

After the sweep of American Moon missions, the Russians did something it would have been hard to imagine Americans doing had the situations been reversed. They gave up. They had a rocket, the N1, that was designed to have the power to go to the Moon, but with its crazy complexity—its first stage was powered by 30 individual rocket engines—they never got it to work. It was launched four times and failed four times. The Soviets had already designed and fabricated a spacesuit for cosmonauts to wear on the Moon. The Smithsonian National Air and Space Museum, in Washington, D.C., has one of the Soviet suits in an exhibit about the U.S.-U.S.S.R. rivalry. With a wryness rare from the Smithsonian, the suit is displayed with a placard that reads, "All dressed up but no way to go."[36]

Apollo was both a victory and a success.

It was followed by a U.S. spaceflight letdown—really, a spaceflight hangover—that has lasted almost 50 years. That's the complaint of the historians about Apollo. That's what they mean when they say, as Roger Launius did, that it didn't open the solar system for human exploration and settlement; when they say, as John Logsdon did, that on balance Apollo did more harm than good to the cause of human spaceflight. No human being has left low Earth orbit—from the U.S. or any country—since December 7, 1972, when Ron Evans, Harrison Schmitt, and Gene Cernan fired the engine of their Apollo 17 ship *America* to head for the Moon.

No less a figure than Wernher von Braun, in an interview with Logsdon after the Moon landings were concluded, said, "The legacy of Apollo has spoiled the people at NASA. They believe that we are entitled to this kind of a thing forever, which I gravely doubt. I believe that there may be too many people in NASA who at the moment are waiting for a miracle, just waiting for another man on a white horse to come and offer us another planet, like President Kennedy."[37]

Inside those critiques is a subtler, more technical criticism: it wasn't going to the Moon that derailed human exploration; it was the way we did it. The leap required a particular package of technology: the big

Saturn V booster, the three-person command module, the very mission-specific lunar module, all produced with urgency. Those were, clearly, the right choices for getting American astronauts to the Moon, safely, in just eight years. But if you were imagining a 50-year arc of human space travel—without the need to dash to the Moon fast, without the need to beat the Russians—would you have made those choices? Apollo left us with a big investment in space hardware that didn't seem quite right for whatever the next steps were.

That's a reasonable point—in the abstract. But it's a completely ahistorical criticism. It ignores reality. NASA and Apollo didn't exist in a world where we could lay out a thoughtful, methodical, half-century-long plan of spaceships and space exploration, and then execute it. Nor has the U.S. government ever shown any talent for that kind of forward thinking. It might be true that the Saturn V or the lunar module weren't the right specific space technologies for whatever the post-Apollo step should have been. But the skills required to build them, the insights and lessons learned about how to travel in space—those were unquestionably transferable to the next step.

In fact perhaps the real problem isn't that Apollo was a success, but that the leadership of both NASA and the U.S. after Apollo didn't see the world clearly and figure out the right strategy for space exploration, and human exploration, with clarity and persuasive power. NASA leaders, presidents, and Congress all made bad space policy decisions in the 1970s and 1980s. But that's their fault, not the fault of the people who got America to the Moon in the 1960s.

Apollo was an unqualified success, and it wasn't—judged on its performance—a waste of money, nor was it a use of money that the United States simply couldn't afford. If we could afford the disaster of Vietnam, we could certainly afford the success of going to the Moon.

Whether it was the right expenditure of money is a different question. The early successes of the Soviet space program lasted for years, but they were built on the big rockets that the Russians had initially designed and built to help them launch their nuclear weapons. How far would the Soviets have gone if we hadn't started chasing them?

The second big critique of the race to the Moon is that manned spaceflight is, by its very nature, a waste of money. It costs so much to fly people in space precisely because of the people: you need spaceships that are sealed, that re-create the environment human beings need, that are fail-safe, and you need to supply all the things people need: food, water, oxygen, spacesuits. And the more provisions and equipment you stack up, the more fuel you need, and the rockets and the spaceships just keep getting bigger and more complicated, more prone to potential failure, and also more expensive. And yet, once the people arrive on the Moon, the argument goes, they can't do that much more science than a well-designed robotic lander could—at least, not enough additional science to justify all that effort. What's more, if anything goes wrong, it's not just a disappointment, it's a tragedy.

That's a much more powerful argument. In fact it's not really arguable: it's true. If all you care about is actual scientific research accomplished per million dollars spent, send unmanned probes. Here too, though, there's often some muddling of what's likely to happen. For what we spent on Apollo, we could have sent a dozen roving probes to the Moon and a dozen more to Mars. But those kinds of missions, of course, never would have gotten that kind of funding, any more than the spending on food stamps goes up when you trim a bomber program. One of NASA's consistent demonstrations of brilliance, one of its steady sources of new knowledge for the world, has been the work of its scientists and engineers in using all kinds of uncrewed technology, including the Hubble Space Telescope and the Mars rovers, to dramatically expand our understanding of the universe.

The serendipitous discovery that human beings in space provide—that's always the rebuttal to uncrewed probes—that serendipity is impossible by remote control. The "Genesis Rock" that Dave Scott and Jim Irwin stumbled on during the second excursion on Apollo 15 was a purely human discovery, and also a wonderfully human discovery.

And there's one other thing that the anti–human spaceflight crew either doesn't know or doesn't bring up: human spaceflight creates a halo of interest, support, and funding for all kinds of other space research, including remote space probes. In a study of the impact of the expanded U.S. space program, Rutgers University professor Jerome Schnee wrote that, heading into the space race in the late 1950s, "astronomy was a small science growing at a modest pace." The U.S. had what he described as "hundreds" of astronomers, and their ranks grew by 4 to 5 percent a year. By the end of the 1960s, the number of U.S. scientists studying astronomy had at least tripled, to 2,500, and by the time of the Moon landings, their ranks were growing by 15 percent a year. (By 2018 the number was roughly 7,600. The number of astronomers grew at twice the pace of population growth.)[38]

But human space exploration and robotic space exploration serve different purposes. You can read a lot about Rome; you can see pictures of Rome; you can watch videos about Rome; you can take virtual walk-through tours of Rome. But eventually you want to go to Rome—because that's how you really experience Rome. No robotic lander could have told us what the Moon smelled like or what the Moon's regolith felt like, that odd combination of powdery but also gritty. No robot could have taken a picture that inspired the line Buzz Aldrin immediately came out with to describe the sweep of the Moon before him, his first two words as he set foot on the Moon: "Magnificent desolation."[39]

In the report they coauthored in early May 1961, in advance of Kennedy's decision to go to the Moon, about whether the United States should embark on a dramatically expanded space program, NASA chief James Webb and Defense Secretary Robert McNamara wrote, "It is man, not merely machines, in space that captures the imagination of the world."[40] People in space inspired the software programmers at MIT and the parachute packers in California. People in space inspired the audience watching on TV, from New York to New Zealand. And being in space also inspired the people in Mission Control and the people who actually went—the astronauts themselves.

Many things in the world need doing. Flying to the Moon is nowhere on the list of national necessities. But if it is not precisely a necessity, it is still essential—in the way, for instance, art and music and storytelling are essential; in the way of scientists trying to solve the mysteries of the universe.

We can get up close to the tools we used to fly to the Moon: white spacesuits worn by the Moon walkers, still covered with the smudges and dust of lunar ground; lunar modules built for testing that never left Earth but that give a sense of the scale and complexity of the vehicle; Apollo capsules that came blazing back to Earth at 25,000 mph, carrying home both the astronauts and the scorch marks of that journey; and the Saturn V rocket itself, laid out horizontally on the ground, so you can take in its scale—so long, at 363 feet, that, lying on its side, it wouldn't fit on an ordinary football field, including the end zones.

What's sad is not the items themselves, which have a vividness that underscores how different an era that was and how risky the journey was. What's sad is that, in part just because of how we've come to think about Apollo, and talk about it, we think those are artifacts from a different America, from an America with a greater spirit of adventure than we have today, that we somehow imagine they represent a better America.

---

At the start of the space race, the Soviets weren't intimidated by the Americans. Up to the moment of President Kennedy's "go to the Moon" speech, they had done everything significant in space first: the first satellite in orbit; the first live creature in orbit, the dog Laika; the first spacecraft to reach the Moon, and then photograph the never-before-seen dark side of the Moon, and then radio those photographs back to Earth; the first living creatures launched to orbit and returned safely to Earth, the dogs Strelka and Belka. And they had done all that while Eisenhower was president.

In 1961 their spaceflight achievements gave the Soviet Union—and its global reputation—a reason to swagger. Russia's good-humored confidence was perfectly captured by a gift Khrushchev gave the new American president. Four weeks after Kennedy's "go to the Moon" speech, three Russian diplomats visited the White House with a fluffy, white-haired puppy for the Kennedys' daughter, Caroline. The puppy, named Pushinka, was a daughter of Strelka, one of the dogs that had spent 24 hours orbiting the Earth and returned safely to the Soviet Union. In a letter to Kennedy, Khrushchev referred to Pushinka as "a direct offspring of the well-known cosmic traveler, Strelka." Pushinka was a gift, but also a message. Give all the speeches you want: the Americans hadn't beaten the Russians to a single big moment in space.[41]

As careful as Kennedy had been to make sure NASA thought the Moon was achievable, that one speech, that one idea, to have America land people on the Moon and bring them back, immediately fired the imaginations of Americans. There are all kinds of signs of that from throughout the decade—just the fact that the man who designed the lunar rover, Ferenc Pavlics, from GM, could look around his house and find the perfect driver for his model lunar rover among his son's toys: Astronaut GI Joe, already outfitted in a spacesuit. That's a reminder of how quickly and thoroughly the Space Age infused the culture of the 1960s.

Perhaps the most revealing illustration is how quickly we started using going to the Moon as a shorthand way of talking about what Americans were capable of in the transformative age of the 1960s.

Not even a year after Kennedy's speech, the agriculture commissioner of Montana was angry about federal farm policy and the impact growing too much wheat was going to have on the livelihoods of Montana farmers. Lowell Purdy invoked Kennedy's Moon mission to criticize the president and his farm program. "Nothing is impossible in this age of miracles," Purdy said. "If we can put a man on the Moon, we surely are capable of seeing that our temporary surplus agricultural products are placed in many hungry stomachs of the world."[42]

Purdy was the first public official to be recorded using that phrase "If we can put a man on the Moon." He said it on May 14, 1962.[43] At that point the U.S. had managed to orbit a single man, John Glenn, alone in a tiny capsule, for three laps around the Earth. NASA hadn't even figured out what a Moon rocket would look like.

But Purdy had perfectly captured his frustration with farm policy: if we can manage the logistics to fly to the Moon, surely we can figure out how to get surplus wheat, grown right here on planet Earth, into the stomachs of hungry people. The fact that we *couldn't* go to the Moon didn't spoil his metaphor.

The next use of the phrase came just three days later, at the opposite end of the country, in the *St. Petersburg (Florida) Times*. Columnist Ann Waldron was writing about the immaculate homes presented in home design magazines, and how silly they look to anyone with a real family and real children. "I have to laugh when I look at those glorious, glossy color pictures in the fancy home magazines," she wrote. One of Waldron's fantasies for combining easy decor with realistic housekeeping turned out to be carpets made of paper that you could simply wad up and throw away. "If we can send a man to the moon," she wrote, "why can't we have paper rugs?"[44] Waldron was using the idea of the Moon in a different way than Purdy: if we can create the technology to fly to the Moon, why can't we do something down-to-Earth, like invent easy-to-clean carpets?

A year later a well-known hero of late 1950s America was testifying before Congress. Captain William R. Anderson, skipper of the first nuclear submarine, the USS *Nautilus*, and his crew had been the first to sail underneath the North Pole in August 1958, cruising 400 feet beneath Arctic ice nonstop for 1,830 miles, from the Pacific Ocean to the Atlantic.[45] Anderson had retired from the navy in May 1963 and was asked by President Kennedy to lead an effort to create a domestic version of the Peace Corps, to put volunteers into the most impoverished parts of the United States. Testifying to a House subcommittee on behalf of the Kennedy plan, Anderson said, "If we can send a man to the Moon, we can do something about the distress of people left to orbit helplessly in the vacuum of despair." Congress didn't fund Kennedy's domestic Peace Corps (Johnson would revive the idea as VISTA), but that line from Anderson appeared in dozens of U.S. newspapers as part of a syndicated "notable quotes" feature.[46]

The *idea* of landing people on the Moon was so persuasive, so vivid, so easy to understand but also so daring, that it leaped far ahead of the actual effort to put people on the Moon. Going to the Moon became the all-purpose yardstick not for accomplishment but for failure on Earth. A Massachusetts state representative complained in 1965, "We can send a man to the Moon but we can't get rid of our garbage and rubbish." After a mysterious and dramatic drop in the population of wild salmon in Idaho's rivers in 1965, the state's director of fish and game remarked, "If we can put a man on the Moon, we certainly can find out where the fish went."[47] Going to the Moon was such an extraordinary leap that it created the space in which we surely ought to be able to perform every routine terrestrial task—even though we hadn't gone to the Moon.

The phrase became a standard trope in the speeches of politicians. Senator Robert Kennedy used it to describe our inability to improve the miserable living conditions of migrant farm workers: "If we can put a man on the Moon before the end of the 1960s, it seems we should be able to work out such a simple problem for farm workers after 30 years of talking about it." Ronald Reagan, then the governor of California, used it campaigning for Richard Nixon's law-and-order presidential bid in 1968 to attack Democrats' "dovishness" on law enforcement: "We can send a man to the Moon, but we cannot guarantee his safety in walking across the street." Nixon's Democratic opponent, Vice President Hubert Humphrey, used the phrase in his standard stump speech: "If we can put a man on the Moon, certainly we can afford to put man on his feet on Earth." Humphrey was particularly fond of the comparison. As vice president, he was the keynote speaker at the 1967 Westinghouse high school science awards in Washington. "If we can put a man on the Moon," he told the winners and their families, "we can surely design a bus that doesn't belch nauseous and poisonous fumes in our faces."[48]

Sometimes people used the trip to the Moon in a simple burst of frustration. A Texas attorney spent two days trying to telephone his injured stepson, a soldier in Vietnam, at a military hospital in Saigon. "You know, it's funny, we can send a man to the Moon," the father told a newspaper reporter, "but we can't get a telephone call to Saigon." And in South Carolina, State Senator James Waddell was furious at the inability of a federal program to provide basic sanitation for poor people in his district. "We can send a man to the Moon," he declared on the floor of the South Carolina Senate, "but we can't build an outhouse."[49]

The wide use and wildfire spread of the phrase isn't just a curiosity or a bit of faddish 1960s slang. It illustrates three important things about America's race to the Moon. First, it shows the sheer power of the idea, which planted itself in Americans' psychology so quickly that "going to the Moon" became a way of thinking about the world. Second, it was, almost instantly, a fresh way of saying "Anything is possible." And third, when people were frustrated with a lack of progress, when they were reaching for inspiration, they immediately thought, *If we can put a man on the Moon . . .*

The work necessary to go to the Moon was mostly invisible. But the phrase shows that Americans absorbed something critical about the journey: it was a stretch. Even for the country that won World War II, that invented the atom bomb, that had 22,500 computers, going to the Moon required us to harness every ounce not just of energy but of imagination and technological innovation.

The point of using the phrase "If we can put a man on the Moon" is precisely that you've chosen the hardest thing you can reach for. That's how you make clear that the problem at hand—paper carpets, nonpolluting city buses, outhouses—should be easy by comparison.

But the most revealing thing about the phrase is how Americans used it, from the beginning, as if we had already done it. In fact the dozens of references from 1963 to the summer of 1969 make absolutely no rhetorical or rational sense because we hadn't actually shown that we *could* go to the Moon. Whether it's being used flippantly by columnists or seriously by the vice president of the United States, the phrase is literally nonsensical. What is the point of comparing something we aren't doing to something we haven't done yet?

But no one ever makes that point. The phrase was used more and more frequently as the sixties proceeded, and people clearly felt it had impact and persuasive power. In fact, the point of saying "If we can put a man on the Moon" is to conclude the conversation on a subject. It's a way not just of finishing but of winning an argument and declaring victory. It must be inarguable that if we can send a man to the Moon, we can deal with our garbage, our racism, our poverty, our missing salmon. Case closed.

We knew we were going to make it. Embodied in the phrase, in the speed with which we adopted it and the way we used it, is the clear sense that Americans considered putting astronauts on the Moon to be simply the latest inspired form of manifest destiny. We had announced we were doing it, and it was as good as done. That attitude seems all the more remarkable as other things unraveled during the sixties—our politics, our cities, our race relations, our ability to figure out how to win in Vietnam.

One writer was wise to the "man on the Moon" construct in a way no one else seemed to be. Matt Weinstock wrote a daily column in the *Los Angeles Times*. In September 1967 he wrote a piece headlined "Found at Last—Flexible Cliché for All Occasions."

"People wishing to show disdain for certain glaring flaws in our civilization appear to have settled on a cliché that could become the symbol of our era," Weinstock wrote. He offered a handy list of his own comparisons, including "We can put a man on the Moon but we can't make hippies take a bath."[50]

In less time than it had taken to go to the Moon, talking about going to the Moon had gone from potent metaphor to platitude. Indeed Weinstock's observation in 1967 that "putting a man on the Moon" had become hackneyed is all the more remarkable because he wrote it in the middle of a period when there was no visible progress on the journey to the Moon, during the two-year stand-down on manned spaceflights that followed the aftermath of the Apollo 1 fire. No Americans were actually going to space, let alone the Moon, but the phrase soared onward.

Weinstock stayed on "man on the Moon" watch. The frequency with which the expression was deployed clearly got under his skin. Twenty months later, in another column, he concluded the situation had become intolerable. The phrase was being used not to inspire but, said Weinstock, "in a nagging tone." Writing in what was then by far the largest newspaper west of the Mississippi River, Weinstock issued a call to boycott use of the phrase, which he said had become "obnoxious." Sadly, he added, "perhaps it's already too late." Weinstock's second column on the "If we can put a man on the Moon" phenomenon was published on June 2, 1969. The lunar module wouldn't land in the Sea of Tranquility for another seven weeks, but we'd already exhausted the idea of its doing so, at least linguistically.[51]

The people inside the space race, the people who knew just how hard going to the Moon was, seemed to think using their work in such comparisons was misplaced, or at least overstated. In that meeting between the anti-poverty protesters, led by Reverend Abernathy, and NASA administrator Paine the day before the launch of Apollo 11, Paine went on to tell Abernathy and the group that

> the great technological advances of NASA were child's play compared to the tremendously difficult human problems with which he and his people were concerned. I said that [Abernathy] should regard the space program, however, as an encouraging demonstration of what the American people could accomplish when they had vision, leadership and adequate resources of competent people and money to overcome obstacles. I said I hoped that he would hitch his wagons to our rocket, using the space program as a spur to the nation to tackle problems boldly in other areas.

Paine's detailed recollection comes from a memo he wrote for his files two days later, as Apollo 11 raced for the Moon, a memo tracked down by the historian Roger Launius. At the end of the meeting that afternoon, Paine asked Abernathy and his fellow protesters to include the Apollo 11 astronauts in their prayers when they held a prayer service later in the day. Abernathy, wrote Paine, "responded with emotion that they would certainly pray for the safety and success of the astronauts, and that as Americans they were as proud of our space achievements as anybody in the country."[52]

The problems that NASA and the vast army of 400,000 people overcame to get Apollo to the Moon were daunting—and are often minimized as requiring no real "technological breakthroughs," just some smart and persistent engineering. That's both dismissive and an odd understanding of the term "breakthrough," given the pioneering work that got done on everything from heat shields and parachutes to the art and science of rendezvous in space.

But it is true that every problem proved solvable, and that's in part because almost none of them involved human behavior or the social systems in which human beings live. Indeed part of the genius of NASA and Apollo in the Webb era was that Webb created his own social system to get the work done.

The problems of inadequate schools, of poverty, of hunger, of health care, aren't susceptible to a "Moon race" fix because they are part of the whole social, cultural, and economic system in which we live. Even students attending the very same school have very different experiences, because they are different children with different teachers. Every lunar module in the same orbit, with the same equipment, with the same duration of rocket firing, heads for the Moon in exactly the same way.

Once you solve the problems of flying to the Moon, you don't wake up the next morning and find those solutions have unraveled overnight. The problems of poverty and neglect and education don't get solved in the same way; they need fresh energy, fresh perspective, fresh attention every day.

The really incredible thing, though, is that the complaint embodied in the phrase "If we can put a man on the Moon, why can't we . . ." is wrong, especially as it applied to America in the 1960s and early 1970s.

Right alongside the race to the Moon, President Kennedy and President Johnson and Congress, and even to some degree President Nixon, applied themselves to exactly those problems. That's what the passage of the Civil Rights Act (1964), the Voting Rights Act (1965), the Great Society programs (1964–65), and the Clean Air (1963) and Clean Water (1972) acts tackled.

And they transformed American society and culture.

Black voter registration across the South soared after the passage of the Voting Rights Act; in two years, it increased from 20 percent to 50 percent of eligible adults in Alabama; from 7 percent to 60 percent in Mississippi; from 25 percent to 50 percent in Georgia. The number of black Americans who voted in the 1964 presidential election jumped to 12 million, from 5 million in 1960.

The proportion of Americans who lived in poverty fell by 40 percent from 1964 to 1973 (and the absolute number of Americans fell by 36 percent). Poverty among senior citizens was cut in half between 1967 and 1977.

Median income for Americans, in constant dollars, rose by almost 40 percent between 1960 and 1975. The GDP of the United States, in constant dollars, rose by 50 percent just between 1960 and 1970.

University enrollment doubled during the 1960s, and the number of women enrolled in universities rose 145 percent.

The number of women in the U.S. workforce grew at twice the rate of men in the workforce, and the number of women in white-collar jobs grew even faster than that.

And half a century later, the air and water everywhere across the United States are cleaner than they were in 1965, because of the laws passed then.[53]

Were the problems—of poverty, opportunity, health, equity—solved? No. And fifty years later, not only do many of them remain unsolved but some have gotten worse. But the big problems that Americans had on January 20, 1961, when President Kennedy took office—every one of those problems had improved when the last Apollo astronauts returned from the Moon in December 1972. In many cases, dramatically improved.

If we can land a man on the Moon, we should be able to tackle our hardest problems right here on Earth. And we did. The very same people did, in the very same decade. As we have lost track of (or never fully appreciated) the impact of the race to the Moon across the economy and culture of the U.S., we also don't credit everything else American society accomplished during that very same time.

---

What could have stopped Apollo?

There were technical problems everywhere—the engines for the Saturn V, the software for the flight computers, the 100 things wrong with the first lunar module delivered to Cape Kennedy—but it's hard to imagine the whole project coming to a halt because of a single or even a fistful of technical or management problems. When things went wrong, no one shrugged or waited for instructions. Part of the culture of Apollo, especially after the fire, involved two core principles: there are no small problems, and every problem can be solved because we aren't going to be the ones who prevent America from going to the Moon.

The Apollo fire could have stopped Apollo, especially if it had happened at a different moment than it did, a year further along, and in space. If fire had killed three astronauts out in space, it might have seemed not just as if space were risky but as if spaceflight were beyond our current competence. At least, it might have seemed that way to Congress. But maybe the response to the fire would always have been, *The only way to honor the astronauts who died is to fly to the Moon, because that is unquestionably what they would want us to do in the wake of their deaths.*

John Kennedy wouldn't have stopped Apollo, but he could have slowed it to the point that, in the presidency after his, the momentum would have dissipated and there would simply not have been enough money and energy to make it happen.

In a project like going to the Moon, momentum isn't just your friend; it's indispensable. It's what keeps you hurdling problems that might otherwise stop you in your tracks.

Just as important, it's worth remembering the things that didn't end up crippling Apollo.

Bill Tindall rescued the computer, which so often rescued Apollo.

The changes to Apollo made after the fire, from the way control panels were wired to the fabric in the spacesuits to how much Velcro was allowed in the spacecraft, were essential and saved the program, and so the fire itself showed NASA how off-course the Moon program was, even if everything looked great on the surface.

Lyndon Johnson had much more authentic passion for the race to the Moon than John Kennedy did. Between his enthusiasm and his mastery of Congress, he wasn't going to let Apollo falter, and the peak funding appropriations for Apollo came as he finished out the final year of Kennedy's term and the year after that, the first year after his landslide victory over Barry Goldwater. Johnson got 61 percent of the popular vote when he ran a year after Kennedy's assassination, the highest percentage of any president in history, before or since.

Upon reflection, it's remarkable that on the fifth day after her husband's murder—not a week later, or a month—during a 15-minute meeting with President Johnson, Jacqueline Kennedy thought to have the nation's spaceport renamed for her husband. And it's remarkable that Johnson not only agreed but picked up the phone and called the governor of Florida to clear the way for the change to happen. Whatever her husband's persistent concerns about the race to the Moon, she didn't share them. She wasn't imagining Cape Kennedy as the site of America's failure to win the space race. By imprinting the spaceport permanently with Kennedy's name (at the time, reporters often referred to it as "America's moonport"), she was quietly but insistently defining the future Moon landings as the most important, or at least most dramatic, tribute to the legacy of John Kennedy. Thousands of stories about Moon missions began with the dateline "Cape Kennedy, FL." (Floridians never appreciated having their cape renamed for the 35th president. It had been called Cape Canaveral for 400 years, well back into the Spanish era. As soon as the Apollo Moon missions were over, 10 years after the assassination, in 1973, the legislature of Florida changed the name of the physical peninsula back to Cape Canaveral. The space facility itself still bears Kennedy's name.)[54]

As Philip Abelson, the physicist and editor of the journal *Science,* had predicted way back in April 1963, in his snarky editorial about the Moon missions, "The first lunar landing will be a great occasion; subsequent boredom is inevitable."[55] After the novelty of Apollo 11 and Apollo 12, the first two landings, and the days-long, white-knuckle rescue of the Apollo 13 astronauts, Americans stopped paying as much attention to the Moon landings. TV viewership fell off dramatically, and the big three TV networks responded in part by scaling back their coverage. Eventually, for the later Moon missions, when the astronauts were in orbit around the Moon for four days or more, when they were on the Moon for three days at a stretch, even the *New York Times* scaled back coverage to just a single, full inside page, or just part of a page, each day.[56]

Even the boredom became front-page news. The day before the January 31 launch of Apollo 14 in 1971—the first flight after the explosion and near disaster of Apollo 13—Norman Mailer held a press conference in San Francisco to promote his newly published book about Apollo, *Of a Fire on the Moon*, which had been serialized in *Life* magazine. Mailer was at the height of his influence as a journalist and cultural commentator, and he was scorching. Americans, he said, were about as interested in the next day's Apollo launch "as in a border war in Bolivia." NASA had drained all the drama, passion, and humanity from the Moon missions. Of the Apollo 11 Moon landing, which was the subject of his book, he said NASA had "succeeded in making the most transcendental event of the 20th century boring." Mailer's press conference was perfectly timed: he made front pages across the country, right alongside the pictures of the Apollo 14 astronauts getting ready for their launch.[57]

The idea that the Moon missions were significant or valuable only to the degree that Americans paid immediate, rapt attention to them was common as the program wrapped up at the end of 1972, and remains a critique in the decades since—about Apollo, and about Americans' disinterest in the Space Shuttle and the Space Station as well.

The fading public interest was itself commented on, giving Apollo 17's landing a valedictory feel. "What even a few years ago seemed like an incredible and impossible task has been done so often and so well that the average citizen now regards a voyage to the Moon as hardly more uncertain than a plane journey to another continent," the *New York Times* editorialist wrote the day after Gene Cernan and Harrison Schmitt blasted off from the lunar surface. The editorial was headlined "Farewell to the Moon" and concluded, "Man evolved on the Earth, but he is no longer chained to it. He has walked on another planet and returned to tell the tale. The impact on the future must be enormous."[58]

The *New York Times* had it exactly right, and Mailer had it exactly wrong: the boredom was a measure of NASA's success. The first splitting of the atom is a milestone event in human history, but every minute of every day, nuclear power runs everything from alarm clocks and coffeepots to aircraft carriers, and the only people watching are the people in charge of the nuclear reactors. The first time a Boeing 747 took off and landed successfully, carrying 350 people, having been christened by the first lady of the United States, it was front-page news. But every day for decades there were hundreds of 747s in routine service, and their comings and goings attracted notice only if something went wrong.[59]

If the point of space travel is to be a spectacle, a show, then lack of public fascination is a problem. If the point of space travel is to travel in space, then lack of public interest is a sign that the operation has matured. That's nowhere so clear as in NASA's robotic probes, which get bursts of attention when they attempt a dramatic landing maneuver or make a particularly interesting discovery or unleash a series of breathtaking photos, but which otherwise operate under the constant tending of their scientists and engineers without the public looking over their shoulder, any more than the public is watching the daily progress of important archaeological digs or cancer research.

Part of the point of Apollo, of course, *was* to be a show. We were showing the Russians, and the world, that we could master space travel. But that wasn't the only point, and Apollo didn't evolve very effectively—or even very consciously—from the spectacular to the operational.

And that critique brings us to a deeper problem with the manned spaceflight effort after Apollo: NASA and its leaders never did a particularly good job of explaining what the point of sending people to space was after they landed on the Moon.

The really big, nonmilitary projects that have been comparable to Apollo are the building of the transcontinental railroad, the building of the Panama Canal, and the construction of the nation's interstate system. Those were huge, challenging undertakings; they were all successful; and they all operate today without constant public attention. But they also had an economic purpose: we weren't just building something awesome; we were laying infrastructure that we needed as a nation. All three changed the economics of life in America and around the world. Even the expedition of Lewis and Clark, a government-funded journey designed to scout the vast reaches of North America, has more in common with the interstate highway system than with Apollo. Lewis and Clark were opening that vast terrain for future settlement, recording its hazards and opportunities.

Indeed most of the legendary explorers of the sailing ship era were, ultimately, on economic missions. They weren't terrestrial astronauts. They were trying to understand the dimensions of the world, yes, but they were also claiming territory and resources.

The race to the Moon had an enormous economic impact; as defenders were fond of pointing out, none of the $24 billion it cost to go to the Moon actually got spent in space; it was spent right here on Earth, and almost all of it in the United States. The economic impact is magnified many times when you account for the power the Moon race had in accelerating the digital revolution. It clearly had enormous political impact in reaffirming that the United States was the world's unrivaled technological leader. But Apollo had no economic purpose at the time. We weren't trying to open the Moon for settlement and economic exploitation. We were just checking it out. And that's why the final Apollo missions—which were, in fact, rich with scientific inquiry—felt a little aimless to ordinary Americans. Hadn't we already been to the Moon? What were we going back for?

That's also why the current burst of commercial space development—the reusable rockets of Elon Musk's SpaceX and Jeff Bezos's Blue Origin, the inflatable habitats of Robert Bigelow—is completely different from what NASA has done in space. It's driven by economic imperatives: there's money to be made in space, and if you can establish the value of a zero-gravity economy, then coming and going to space will become routine, regardless of how hazardous it is.

We are puzzled by Apollo not because it was a failure or a waste of money and effort; we're puzzled because we don't take it on its own terms. We don't appreciate exactly how hard it was to fly to the Moon, and why we did it. And we can't understand why Apollo alone didn't provide the momentum to keep going. But that was never its purpose.

One of the critiques in the mid-1960s, as the cost and time of going to the Moon were really settling in, was that the pace of the leap to the Moon was the foolish and expensive part. We *should* go to the Moon, this argument went, but we should do it in a calm, rational, stepwise program. We shouldn't race the Russians. With that more reasoned pace, the costs would be spread out so they were easier to manage.

But that type of program never would have happened. If we'd slowed to a "rational" pace, we never would have gone. Since the last

Apollo flight, in December 1972, we've had four decades of stepwise plans from presidents, NASA administrators, and Congress on how to take the next steps in space. None of the grandest plans has ever amounted to anything. And the more routine, stepwise plans got us the Space Shuttle and the International Space Station, both in many ways the opposite of Apollo: they didn't deliver what was promised, while costing sums that make Apollo look like a bargain, and despite decades of operation for both, it's never been quite clear what the actual purpose of either was.

Here's another thing we don't give Apollo credit for, then: it was dramatically ahead of its time. Part of why it has left so much space behind it is that it wasn't just a leap to the Moon; it was a leap that took the technology and the people to a place we weren't otherwise ready to go. We haven't spent 50 years neglecting space; we've spent 50 years catching up.

---

Americans love space.

Even as the Moon trips started to feel routine, they always provided a spark of connection and pride. During Apollo 17, mission commander Gene Cernan accidentally dropped a geology hammer, which hit the right rear fender of the lunar rover, knocking it loose. The astronauts tried to use the rover without the fender, but the dust flying everywhere was too disruptive. So, with some help and instructions from Mission Control, Cernan and crewmate Harrison Schmitt constructed a replacement fender inside the lunar module, using plastic maps of the Moon and duct tape, and secured it to the rover using clips from the lunar module cabin.[60]

The president of the Auto Body Association of America, Reg Predham, was so impressed that immediately—before the astronauts had even left lunar orbit—he conferred official status on Cernan and Schmitt as lifetime members of the Auto Body Association. "We're delighted to see that when something like this happens on the Moon," Predham said from his auto body repair shop in Neptune City, New Jersey, "that they had the ingenuity to put it all back together. Those astronauts: College graduates. Pilots and geologists. They make damn good body and fender men."[61]

We use the phrase "If we can put a man on the Moon" as often in the 2010s as we did in the 1980s and 1990s, although more often now with a sense of wryness. Does using the expression 50 years after the fact give it more punch or more irony?

Way back in 1986, the *New York Times* joined the *Los Angeles Times* in calling for a halt to the phrase. "We can send a man to the Moon," the *Times* editorialist wrote, "but we cannot stop public speakers from saying, 'We can send a man to the Moon, but we cannot. . . .' So awesome was Neil Armstrong's giant leap for mankind that it has created the cliché standard for a whole generation."

On January 1, 2018, the *Wall Street Journal* used it in what should, rightly, be its final use ever, about NASA's sluggish efforts to return to the Moon. The headline of the story: "If We Can Put a Man on the Moon, Why Can't We Put a Man on the Moon?"[62]

We still use the phrase in 2018 for the same reason we did in 1968: going to the Moon remains one of the most amazing things ever accomplished.

Seventy percent of Americans today weren't born, or were younger than five, when we first went to the Moon—which is to say, for 70 percent of Americans, the Moon landings are something to find on YouTube or in books. By the end of 2018, eight of the twelve men who walked on the Moon had died. Most of those who led the effort have died, as have most of the hundreds of thousands of Americans who worked to make it possible. But the appeal of the accomplishment—which 50 years later is often separated from both the politics that inspired it and what it cost—retains powerful allure.

We love space. We love tales of space: *Star Trek* and *Star Wars*, *Alien* and *Avatar*, *Gravity* and *The Martian*. *Apollo 13*, which was nominated for nine Academy Awards and won two. The original movie for the Space Age, *2001: A Space Odyssey*, which came out eight months before Apollo 8 flew to the Moon in 1968 with its own impressive computer, albeit not one that could talk.

We are not, in fact, bored by the romance and adventure of our own space travel. The Smithsonian Air and Space Museum has a position of prominence on the Mall in Washington, and it has a second, even bigger set of buildings out by Dulles International Airport. (The Dulles facility is so large you could put the original Air and Space Museum inside it.) Between those two locations, the Air and Space Museum is the most-visited museum anywhere in the world, with 24,000 visitors a day, 364 days a year, more even than the Louvre with its 8 million visitors a year.[63]

We visit for all kinds of reasons—spaceflight hardware is, quite simply, cool and amazing. But one reason we visit is to step out of our daily lives for a few moments and connect with the spirit of adventure and daring that flying in space requires. Apollo spacesuits and Gemini capsules look accessible; you can easily imagine yourself in them. That's where the wistfulness about Apollo comes from, as well.

What has become of the America that planted a flag on the Moon? We used to do things like that. Why don't we anymore?

That spirit of America is just fine. It's alive and well. In the halo after Apollo, it created Microsoft and Intel, Apple and Google. Have you noticed that all of human knowledge is accessible from a device that fits in your hand? Did not creating that world—the world we have so quickly come to take for granted—require spirit and determination, vision and daring? Of course it did. It didn't always require physical courage, but it required intellectual courage and relentless determination and boldness of imagination.

Americans created the internet. Americans decoded the genome. American spaceships leap the solar system to unlock the mysteries of Mars, Jupiter, Saturn, and all kinds of quirky asteroids and comets and moons.

When you talk to the people who took America to the Moon, when you read what they said at the time and how they reflected on it decades later, those people will tell you that in working on Apollo, they did something extraordinary—that it was the greatest experience of their lives, whether they were 24 when they worked on it or 54. Those folks never diminish the accomplishment, or the commitment it required.

But they always say two other things: they didn't do it alone, and they do not consider themselves extraordinary. The task inspired them and motivated them and brought out of them work they might not have been able to do in other circumstances.

That is the spirit of America, and it is the essence of the American dream: to imagine something that is out of reach, and then do what's necessary to make it happen, to prove that it wasn't out of reach after all.

John F. Kennedy was our poet of space and also our philosopher of space. Put aside for the moment the fact that in his own heart he was apparently never quite captured by space. He nonetheless gave voice to the value space travel had for us in his era and the role it can have for us in the future.

In that first speech about the Moon, Kennedy started by pointing out "the impact of this adventure on the minds of men everywhere." Space, he said, is a way for America "to take longer strides"; it is "a great new American enterprise."

Kennedy worried that bureaucracy or labor issues would somehow hobble an effort that wasn't even under way yet. Without realizing it, he imagined the culture that NASA would go on to create, a culture that owed much to Kennedy's call itself—something those who worked on Apollo also mention. Kennedy said in that first speech, "Every scientist, every engineer, every serviceman, every technician, contractor and civil servant [must] give his personal pledge that this nation will move forward, with the full speed of freedom in the exciting adventure of space."

To go to the Moon, with the full speed of freedom. A single phrase that manages to capture all the complexity of motivation and politics involved in the race to the Moon.

But most important, Kennedy didn't say the astronauts would be going to the Moon. He didn't say NASA's scientists and engineers would take us to the Moon. In that first speech he said, "In a very real sense, it will not be one man going to the moon. We make this judgment affirmatively: It will be an entire nation. For all of us must work to put him there."[64]

The next year, at Rice University, Kennedy came back to the same idea: "This country was not built by those who waited and rested and wished to look behind them. This country was conquered by those who moved forward." Complacency is not the American personality, courage before a challenge is.

"We choose to go to the Moon"—he said it three times in a row—"because the challenge is one that we are willing to accept, one we are unwilling to postpone, and one which we intend to win."

We're going to the Moon, all of us, because going to the Moon is a uniquely American challenge. And that too has proven true.

Americans like to be reminded of the best of their national character.

And Americans want to be asked to do hard things. We want a mission, and we will do it. That's part of the mythology we tell ourselves, of course, but it's the reality as well. It is one of the invisible bonds, across education and wealth and opportunity. We rise to the occasion.

Those words that NASA chief Thomas Paine had for the protesters the day before Apollo 11, they were an echo of Kennedy's own eloquence, and they contain the key lesson of the race to the Moon, the lesson that is so often misunderstood. No, the leap to the Moon is not the perfect model for solving the problems of poverty or any of the other problems of American society on Earth. But it does contain a wider truth: with inspired leadership, with resources, and, most important, with clarity of purpose, with an explanation of the need, Americans will solve the hardest problems they are asked to tackle.

But we have to be asked. We have to be rallied to the cause. Nothing has faded of the American spirit, or the spirit of Americans themselves in the past 50 years. What has shifted is the way we talk about our relationship to our country. Kennedy, sensing that after eight years of the relative quiet of the Eisenhower administration, energy and initiative and change would mark the 1960s, concluded his inaugural address with what has become a famous reminder to Americans that they are their government, they are their country. "My fellow Americans," he said, "ask not what your country can do for you, ask what you can do for your country." And he did not hesitate to ask.

The big problems that shadow us in the early 21st century—crumbling and dated infrastructure, a fading sense of opportunity in the American economy, climate change—we know how to solve those problems. As the folks who flew us to the Moon came to appreciate, the hard part is not the actual solutions. The hard part is the human part: motivation, giving people a role and a goal.

When the country was attacked on September 11, 2001, President George W. Bush asked Americans to keep shopping, to "get down to Disney World in Florida. Take your families and enjoy life."[65] And remarkably, we did just that. In the month after the attacks American consumers—in a recession—spent more money at retail than they ever had before. If shopping would help fight the terrorists, recession notwithstanding, we would shop.[66]

If we want to tackle climate change, we can. It can't be solved with "a moonshot," in the sense that Apollo was solved with a series of brilliant technical, engineering, and management efforts. But it can be solved with a moonshot in the sense of rallying Americans to a purpose, to a mission, to something that takes incredible effort. With leadership and clarity of purpose. We just need to be asked.

It is a revealing affirmation of that American spirit that the sense of wonder we feel about Apollo, and any disappointment connected to it, isn't about our failure to exploit the Moon or our geopolitical advantage in space. That wistfulness is aspirational, not nationalistic. Why haven't we built on that achievement to create the next one?

In the speech he gave on Thursday afternoon, November 21, 1963, in San Antonio, 20 hours before he would be killed, Kennedy was dedicating a facility for researching the medical implications of spaceflight for space travelers and also for understanding how what we learn in space could be of value for tackling medical problems back on Earth. "The conquest of space must and will go ahead," he said. "That much we know. That much we can say with confidence and conviction." Kennedy concluded the speech this way:

> Frank O'Connor, the Irish writer, tells in one of his books how, as a boy, he and his friends would make their way across the countryside, and when they came to an orchard wall that seemed too high and too doubtful to try, and too difficult to permit their voyage to continue, they took off their hats and tossed them over the wall—and then they had no choice but to follow them.
>
> This nation has tossed its cap over the wall of space, and we have no choice but to follow it. Whatever the difficulties, they will be overcome. Whatever the hazards, they must be guarded against. With the vital help of this Aerospace Medical Center, with the help of all those who labor in the space endeavor, with the help and support of all Americans, we will climb this wall with safety and with speed—and we shall then explore the wonders on the other side.[67]

# ACKNOWLEDGMENTS

In the spring of 1986, I was a *Washington Post* reporter, assigned full-time to the team investigating the aftermath of the space shuttle *Challenger* disaster. I spent weeks at a time working out of an improvised press center at the Johnson Space Center, south of Houston.

The temporary work area for reporters had been set up in what was then the visitors' center, and my worktable was across the room from the full-scale lunar module on display there. That lunar module (LTA-8) was not a flight-ready model, but it was a full-scale, human-rated version sent to Houston for testing and training astronauts.

My worktable was alongside that of Howard Witt, a reporter for the *Chicago Tribune*, and for both of us, the lunar module held a magnetic allure. Every evening, the visitor center would close at 6 p.m., and Howard and I would be alone writing our stories, and waiting to be cleared by our editors, in the silent visitor center in the shadow of that lunar module.

The JSC lunar module was "protected" by a little barrier—a single strand of chain-links made of plastic, about knee high. We desperately wanted to climb up and look inside—a real spaceship.

We debated two questions endlessly: Was the LM alarmed in some unseen way? With lasers surrounding it or pressure sensors underneath the footpads?

And the second question: how to get a look inside. Should we simply climb the ladder one quiet evening (the hatch was open), and look, without permission? Or should we ask permission?

If we climbed up without permission and got busted, we ran the risk of having our press credentials yanked, which would make doing our jobs almost impossible. And NASA was so skittish in the aftermath of *Challenger* that for weeks a media representative went with us to the bathroom. But if we asked permission and were told no, then we couldn't sneak up—we'd never see the inside.

One quiet afternoon, on the spur of the moment, I decided for both of us. I told a friendly media relations guy about our nightly temptation. I'm sad to say I don't remember who it was. He just laughed and said, "C'mon, let's go look inside right now."

At that moment, the visitor center was swarmed with a school group of third graders. I said, "Are you sure you want to do this now? We could have forty rioting third graders if you let us climb up in front of them."

He laughed again and said, "I decide who gets to see the inside of the lunar module."

I vividly remember stepping over the plastic chain. Howard and I went up one at a time. This was more than thirty years ago now, and just two things stick in my mind. The lunar module cabin is very small—about the size of two phone booths. And the interior, lined with control panels, gauges, and instruments, looked very much like the cockpit of an older military aircraft, or the bridge of a navy ship. Gray paint, rows of indicator lights and black switches, a joystick. Very well organized, very busy, not much romance.

The lunar module's design was already twenty years old at that moment. And we'd seen spaceship cockpits not just in *Star Trek*, but also in the first three *Star Wars* movies. The lunar module cabin did not have a "Space Age" feel at all. But it was intimidating, or at least humbling. You could not simply have hopped in and flown away.

I was eight years old when Armstrong and Aldrin stepped onto the Moon, and I followed the missions intensely, building Apollo rocket and spaceship models between flights. But if this book had a spark of inspiration, it was when that good-humored NASA media person let Howard Witt and me climb into the cockpit of that spaceship.

*One Giant Leap* is different from my previous books because it is a blend of journalism and history.

Raphael Sagalyn, my agent and my friend, knew what to say to get this book done, at the beginning, in the middle, and at the end. It wouldn't have happened without him.

Jonathan Karp and Ben Loehnen, my editors at Simon & Schuster, were fabulous—enthusiastic, and also unwavering in their support, their confidence, and their faith.

In June 2016, the space historian Roger Launius sat down for lunch with me. His challenging questions and well-placed skepticism helped me frame a new way of thinking about the importance of Apollo and the race to the Moon.

Arthur Applbaum, a friend who is a Harvard professor, gave me great advice about how to turn that skepticism, and that fresh perspective, into actual book chapters.

At the very beginning of this project, John Buzbee provided friendship, good humor, and absolute confidence in the value of the story, and he bought me the first container of Tang I ever mixed up and drank.

I am grateful to a battalion of NASA staff over the years—those who documented their work, those who preserved that documentation, and the recent NASA staff who have made all those reports and histories available on the internet.

The staff at NASA headquarters' history office knows the history of their agency, knows the stories that have been told and those that haven't, and knows where to find both. I'm grateful in particular for the help, patience, and insight of Colin Fries and Elizabeth Suckow.

Richard Orloff, who worked on the business side at the *Asbury Park Press*, was also a part-time NASA historian who wrote a book called *Apollo by the Numbers.* That book has been so valuable—an almanac of Apollo chronologies and data—that I photocopied the pages I used the most from it and kept them on my desk; other important pages were flagged with Post-It notes. Not a day of work on *One Giant Leap* went by without my consulting Orloff. I so appreciate the diligence, determination, and imagination he brought to *Apollo by the Numbers.*

Scott DeCarlo, of *Fortune*, Leslie Berlin, from Stanford University, and Pete Pranica, from Memphis, provided indispensable professional help at critical moments.

Anne Platoff shared with me all of her original research into how NASA managed, at the last minute, to get a flag onto the Moon. Mike Marcucci shared with me audio files of interviews about the lunar module that he had done, but not used. In both cases, the material proved invaluable.

Many people stepped up to help with photos. Tiki Tindall Williams, a daughter of legendary NASA engineer and manager Bill Tindall, was excited at the prospect of seeing some fresh photos of her dad in print, and provided a wide selection. Drew Crete, at the Charles Stark Draper Laboratory, did the same for Doc Draper, finding rarely seen photos of the aeronautical pioneer. Edward Anderson, Vic Beck, and Dianne Baumert-Moyik, of Northrop Grumman Corporation, provided dozens of behind-the-scenes photos of the lunar module under construction in Bethpage, Long Island; they vividly brought the LM's creation to life. Eric Jones and Jon Hancock, of the *Apollo Lunar Surface Journal*, were generous with both information and help for the panoramic photos from NASA missions that some *ALSJ* staff had created. At NASA headquarters, photo researcher Connie Moore was able to track down every photo I couldn't find on my own.

Creating a real-world book that has heft and an irresistible cover, that is free of infelicities, grammatical flubs and typos, and that people learn about—that all requires a large cast. At Simon & Schuster, I'm grateful for the careful copyediting of Jonathan Evans and Judith Hoover; for Lewelin Polanco, Alison Forner, and Ryan Raphael, who did *One Giant Leap*'s beautiful and arresting design, inside and out; and for the work of Larry Hughes, Christine Calella, Stephen Bedford, Carolyn Kelly, Julianna Haubner; and of Kimberly Goldstein and Annie Craig; and thanks to Laura Ogar for a savvy, thorough, and useful index.

Myrtle Kearse, who has been with me for every book, retired during this one, and then came back when it turned out I couldn't do it without her. Myrtle has been a blessing for two decades. Heather Craige, too, has been with me for every book, but she decided to let me do this one on my own.

I lean on my friends—especially Kevin Spear, Chuck Salter, and Ruth Sheehan—for storytelling advice, and for encouragement at moments when the work sometimes flags.

My parents, Suzanne and Lawrence Fishman, are my biggest supporters and among my first readers. The fact that they can't wait to read the next book (and tell everyone they know about it) is a joy.

I have great siblings—smart, appreciative, supportive: Matthew Fishman, Betsy Rosenfeld, Andrew Fishman. They are also funny, which comes in handy.

Nicolas Fishman and Maya Wilson have never known a home in

which a book wasn't being written. Nicolas helped do the math for this one, and tweaked passages delivered the way he likes—by text message. Maya read many chapters with a sense of wonder, a red pen, and an editor's judgment. Her joie de vivre is as indispensable as gravity itself. Together, they have created the world we live in; they also never let me get lost in the work.

Two people, as ever, deserve special mention.

The books keep getting longer, but Geoff Calkins has heard every word of this one, too, read aloud, many parts more than once. Geoff knows how to uncrinkle tangled syntax; is unafraid to say some poorly explained aspect of space travel doesn't make sense; he knows the boggy parts, and also the parts that work. Geoff doesn't think of himself as an editor, but he's a great one. And he's willing to hear the next chunk no matter what time it is, no matter how busy he is, whether he's running or driving or sitting in a press box. He has certainly absorbed more space science during the last three years than he ever expected to have to. He remains that rarest of companions, a best friend of thirty-seven years.

And this book would, quite simply, not have happened without my wife, Trish Wilson. She knows the right word when the wrong one slips in; she knows how to reorganize a section or a chapter when the pacing flags or the point is lost. Her editing is both unerring and enthusiastic. Her sharp eye, her sharp ear, made every page better. Her astonishing confidence, greater often than my own, is what quietly pulled the book to completion. She is my partner, my best editor, and my heartbeat.

# NOTES

A note on the notes: It is the rare source note on the pages that follow that does not contain a web address for a report, document, newspaper article, or video. To make the notes more useful, they will be posted online, at the website of the book. Go to www.onegiantleap.space.

## PREFACE: The Mystery of Moondust

1 Eric M. Jones, "A Visit to the Snowman," *Apollo 12: Lunar Surface Journal*, September 14, 2017, https://www.hq.nasa.gov/alsj/a12/a12.landing.html, 109:51:50.

2 Astronauts describing the smell of lunar dust: Armstrong (Apollo 11): James R. Hansen, *First Man* (New York: Simon & Schuster, 2005), pp. 531–32. Aldrin (Apollo 11): Buzz Aldrin with Ken Abraham, *Magnificent Desolation* (New York: Three Rivers Press, 2009), p. 45. Schmitt (Apollo 17): Eric M. Jones, "Ending the Second Day," *Apollo 17: Lunar Surface Journal*, April 34, 2015, https://www.hq.nasa.gov/alsj/a17/a17.eva2post.html, 148:23:10. Irwin and Scott (Apollo 15): Eric M. Jones, "Post-EVA-1 Activities," *Apollo 15: Lunar Surface Journal*, May 29, 2012, https://www.hq.nasa.gov/alsj/a15/a15.eva1post.html, 126:16:16ff.

3 Leonard David, "The Moon Smells: Apollo Astronauts Describe Lunar Aroma," Space.com, August 25, 2014, https://www.space.com/26932-moon-smell-apollo-lunar-aroma.html; Jeremy Pearce, "Thomas Gold, Astrophysicist and Innovator, Is Dead at 84," *New York Times*, June 24, 2004, https://www.nytimes.com/2004/06/24/us/thomas-gold-astrophysicist-and-innovator-is-dead-at-84.html.

4 David, "The Moon Smells."

5 Hansen, *First Man*, pp. 532–33.

6 Tony Phillips, "The Mysterious Smell of Moondust," *Science@NASA: Apollo Chronicles*, January 30, 2006, https://www.nasa.gov/exploration/home/30jan_smellofmoondust.html; Conrad quote: Eric M. Jones, "Post EVA Activities in the LM," *Apollo 12: Lunar Surface Journal*, June 19, 2014, https://www.hq.nasa.gov/alsj/a12/a12.posteva1.html, 121:31:45.

7 Arlene Levinson, "Top 100 News Stories of the Century," Associated Press, *Deseret News* (Salt Lake City), April 15, 1999, https://www.deseretnews.com/article/691495/Top-100-news-stories-of-the-century.html.

8 Donna Scheibe, "Woman Chute Rigger Helps Bring Back Astronauts Alive," *Los Angeles Times*, July 11, 1971, pp. J1, J2, https://www.newspapers.com/image/384778395/.

9 Bernard Weinraub, "Hundreds of Thousands Flock to Be 'There,'" *New York Times*, July 16, 1969, p. 22, https://timesmachine.nytimes.com/timesmachine/1969/07/16/78356035.html; John F. Kennedy, "Address at Rice University in Houston on the Nation's Space Effort," September 12, 1962, *American Presidency Project*, https://www.presidency.ucsb.edu/documents/address-rice-university-houston-the-nations-space-effort.

## 1: Tranquility Base & the World We All Live In

1 These are lines from the speech William Safire prepared for President Nixon in case the Apollo 11 astronauts encountered disaster on the Moon, and were unable to return to Earth. The typescript of the text is imaged at the National Archives, https://www.archives.gov/files/presidential-libraries/events/centennials/nixon/images/exhibit/rn100-6-1-2.pdf.

Safire, who went on to be a columnist for the *New York Times*, wrote about having to write the speech on the 30th anniversary of the Moon landing. William Safire, "Disaster Never Came," *New York Times*, July 12, 1999, http://www.nytimes.com/1999/07/12/opinion/essay-disaster-never-came.html.

2 Playtex was born in 1932 as International Latex Corporation (ILC). The company started using latex to design and manufacture various products, including gloves and women's lingerie, especially girdles and bras. The Playtex division—named by combining the words "play" and "latex"—was formed in 1947, in part to produce rubberized diaper pants for babies. During most of the 1960s, the division that worked on the spacesuits was considered the industrial division of Playtex, and the company was known by its best-known consumer brand. "Even at the time, people in NASA

called the company Playtex, as opposed to ILC, partially as you would call someone by their nickname, partly as a kind of, can you believe we're dealing with Playtex here?" said Nicholas de Monchaux, who wrote a book about the history and design of the Apollo spacesuit called *Spacesuit: Fashioning Apollo*. (De Monchauz is quoted here from the documentary "Putting Man on the Moon," at minute 24:30, https://www.aljazeera.com/programmes/aljazeeracorrespondent/2015/10/putting-men-moon-151013082436203.html.)

Located in Delaware, the industrial division of Playtex was known formally as ILC Dover, and its address through the 1960s was: Playtex Park, Dover, DE. ILC Dover still produces space suits for NASA. The address is now: One Moonwalker Drive, Frederica, DE. ILC Dover is now an independent company. The underwear portion of Playtex is now part of Hanes.

"Cross Your Heart" bra introduced in 1954: "History of the Plaxtex Brand," https://www.playtex.co.uk/c/brand-history-70100/.

The observation "form-fitting and flexible": *Moon Machines*, Episode 5: "The Space Suit," directed by Nick Davidson, 2008, Dox Productions for The Science Channel, Discovery Communications, Playtex's pitch to NASA starting at 8:00.

3 Eric M. Jones, "Mobility and Photography," *Apollo 11: Lunar Surface Journal*, November 9, 2016, https://www.hq.nasa.gov/alsj/a11/a11.mobility.html, 110:13:42.

The video of this part of Armstrong and Aldrin's Moon walk runs from 52:40 to 55:00 on "Apollo 11: 'First Moonwalk on TV' (Restored)," YouTube, posted February 17, 2015, by Dan Beaumont Space Museum, https://www.youtube.com/watch?v=L9Go_j_i6o8. Aldrin comment: Jones, ed., *Apollo 11 Lunar Surface Journal*, "Mobility and Photography," 110:13:42.

4 Reihm observation: Eric Ruth, "Man on the Moon," *University of Delaware* Messenger, v. 25, n. 3 (Dec. 2017), p. 40, http://www1.udel.edu/udmessenger/vol25no3/stories/alumni-reihm.html.

5 Jones, ed., *Apollo 11 Lunar Surface Journal*, "Mobility and Photography," 110:13:42.

6 *Moon Machines*, Episode 5: "The Space Suit," Eleanor Foraker comments at 41:40, Joe Kosmo comments at 41:30.

7 Eric M. Jones, ed., *Apollo 11 Lunar Surface Journal*, "Post-landing activities," https://www.hq.nasa.gov/alsj/a11/a11.postland.html, 104:39:14.

8 *Moon Machines*, Episode 5: "The Space Suit," Sonny Reihm comments at 42:00.

9 There is a three-minute compilation video of astronauts on the Moon skipping, running, falling, and singing. It's astronauts as you've never seen them: Childlike with joy. (Unfortunately the astronauts in each scene are not identified.) Available on "Astronauts Tripping on the Surface of the Moon," posted September 20, 2013, by amovees, YouTube: https://www.youtube.com/watch?v=x2adl6LszcE. It appears to be an excerpt from Al Reinert, *For All Mankind*, The Criterion Collection, 1989, available from Amazon: https://www.amazon.com/All-Mankind-Neil-Armstrong/dp/B004BQTEGA.

10 Alex Wellerstein, "How Many People Worked on the Manhattan Project?," *The Nuclear Secrecy Blog*, November 1, 2013, http://blog.nuclearsecrecy.com/2013/11/01/many-people-worked-manhattan-project/.

Richard W. Orloff, "Apollo Program Budget Appropriations," in *Apollo by the Numbers: A Statistical Reference* (Washington, D.C.: NASA History Division, 2000), p. 281.

11 Christopher Kraft, interviewed in the video "Human Spaceflight: The Kennedy Legacy," published May 25, 2011, by NASA, starting at 01:15, YouTube, https://www.youtube.com/watch?v=biSdeqwcGMk.

12 Roger D. Launius, "Public Opinion Polls and Perceptions of U.S. Human Spaceflight," *Space Policy*, vol. 19, no. 3, August 2003, pp. 163–75.

Frank Newport, "Despite Recent High Visibility, Americans Not Enthusiastic About Spending More Money on Space Program," *Gallup*, 28 July 1999, http://www.gallup.com/poll/3688/Despite-Recent-High-Visibility-Americans-Enthusiastic-About-Spend.aspx.

13 Earthrise photo from "100 Photos: The Most Influential Images of All Time," *Time*, undated, http://100photos.time.com/photos/nasa-earthrise-apollo-8.

14 "Men of the Year," *Time*, January 3, 1969, pp. 9ff., http://content.time.com/time/magazine/article/0,9171,900486,00.html.

15 Ibid.

16 Polling in the U.S. about enthusiasm for the Moon landing and the space program: Louis Harris, "Moon Landings, Space Exploring Are Unpopular," *Burlington (VT) Free Press*, February 18, 1969, p. 7, https://www.newspapers.com/image/200546246/.

The Gallup organization didn't poll Americans with exactly the same precision about attitudes toward the Moon landing as the Harris Poll did. In January 1969, Gallup polled the question of spending more generically—"many billions of dollars on space research"—then asked if Americans favored increasing space spending, holding it the same, or decreasing it.

The result: increase spending, 14%; keep spending the same, 41%; rreduce spending, 40%; no opinion, 5%. So just four weeks after Apollo 8, according to Gallup, 40% of Americans thought spending for space should be reduced. George Gallup, "55 Percent Favor Space Spending, " *Star Tribune* (Minneapolis), February 27, 1969, p. 30, https://www.newspapers.com/image/185072275/.

Vietnam spending: "Vietnam Statistics—War Costs: Complete Picture Impossible," *CQ Almanac 1975*, 31st ed., pp. 301–5, 1976, http://library.cqpress.com/cqalmanac/cqal75-1213988.

Vietnam casualties: "Vietnam War U.S. Military Fatal Casualty Statistics," National Archives, https://www.archives.gov/research/military/vietnam-war/casualty-statistics.

17 Roger D. Launius, former chief historian for NASA, in-person interview, June 9, 2016.

18 For the history of Tang: Matt Blitz, "How NASA Made Tang Cool," *Food & Wine*, May 18, 2017, https://www.foodandwine.com/lifestyle/how-nasa-made-tang-cool.

ABC News coverage of the Apollo 8 mission, with Jules Bergman and the Tang logo, can be seen in "Apollo 8: 'Enters Moon's Gravity,' ABC News, December 23, 1968," posted December 24, 2011, by Dan Beaumont Space Museum, YouTube, https://www.youtube.com/watch?v=rbYW54FUEPY.

For the Apollo 11 crew taste-testing and rejecting Tang: Hansen, *First Man*, pp. 417–18. For the history of Velcro: Claire Suddath, "A Brief History of: Velcro," *Time*, June 15, 2010, http://content.time.com/time/nation/article/0,8599,1996883,00.html.

NASA's webpage clarifying that it did not invent Tang or Velcro: "Are Tang, Teflon, and Velcro NASA Spinoffs?," NASA, undated, https://www.nasa.gov/offices/ipp/home/myth_tang.html.

19 "TV Roundup: 53.5 Million Homes Viewed Apollo Flight," unbylined, *The Philadelphia Inquirer*, August 26, 1969, p. 23, https://www.newspapers.com/image/179913144/.

"Latest Nielsens: All in Family Keeps Top Rating," *Los Angeles Times*, January 1, 1973, part IV, p. 24, https://www.newspapers.com/image/385960836/.

20 Richard C. Levin, "The Semiconductor Industry," in Richard R. Nelson, ed., *Government and Technical Progress* (Elmsford, NY: Pergamon Press, 1982), p. 36 (integrated circuit prices), pp. 63–64 (NASA purchasing).

21 David A. Mindell, Frances and David Dibner Professor of the History of Engineering and Manufacturing and professor of aeronautics and astronautics, Massachusetts Institute of Technology, telephone interview, October 29, 2017.

22 James E. Tomayko, "Computers in Spaceflight: The NASA Experience," NASA, Contractor Report 182505 (Washington, D.C., 1988), pp. 251, 255.

23 Table No. 1098: Homes With Selected Electrical Appliances: 1953 to 1970, *Statistical Abstract of the United States* (Washington, D.C.: U.S. Department of Commerce, Bureau of the Census, 1970), p. 687.

24 The electric can opener is one of the true oddities that the *Statistical Abstract of the United States* tracked through the years. Sales data first appear in 1960 (1.2 million can openers sold). By 1965, 20% of U.S. homes had one; by 1970, 43% of U.S. homes had one. The *Statistical Abstract* last reports data for can opener ownership in its combined 1982–83 edition, by which point 51 million U.S. homes had one—64% nationwide ownership of electric can openers.

Handheld can openers that just peeled open the lid were apparently considered dangerous, helping explain the boom in electric versions. In 1950 the *New York Times* ran a story saying, "The old-fashioned can opener that leaves sharp jagged edges is the No. 1 hazard in the average New York home, the Greater New York Safety Council reported yesterday," ahead of loose rugs on the floor and smoking in bed. "Hazards in Home Listed," *New York Times*, March 5, 1950, p. 15, https://timesmachine.nytimes.com/timesmachine/1950/03/05/98603797.html.

25 "Out of Order," *Time*, October 14, 1957, pp. 102–4.

26 "Just Heat and Serve," *Time*, December 7, 1959, pp. 92–93.

27 Table 1167: Manufacturers Sales and Retail Value of Home Appliances, 1955–1969, *Statistical Abstract of the United States*, 1970, pp. 728–29.

28 "The Selectric Typewriter," IBM, http://www-03.ibm.com/ibm/history/ibm100/us/en/icons/selectric/. "Sony Design: History: 8FC-59," Sony, https://www.sony.net/Fun/design/history/product/1960/8fc-59.html.

29 Touch-tone telephone service sources: AT&T introduces at the 1962 World's Fair: "AT&T Archives: Century 21 Calling . . . ," AT&T, company video, accessed July 8, 2018, http://techchannel.att.com/play-video.cfm/2013/7/10/AA11167-Century-21-Calling, 14:11. Washington, D.C., gets touch-tone service: Willard Clopton, "D.C. Suburbs Using Push-Button Phones: Halves Calling Time," *Washington Post*, January 2, 1965, p. A10.

Four million touch-tone phones in use in U.S.: "Telephone Dials," *Engineering and Technology History Wiki*, accessed July 8, 2018, ethw.org/Telephone_Dials; 109 million phones total in U.S.: *Statistical Abstract of the United States*, 1974, Table 826: Telephones, Calls, and Rates: 1950 to 1973, p. 500.

30 Phil Hevener, "Smiling, Wide-Eyed Jeannie Wows 'Em in Cocoa Beach," *Florida Today*, June 28, 1969, pp. 1A, 6A, https://www.newspapers.com/image/124882970/.

31 The lunar module briefing document describes the ladder as having nine rungs, nine inches apart, with the lowest rung being 30 inches from the Moon's surface. "Lunar Module: Quick Reference Data," Grumman Corporation, https://www.hq.nasa.gov/alsj/LM04_Lunar_Module_ppLV1-17.pdf, p. LV-12.

32 As President Kennedy spoke on May 25, 1961, the U.S. had launched only a single manned spaceflight: Alan Shepard's suborbital arc in the tiny Mercury capsule, *Freedom 7*, three weeks earlier on May 5, 1961. The flight lasted 15 minutes from blastoff to splashdown, Shepard went 102 miles high, was weightless for just 5 minutes, and landed only 302 miles southeast of his launch pad at Cape Canaveral. See "Mercury-Redstone 3," https://www.nasa.gov/mission_pages/mercury/missions/freedom7.html.

33 Hansen, *First Man*, p. 304.

"Senate Report Assails NASA on Apollo Deaths," United Press International, *The New York Times*, February 1, 1968, p. 8, https://timesmachine.nytimes.com/timesmachine/1968/02/01/76867119.html.

34 "Legal History of Contraceptives," *Jurist: Legal News & Research*, January 28, 2014, https://www.jurist.org/archives/feature/legal-history-of-contraceptives-in-the-us/; Martha J. Bailey, "How Contraception Transformed the American Family," *Atlantic*, June 16, 2015, https://www.theatlantic.com/politics/archive/2015/06/griswold-50th-anniversary/395867/.

35 U.S. births during the 1960s falling: *Statistical Abstract of the United States*, 1972, Table 8, p. 10; *Statistical Abstract of the United States*, 1974, Table 68, p. 53. Women 18 and over: *Statistical Abstract of the United States*, 1972, Table 33, p. 30. College enrollment: *Statistical Abstract of the United States*, 1972, Table 33, p. 108. Women and men in the workforce: *Statistical Abstract of the United States*, 1971, Table 349, p. 223. Women in various work categories: *Statistical Abstract of the United States*, 1971, Table 347, p. 222.

36 Top TV shows: "Top-Rated TV Shows of Each Season, 1950–51 to 1999–2000," in *World Almanac and Book of Facts 2001* (Mahwah, NJ: World

Almanac Books, 2001), p. 317. Top music 1960 and 1969: "Hot 100 55th Anniversary: Every No. 1 Song (1955–2013)," *Billboard*, August 2, 2013, https://www.billboard.com/articles/columns/chart-beat/5149230/hot-100-55th-anniversary-every-no-1-song-1958-2013.

37 Decline of the bald eagle: "Bald Eagle: Fact Sheet: Natural History, Ecology, and History of Recovery," U.S. Fish & Wildlife Service, June 2007, https://www.fws.gov/midwest/eagle/recovery/biologue.html.

Smog pollution, New York City: Jim Dwyer, "Remembering a City Where the Smog Could Kill," *New York Times,* March 1, 2017, p. A22, https://www.nytimes.com/2017/02/28/nyregion/new-york-city-smog.html. Smog pollution, Los Angeles: Douglas Smith, "Fifty Years of Clearing the Skies," California Institute of Technology, Caltech Media Relations, April 25, 2013, http://www.caltech.edu/news/fifty-years-clearing-skies-39248.

38 "The Modern Environmental Movement: Timeline," *American Experience*, PBS, accessed June 23, 2018, https://www.pbs.org/wgbh/americanexperience/features/earth-days-modern-environmental-movement/.

39 It is hard to compare large spending projects across decades and centuries; what's the labor of Chinese immigrants building the transcontinental railroad worth compared to the technicians who assembled the lunar module? But the scale of Apollo is dramatically larger than those earlier projects.

As of 1974, NASA told Congress that Apollo cost $24.5 billion. Apollo employed 411,000 people at its peak in 1965.

The Manhattan Project cost $2 billion in 1945 dollars ($5.7 billion in 1974 dollars) and employed 125,000 people at its peak.

The Panama Canal cost $375 million in 1914 dollars ($1.9 billion in 1974 dollars) and employed 45,000 people during the peak of the U.S. effort.

Estimates of the cost and the laborers for the transcontinental railroad, built by three private railroad companies, are harder to nail down. The cost seems to have been between $110 million and $124 million in 1869 dollars. Using the Panama Canal inflation figure from 45 years later, that's at least $500 million in 1974 dollars. Even doubling that to account for the years for which there is no CPI data gets to only $1 billion in modern dollars. Similarly, employment figures are imprecise, but at peak employment the transcontinental railroad had 30,000 or more workers at once.

One more contemporary effort that isn't really a "project" was the Marshall Plan to provide aid to help rebuild Western Europe after World War II. Marshall Plan funding totaled $13 billion from 1947 to 1951. In 1972 dollars, that is $21 billion—the scale of Apollo, but for a much different style of effort.

Manhattan Project sources: Alex Wellerstein, "The Price of the Manhattan Project," *Restricted Data: The Nuclear Secrecy Blog,* May 17, 2013, http://blog.nuclearsecrecy.com/2013/05/17/the-price-of-the-manhattan-project/; Alex Wellerstein, "How Many People Worked on the Manhattan Project," *Restricted Data: The Nuclear Secrecy Blog,* November 1, 2013, http://blog.nuclearsecrecy.com/2013/11/01/many-people-worked-manhattan-project/.

Panama Canal sources: "End of Construction," *A History of the Panama Canal: Prepared by the Panama Canal Authority,* accessed June 25, 2019, https://www.pancanal.com/eng/history/history/end.html; Azad Abdulhafedh, "The Panama Canal: A Man-Made Engineering Marvel," *International Journal of Society Science and Humanities Research* 5, no. 1 (January–March 2017), p. 330, www.researchpublish.com/download.php?file=The%20Panama%20Canal-4402.pdf&act=book.

Transcontinental Railroad sources: Maury Klein, "Financing the Transcontinental Railroad," Gilder Lehrman Institute of American History: AP US History Study Guide, accessed June 25, 2018, https://ap.gilderlehrman.org/essays/financing-transcontinental-railroad; "Construction Cost of the Transcontinental Railroad: CPRR Discussion Group," *Central Pacific Railroad Photographic History Museum,* June 1, 2008, http://discussion.cprr.net/2008/06/construction-cost-of-transcontinental.html; "How Many Workers: CPRR Discussion Group," *Central Pacific Railroad Photographic History Museum,* September 10, 2006, http://discussion.cprr.net/2006/09/how-many-workers.html.

Marshall Plan source: "George C. Marshall: The Marshall Plan: History of the Marshall Plan," George C. Marshall Foundation, accessed July 2, 2018, https://www.marshallfoundation.org/marshall/the-marshall-plan/history-marshall-plan/.

40 Total forces in Vietnam: "Vietnam Conflict: U.S. Military Forces in Vietnam and Casualties Incurred: 1961 to 1975," Table 590, *Statistical Abstract of the United States,* 1977, p. 369.

Total Apollo staff: NASA, *NASA Historical Data Book, 1958–1968,* vol. 1, Washington, D.C.: NASA SP-4012, 1976, Table 3-26: "Total NASA Employment, Selected Characteristics," p. 106.

41 Numbers for employees at the largest U.S. firms come from the *Fortune 500* list of the largest U.S. corporations for the years 1963 to 1970. Data

provided by Scott DeCarlo, list editor, *Fortune* magazine, via spreadsheet, May 2018.

42 Total Saturn V rockets: Tim Sharp, "Saturn V Rockets & Apollo Spacecraft," Space.com, October 17, 2018, https://www.space.com/16698-apollo-spacecraft.html.

Total Apollo command modules and their disposition: "Location of Apollo Command Modules," Smithsonian National Air and Space Museum: Spacecraft & Vehicles, accessed January 10, 2019, https://airandspace.si.edu/explore-and-learn/topics/apollo/apollo-program/spacecraft/location/cm.cfm.

Total lunar modules and their disposition: "Location of Apollo Lunar Modules," Smithsonian National Air and Space Museum: Spacecraft & Vehicles, accessed January 10, 2019, https://airandspace.si.edu/explore-and-learn/topics/apollo/apollo-program/spacecraft/location/lm.cfm.

43 MIT staffing for Apollo during the 1960s: *MIT's Role in Project Apollo: Final Report on Contracts NAS 9-153 and NAS 9-4065* (Cambridge, MA: Massachusetts Institute of Technology, 1972), p. 7. "20,000 companies": *NASA Historical Data Book, 1958-1968*, vol. 1, "Table 5-1: Total Number of Procurement Actions by Kind of Contractor: FY 1960–FY 1968," p. 164.

44 NASA staffing and contractor staffing: *NASA Historical Data Book, 1958-1968*, vol. 1, "Table 3-26: Total NASA Employment, Selected Characteristics," p. 106.

State-by-state Apollo contracting dollars from NASA: *NASA Historical Data Book, 1958–1968*, vol. 1, "Table 5-17: Distribution of NASA Prime Contract Awards by States: FY 1961–FY 1968," p. 182.

45 Vietnam fatality statistics: "Vietnam Conflict: U.S. Military Forces in Vietnam and Casualties Incurred: 1961 to 1975," Table 590, *Statistical Abstract of the United States*, 1977, p. 369. Vietnam cost statistics: "Vietnam Statistics—War Costs: Complete Picture Impossible—Full War Costs" listing, *CQ Almanac 1975*, 31st ed., 301-5, 1976, http://library.cqpress.com/cqalmanac/cqal75-1213988.

46 "Most View the CDC Favorably; VA's Image Slips: Ratings of Government Agencies," Pew Research Center, January 22, 2015, http://assets.pewresearch.org/wp-content/uploads/sites/5/2015/01/1-22-15-Favorability-release.pdf (current data, p. 1; historic data on NASA, p. 10).

47 The author did an analysis of the appearance in print media of the phrase "if we can put a man on the Moon" going back to the earliest citation, before John Kennedy's call to fly to the Moon, on May 25, 1961.The use of

the phrase in the U.S. media is discussed in detail in Chapter 9, starting on page 315.

Research on the phrase was based on searching the U.S. nationwide newspaper database newspapers.com, and the databases of the *New York Times*, the *Washington Post*, the *Wall Street Journal*, the *Chicago Tribune*, and the *Los Angeles Times.*

The collating and citations are actually based on searching for all instances of three phrases with slightly different wording, and without the initial word "if," because the construction in stories was often, "we can send a man to the Moon, but . . ." The three different wordings are: "we can put a man on the Moon," "we can land a man on the Moon," and "we can send a man to the Moon."

The first use of the phrase occurred in the student newspaper of the University of North Carolina at Chapel Hill, *The Daily Tar Heel*, on November 8, 1958. Every instance of each of those phrases from that first occurrence through the launch of the first Moon landing mission, Apollo 11, on July 16, 1969, was read and cataloged. As well, the total occurrences for the phrases for each decade from 1950 forward were tabulated. Totals for all three phrases, by decade, are below. Note that each period is 10 years long—from January 1 of the starting year through December 31 of the ending year—except for 2010 to 2018, which is 9 years.

**1950 to 1959:** 9
**1960 to 1969:** 440
**1970 to 1979:** 1,486
**1980 to 1989:** 1,938
**1990 to 1991:** 1,592
**2000 to 2009:** 771
**2010 to 2018:** 423

48 Tom Alexander, "The Unexpected Payoff of Project Apollo," *Fortune*, July 1969, p. 156.

49 Lyndon Baines Johnson, *The Vantage Point* (New York: Popular Library, 1971), p. 285.

50 John M. Logsdon, *John F. Kennedy and the Race to the Moon* (New York: Palgrave Macmillan, 2010), p. 101. Logsdon says Robert Seamans, the associate administrator of NASA, told McNamara that a Mars mission simply wasn't technically feasible. In his own memoir, *Aiming at Targets: The*

*Autobiography of Robert C. Seamans, Jr.* (Honolulu: University Press of the Pacific, 2004), Seamans says NASA and the country were not "in any position whatsoever to take that on as an objective" and that announcing a mission straight to Mars would have been "foolhardy" (p. 89).

51 Kennedy, "Address at Rice University in Houston on the Nation's Space Effort."

52 The Pew Research Center has been tracking public trust in government since 1958. At the end of 2017, the percentage of Americans who "trust the government in Washington always or most of the time" was at 18, and the moving average was also at 18. The average was lower—15% and 17%—briefly in 2011. Individual polls have put faith in government lower than 18% just five times from 2007 to 2017, out of 34 polls. "Public Trust in Government: 1958–2017," Pew Research Center, December 14, 2017, http://www.people-press.org/2017/12/14/public-trust-in-government-1958-2017/.

53 Northrop Ventura Corp., "Project Apollo: The Last Five Miles Home," November 1966, YouTube, posted by spaceaholic, August 7, 2010, 14:43, https://www.youtube.com/watch?v=gDNDQQlx1JE.

54 "MIT Science Reporter: Computer for Apollo (1965)," hosted by John Fitch, YouTube, posted January 20, 2016, by the Vault of MIT, 29:20, https://www.youtube.com/watch?v=ndvmFlg1WmE.

55 Thomas J. Kelly, *Moon Lander* (Washington, D.C.: Smithsonian Books, 2001), p. 139.

56 "Backgrounder: The Apollo Heat Shield," press kit, Avco Space Systems Division, Lowell, Mass., Apollo 8, Contractors, Kits, Record #012863, NASA History Office, Washington, D.C.

57 The 11 missions had 33 crew members, but four Apollo astronauts from the first four missions—which didn't land—got to fly a second time on Moon-bound missions: Jim Lovell (on Apollo 13, who didn't get to land), David Scott, John Young, and Gene Cernan. So while there were 33 crew members, there were only 29 actual Apollo flight astronauts.

58 The statistic that every hour of Apollo space flight required 1 million hours of work on Earth was calculated using basic data from NASA. The total number of spaceflight hours—2,502—comes from Orloff, *Apollo by the Numbers*, p. 305.

The total number of hours worked on the ground was calculated using the total number of NASA staff members—NASA and contractor employees—for each year. That provided a basic "work years" figure. But that was total NASA and contractor staff for all NASA projects. That total "staff years worked" figure was then adjusted, based on the percent of the NASA budget that year devoted to Apollo. The staffing numbers come from *NASA Historical Data Book, 1958-1968*, vol. 1, p. 106. The percent of the NASA budget devoted to Apollo in each year is from Orloff, p. 281.

Those numbers are below:

| YEAR | STAFF / WORK YEARS | % TO APOLLO | ADJ WORK YEARS |
|---|---|---|---|
| 1960 | 47,000 | 0% | |
| 1961 | 75,000 | 1% | 750 |
| 1962 | 139,000 | 10% | 13,900 |
| 1963 | 248,000 | 17% | 42,160 |
| 1964 | 380,000 | 57% | 216,600 |
| 1965 | 411,000 | 61% | 250,710 |
| 1966 | 396,000 | 66% | 261,360 |
| 1967 | 309,000 | 70% | 216,370 |
| 1968 | 246,000 | 64% | 157,440 |
| 1969 | 218,000 | 63% | 137,340 |
| 1970 | 100,000 | 54% | 54,000 |
| 1971 | 80,000 | 36% | 28,800 |
| **TOTAL WORK YEARS**, Apollo, 1961 to 1971: 1,379,430 | | | |

If you assume a basic work year of 2,000 hours (40 hours a week for 50 weeks), then multiply 2,000 hours by the 1,379,430 work years, that comes to: 2,758,860,000 hours of work—2.8 billion hours. If you divide 2.8 billion hours of work on Earth by 2,502 hours of space flight, you get 1.1 million hours of work on Earth for every hour of space flight.

The calculation is rough in two areas. First, the percent of the NASA budget going to Apollo is only a proxy for the percent of total NASA and contractor staff members working on Apollo, but not a bad proxy. And 2,000 hours of work for each "work year" is, if anything, low. Most NASA and contractor staff members reported working extremely long days, and extremely long weeks, during the Apollo project, often including weekends, for months and years on end. But both those approximations mean that the calculation is, if anything, conservative. Each hour of spaceflight most likely required more than 1 million hours of work on the ground—but certainly required at least that.

## 2: The Moon to the Rescue

1 The account of Gagarin's historic orbital flight comes from several sources: Asif A. Siddiqi, *Challenge to Apollo: The Soviet Union and the Space Race, 1945–1974*, NASA History Division, Office of Policy and Plans, Washington D.C., 2000, https://history.nasa.gov/SP-4408pt1.pdf (full story of Gagarin's flight: pp. 243–98; detail of Gagarin's landing: p. 281). All direct quotes from Gagarin are from this account. A second account, also relying on original Russian documents: Anatoly Zak, "Vostok 1: Vostok Lands Successfully," Russian Space Web, accessed June 22, 2017, http://www.russianspaceweb.com/vostok1_landing.html. Descriptions of the field, "planting potatoes," and the woman and her granddaughter are from Théo Pirard, "Yuri Gagarin, 12 April 1961: 'I come from outer space!' (1)," *Reflexions*, Liège Université, accessed July 10, 2018, http://reflexions.ulg.ac.be/cms/c_33931/youri-gagarine-12-avril-1961-je-viens-du-cosmos-1. On Google Maps, "The landing of Gagarin, Saratov Oblast, Russia," is a designated landmark, including visitor photos from people who have visited the monument there.

2 "Guard parachute," "military men safeguard spacesuit, pistol, watch, handkerchief," all from Zak, "Vostok 1." Russian cosmonauts armed with pistols, and one pair landed in taiga and needed it: Alexander Korolkov, "Pistol-Packing for the Final Frontier: Why Were Cosmonauts Armed?," *Russia Beyond*, July 17, 2014, https://www.rbth.com/defence/2014/07/17/pistol-packing_for_the_final_frontier_why_were_cosmonauts_armed_38279.html.

3 Kennedy being warned of launch in advance: Hugh Sidey, "How the News Hit Washington—With Some Reactions Overseas," *Life*, April 21, 1961, pp. 26–27, https://books.google.com/books?id=9FEEAAAAMBAJ&printsec=frontcover#v=onepage&q&f=false. Clifton asks if President Kennedy wants to be woken up: Hugh Sidey, *John F. Kennedy, President* (New York: Crest Books, 1964), p. 110. Wiesner called by Pentagon: "The Cruise of the Vostok," *Time*, April 21, 1961, p. 49, http://content.time.com/time/subscriber/article/0,33009,895299,00.html. *New York Times* reporter calls Salinger: Sidey, *John F. Kennedy*, p. 111.

4 "Soviet Orbits Man and Recovers Him," United Press International, *New York Times*, April 12, 1961, pp. 1, 22. https://timesmachine.nytimes.com/timesmachine/1961/04/12/issue.html.

5 "The Bay of Pigs," *JFK in History*, John F. Kennedy Presidential Library and Museum, accessed July 10, 2018, https://www.jfklibrary.org/JFK/JFK-in-History/The-Bay-of-Pigs.aspx; Tad Szulc, "Castro Says Attack Is Crushed; Cuba Rebels Give Up Beachhead, Report New Landings on Island," *New York Times*, April 20, 1961, pp. 1, 10, https://timesmachine.nytimes.com/timesmachine/1961/04/20/issue.html.

6 George Gallup, "Most Nations Believe Soviets Will Lead in Science by 1970," *Ogden (UT) Standard-Examiner*, February 12, 1960, p. 5, https://www.newspapers.com/image/28147217/; USIA Polling Data: Warren Unna, "U.S. Prestige Slip Seen in Poll of 5 Major Allies," *The Washington Post*, October 29, 1960, pp. A1, A7.

7 John Kennedy was part of a run of U.S. presidents who believed in the value of press conferences. In the 34 months of his presidency, he had 65 press conferences, one every 16 days. He was the first president to have the press conferences broadcast live, on television. Kennedy had more press conferences in his three years than Reagan had in eight, more than Nixon had in six years. Kennedy averaged 23 a year, more than Obama (21 a year), Carter (15), and Ford (16). But Kennedy's numbers are low compared to the presidents immediately surrounding him: Eisenhower averaged 24 a year, Johnson 26, and Truman met with the press almost every week (42 a year). "Presidential News Conferences," *American Presidency Project*, undated, http://www.presidency.ucsb.edu/data/newsconferences.php.

8 Information on press conference preparation from Pierre Salinger, *With Kennedy* (New York: Doubleday, 1966), pp. 137–38, via JFK Library website, https://www.jfklibrary.org/Education/Students/Presidential-Press-Office.aspx. The book is available on Google Books: https://books.google.com/books?id=vx45mXCc4JoC&pg=PP5&dq=%22pierre+salinger%22+%22with+kennedy%22#v=onepage&q=%22pierre%20salinger%22%20%22with%20kennedy%22&f=false.

9 John F. Kennedy, "Excerpts from a Speech Delivered by Senator John F. Kennedy, Civic Auditorium, Portland, Oregon," September 9, 1960, https://www.presidency.ucsb.edu/documents/excerpts-from-speech-delivered-senator-john-f-kennedy-civic-auditorium-portland-or-advance.

10 President Kennedy used "vigor" and "vigorous" in his speeches regularly—in his speech accepting the Democratic nomination for President on July

15, 1960, in Los Angeles he said, "Leadership is the ability to . . . lead vigorously": "Acceptance speech at Democratic Convention. Nomination for President," July 15, 1960, JFK Presidential Library and Museum, https://www.jfklibrary.org/Asset-Viewer/AS08q5oYz0SFUZg9uOi4iw.aspx.

The *New York Times* sometimes used the word "vigor" in headlines about Kennedy's campaign: Damon Stetson, "A Need for Vigor Cited by Kennedy," *The New York Times*, February 26, 1960, p. 2, https://timesmachine.nytimes.com/timesmachine/1960/02/26/issue.html; Clayton Knowles, "Kennedy's Reply to Truman Asks Young Leaders: Senator Contends 'Strength & Vigor' Are Required in the White House," *The New York Times*, July 5, 1960, p. 1, https://timesmachine.nytimes.com/timesmachine/1960/07/05/issue.html.

Speech references "an America on the march" and "Americans are tired of standing still" from "Speech by Senator John F. Kennedy, Convention Hall, Philadelphia, PA," *American Presidency Project*, October 31, 1960, https://www.presidency.ucsb.edu/documents/speech-senator-john-f-kennedy-convention-hall-philadelphia-pa.

11 Kennedy, "Address of Senator John F. Kennedy Accepting the Democratic Party Nomination for the Presidency of the United States."

12 Coverage of the "Kitchen Debate": Harrison E. Salisbury, "Nixon and Khrushchev Argue in Public as U.S. Exhibit Opens; Accuse Each Other of Threats," *The New York Times*, July 25, 1959, pp. 1, 2, https://timesmachine.nytimes.com/timesmachine/1959/07/25/issue.html; Willard Edwards, "Khrushchev, Nixon Debate," *Chicago Daily Tribune*, July 25, 1959, pp. 1, 4, http://archives.chicagotribune.com/1959/07/25/; John Scali, "Khrushchev and Nixon Trade Gibes," *The Washington Post*, July 25, 1959, pp. A1, A4.

Quote from Kennedy-Nixon presidential debate: "October 21, 1960 Debate Transcript," Commission on Presidential Debates, http://www.debates.org/index.php?page=october-21-1960-debate-transcript.

13 Ted Sorenson, *Kennedy: The Classic Biography* (New York: Harper Perennial, 2009), p. 524.

14 Karl G. Harr, Jr., "Industry and World War II—Embryo to Vigorous Maturity," *Air Force/Space Digest*, September 1965, p. 64, https://www.aia-aerospace.org/wp-content/uploads/2016/06/INDUSTRY-AND-WW-II.pdf.

The Soviet Union was offered U.S. rebuilding assistance through the Marshall Plan, but declined. Some scholars date the start of the Cold War to the creation of the Marshall Plan, and the Soviet Union's suspicion that it was a way of creating an anti-Soviet alliance in Western Europe. See Scott D. Parrish and Mikhail M. Narinsky, "New Evidence on the Soviet Rejection of the Marshall Plan, 1947: Two Reports," Working Paper No. 9, Woodrow Wilson International Center for Scholars, https://www.wilson center.org/sites/default/files/ACFB73.pdf.

15 Ten cabinet members sworn in: "Swearing-in Ceremony and Reception for Cabinet Secretaries, 4:00 PM," JFK Presidential Library and Museum, January 21, 1961, https://www.jfklibrary.org/Asset-Viewer/Archives/JFK WHP-AR6287-E.aspx. Glennan leaves without talking to anyone except for brief conversation with LBJ: *The Birth of NASA: The Diary of T. Keith Glennan*, NASA, https://history.nasa.gov/SP-4105.pdf (conversation with LBJ, at Glennan's insistence, January 9, 1961, p. 300; leave to drive home, p. 309).

16 All quotes on the following pages from President Kennedy's April 12, 1961, press conference, along with the attendance figure, are from the transcript (and audio recording) at the Kennedy Library: John F. Kennedy, "News Conference 9, April 12, 1961," John F. Kennedy Presidential Library and Museum, https://www.jfklibrary.org/archives/other-resources/john-f-ken nedy-press-conferences/news-conference-9.

17 Kennedy was captivated by desalination and the possibilities it offered for reducing poverty and suffering. On June 21, 1961, he used a switch that had been temporarily installed on his desk in the Oval Office to turn on the first commercial desalination plant in the U.S., located in Freeport, Texas, saying it represented "one of the oldest dreams of man—extracting water from the seas."

Kennedy teamed up with Congress to invigorate the Office of Saline Waters, in the Interior Department, including winning funding for more than half a dozen desalination plants, several of which were designed to research the best methods to turn seawater into drinking water. "President Hails Desalting Plant," United Press International, *New York Times*, June 22, 1961, p. 21, https://timesmachine.nytimes.com/timesmachine/1961 /06/22/101467703.html; Robert C. Toth, "Efforts to Desalt Water Are Stepped Up in U.S.," *New York Times*, February 21, 1963, p. 11, https:// timesmachine.nytimes.com/timesmachine/1963/02/21/issue.html.

18 "Eisenhower Notes Soviet Space Feat," Associated Press, *New York Times*, April 15, 1961, p. 2, https://timesmachine.nytimes.com/timesmachine /1961/04/15/118907150.html.

19 "The Cruise of the Vostok," *Time*.

20 "Soviet Lands Man after Orbit of World; K Challenges West to Duplicate Feat," *Washington Post*, April 13, 1961, A1, A24. The name Vostok has some historical resonance for Russians. The 130-foot-long Russian sloop-of-war *Vostok*, captained by Faddey Bellingshausen, discovered the continent of Antarctica on January 28, 1820, and returned to Antarctica the following January. Vostok Station, a Russian research station in Antarctica established in December 1957, is named for the ship that discovered the continent, and Antarctica's largest subglacial lake, Vostok Lake, is in turn named for the Russian research center. The *Washington Post* routinely referred to Soviet premier Khrushchev simply as "K" in headlines.

21 "Gagarin in Space: 'Jules Bergman Announces and Explains the Event,' ABC News Special, April 12, 1961," YouTube, posted by Dan Beaumonth Space Museum, November 14, 2011, 4:33, https://www.youtube.com/watch?v=ZYucEQFYNf0&t=4s. "Russia Reaps Praise for Mighty Deed," *Washington Post*, April 13, 1961, A24.

22 Osgood Caruthers, "Gagarin Is Hailed by All of Moscow," *New York Times*, April 14, 1961, pp. 1, 2, https://timesmachine.nytimes.com/timesmachine/1961/04/15/issue.html. The 2 million figure and Khrushchev's tears, from "Reds Pay Homage to Their Space Hero," Reuters, *Chicago Daily Tribune*, April 15, 1961, pp. 1, 2, https://chicagotribune.newspapers.com/image/374621899.

23 "Text of the Soviet Statement Praising First Space Flight," Reuters, *New York Times*, April 13, 1961, p. 14, https://timesmachine.nytimes.com/timesmachine/1961/04/13/issue.html.

24 TV ownership: "Selected Communications Media: 1920 to 1998," Table No. 1440, *Statistical Abstract of the United States*, 1999, p. 885, https://www2.census.gov/library/publications/1999/compendia/statab/119ed/tables/sec31.pdf.

Popular shows: *The World Almanac 2001*, p. 317.

Car ownership: "Motor-Vehicle Registration (Passenger Cars, Buses, and Trucks) by States: 1920 to 1961," Table No. 766, *Statistical Abstract of the United States*, 1962, p. 563.

Interstate Highway System: Richard F. Weingroff, "The Greatest Decade: Celebrating the 50th Anniversary of the Eisenhower Interstate System," U.S. Department of Transportation: Federal Highway Administration, https://www.fhwa.dot.gov/infrastructure/50interstate.cfm.

Park visits: *Historical Statistics of the United States: Colonial Times to 1957*, Washington, D.C.: U.S. Department of Commerce, Bureau of the Census, 1960, pp. 222–23.

Shopping malls: Thomas W. Hanchett, "U.S. Tax Policy and the Shopping-Center Boom," *American Historical Review*, October 1996, pp. 1082–1110.

Family income: "Money Income of Families—Median Income in Current and Constant (1997) Dollars, by Race and Type of Family: 1947 to 1997," Table No. 1427, *Statistical Abstract of the United States*, 1999, p. 877, https://www2.census.gov/library/publications/1999/compendia/statab/119ed/tables/sec31.pdf.

25 Levittown: Clare Suddath, "The Middle Class," *Time*, February 27, 2009, http://content.time.com/time/nation/article/0,8599,1882147,00.html.

Pace of U.S. home building: Lizabeth Cohen, "A Consumer's Republic: The Politics of Mass Consumption in Postwar America," *Journal of Consumer Research*, vol. 31, no. 1 (June 2004), p. 237, https://dash.harvard.edu/bitstream/handle/1/4699747/cohen_conrepublic.pdf.

Outdoor cooking: "Yards Provide Good Trade in Equipment," *Washington Post*, May 15, 1955, p. G-11.

26 Homer Bigart, "Troops Enforce Peace in Little Rock As Nine Negroes Return to their Classes; President to Meet Southern Governors," *New York Times*, September 26, 1957, pp. 1, 12, https://timesmachine.nytimes.com/timesmachine/1957/09/26/issue.html.

27 "Halfway across Arizona": Gladwin Hill, "2nd Atomic Blast in 24 Hours Jolts Wide Nevada Area," *New York Times*, January 29, 1951, pp. 1, 25, https://timesmachine.nytimes.com/timesmachine/1951/01/29/88430448.html.

"brightened the sky in four states": "Latest Nevada A-Blast Seen in 4 States," *Washington Post*, February 2, 1951, p. A1.

"eerie sun-gold glow": "4th A-Blast 'Like a Quake' in Las Vegas," *Washington Post*, February 3, 1951, p. A1.

28 "Latest Nevada A-Blast Seen in 4 States," *Washington Post*, February 2, 1951, p. A1.

29 "Reds Admit New Test of A-Weapon," *Washington Post*, September 1, 1956, pp. A1, A6.

30 KTLA broadcasts of atomic tests: "KTLA A-Bomb Coverage April 22, 1952," YouTube, posted by Robert Black, August 18, 2015, 10:15, https://www.youtube.com/watch?v=OF3JvVJMtzg. "Atomic / Nuclear Bomb Nevada Test—5/1/1952," posted by TheShottingstar31, June 24, 2012, 9:12, https://www.youtube.com/watch?v=d5JsrCPWCl8. *Washington Post* TV critic's review: Sonia Stein, "Incredibly, 'Bomb' Was Tame on TV," *Washington Post*, March 22, 1953, p. L1.

31 Soldiers near bomb blast: Elton C. Fay, "Men Shaken But Safe 2 Miles from A-Blast," *Washington Post*, March 18, 1953, p. A1. Monkeys and rats: "High A-Blast Seen 1000 Miles Away," *Washington Post*, April 7, 1953, p. A4. Members of Congress: "Big A-Blast Shakes up 14 Congressmen: Knocks off Hats," *Washington Post*, April 26, 1953, p. A1.

Casino atomic-bomb watching parties: "How 1950s Las Vegas Sold Atomic Bomb Tests as Tourism," YouTube, posted by Smithsonian Channel, March 31, 2017, 1:30, https://www.youtube.com/watch?v=FghT80tVFKo.

32 Reporters on B-25: Gladwin Hill, "Air Force Planes Chase and Test Atomic Cloud After Nevada Blast," *New York Times*, March 2, 1955, pp. 1, 12, https://timesmachine.nytimes.com/timesmachine/1955/03/02/issue.html.

Testing "Survival Town": "A-Blast Shatters 'Survival Town,'" *Washington Post*, May 6, 1955, p. A1.

Video of "Survival Town" test: "Nuclear Bomb Tests (1955) Dummy Town," YouTube, posted by WesternWorldHistory, June 13, 2012, 15:11, https://www.youtube.com/watch?v=thPfjOt5WEo.

33 Radioactive snow: "4th A-Blast 'Like a Quake" in Las Vegas," *Washington Post*. "Tea-cup" bombs: "Tiny A-Bombs Rumor Stirs Northwest," *Washington Post*, February 19, 1953, p. A1.

34 "New Film to Help in Bomb Training," *New York Times*, January 25, 1952, p. 7, https://timesmachine.nytimes.com/timesmachine/1952/01/25/84299073.html. Jake Hughes, "Duck and Cover," The Library of Congress: National Film Preservation Board, Film Essays, https://www.loc.gov/programs/static/national-film-preservation-board/documents/duck_cover.pdf.

35 George Gallup, "Many Believe Atomic Tests Caused Tornadoes This Spring," *Washington Post*, June 19, 1953, p. A23. "A-Bomb Link to Tornadoes is Discounted," *Washington Post*, November 18, 1954, p. A13.

36 "The Nuclear Testing Tally," Arms Control Association, September 3, 2017, https://www.armscontrol.org/factsheets/nucleartesttally.

37 Max Frankel, "Russians Announce Firing Intercontinental Missile 'Huge Distance' to Target," *New York Times*, August 27, 1957, pp. 1, 6, https://timesmachine.nytimes.com/timesmachine/1957/08/27/issue.html. "Text of Soviet Statement," *New York Times*, August 27, 1957, p. 6, https://timesmachine.nytimes.com/timesmachine/1957/08/27/issue.html.

"the ultimate weapon": Harry Schwartz, "The Real Threat of Moscow's Missile," *New York Times Magazine*, September 15, 1957, pp. 20ff., https://

timesmachine.nytimes.com/timesmachine/1957/09/15/167870592.html?pageNumber=282.

38 Kathleen Teltsch, "Soviet Jets to Fly Delegates to U.N.," *New York Times*, August 27, 1957, pp. 1, 6, https://timesmachine.nytimes.com/timesmachine/1957/08/27/issue.html.

39 Marquis Childs, "Reds Fire 6 Missiles of Intercontinental Range over Siberia," *Washington Post*, August 30, 1957, p. A1.

40 "5,000- Mile Missile Explodes at Test," *New York Times*, June 12, 1957, pp. 1, 7, https://timesmachine.nytimes.com/timesmachine/1957/06/12/issue.html.

41 Sputnik: William J. Jorden, "Soviet Fires Earth Satellite into Space; It Is Circling the Globe at 18,000 M.P.H.; Sphere Tracked in 4 Crossings Over U.S.," *New York Times*, October 4, 1957, pp. 1, 3, https://timesmachine.nytimes.com/timesmachine/1957/10/05/issue.html. Significance of weight: "Device is 8 Times Heavier Than One Planned by U.S.," *New York Times*, October 5, 1957, pp. 1, 3, https://timesmachine.nytimes.com/timesmachine/1957/10/05/issue.html.

Interrupted network broadcasts: Roy Silver, "Satellite Signal Broadcast Here," *New York Times*, October 5, 1957, pp. 1, 2, https://timesmachine.nytimes.com/timesmachine/1957/10/05/issue.html.

42 "Text of Satellite Report," Reuters, *New York Times*, October 4, 1957, p. 3, https://timesmachine.nytimes.com/timesmachine/1957/10/05/issue.html.

43 Sputnik comment sources: Grumpy scientist: Richard Witkin, "U.S. Delay Draws Scientists' Fire," *New York Times*, October 4, 1957, p. 2, https://timesmachine.nytimes.com/timesmachine/1957/10/05/issue.html. U.S. Senators: "Senators Attack Missile Fund Cut," *New York Times*, October 6, 1957, pp. 1, 43, https://timesmachine.nytimes.com/timesmachine/1957/10/06/issue.html. Eisenhower at his Gettysburg farm: Rutherford Poats, "Soviet Launching Brings Demand for Probe by Congress," *Washington Post*, October 6, 1957, p. A1.

44 NBC News anchor: "Red Moon over the U.S.," *Time*, October 14, 1957, p. 27, http://time.com/vault/issue/1957-10-14/page/29/. "Soviet Claiming Lead in Science," *New York Times*, October 5, 1957, p. 2, https://timesmachine.nytimes.com/timesmachine/1957/10/05/issue.html.

45 "Reaction to the Soviet Satellite—A Preliminary Evaluation," memo prepared for President Eisenhower, October 16, 1957, Dwight D. Eisenhower Presidential Library, Museum and Boyhood Home, Research: Sputnik and

the Space Race, https://www.eisenhower.archives.gov/research/online_documents/sputnik.html.

46 Official Soviet statement on Sputnik: Reuters, "Text of Satellite Report," *New York Times*, October 5, 1957, p. 3, https://timesmachine.nytimes.com/timesmachine/1957/10/05/issue.html. Run on binoculars at New York City department stores: "Sputnik No. 1 on 34th Street," *New York Times*, October 10, 1957, p. 50, https://timesmachine.nytimes.com/timesmachine/1957/10/10/84771386.html. Sputnik cocktail: Associated Press, "Engineers' Cocktail Features Sour Grapes," *Great Falls (MT) Tribune*, October 20, 1957, p. 13, https://www.newspapers.com/image/239205949. Albania: Constance McLaughlin Green and Milton Lomask, *Vanguard: A History*, NASA Historical Series (Washington, D.C.: NASA, 1970), p. 188. *Life* magazine: "Soviet Satellite Sends U.S. into a Tizzy," *Life*, October 14, 1957, pp. 34–37. Vatican: "Vatican Sees Satellite as Infernal Toy," *Chicago Daily Tribune*, October 10, 1957, part 1, p. 5, https://www.newspapers.com/image/371748773.

47 Sputnik's beep can be heard easily on the internet and via YouTube. Here is NASA's version: "Sputnik: Beep," NASA Image and Video Library, March 22, 2017, https://images.nasa.gov/details-578626main_sputnik-beep.html. Arthur C. Clarke is credited with applying the word "beep" to what became known as the electronic computer tone, in his first novel, *The Sands of Mars*, published in 1951. Laurence Cawley, "The Ubiquity of the Modern Beep," *BBC News Magazine*, May 12, 2014, https://www.bbc.com/news/magazine-27308544. Clarke's reference is on p. 82 of *The Sands of Mars*: http://avalonlibrary.net/ebooks/Arthur%20C%20Clarke%20-%20The%20Sands%20Of%20Mars.pdf.

48 Green and Lomask, *Vanguard*, p. 187.

49 Associated Press, "Soviet Fires New Satellite, Carrying Dog; Half-Ton Sphere Is Reported 900 Miles Up," *New York Times*, November 3, 1957, p. 1, https://timesmachine.nytimes.com/timesmachine/1957/11/03/issue.html.

50 October Revolution celebration: William J. Jorden, "Khrushchev Asks East-West Talks to End 'Cold War,'" *New York Times*, November 7, 1957, pp. 1, 11, https://timesmachine.nytimes.com/timesmachine/1957/11/07/issue.html. Possible nuclear "firework" at the Moon: Walter Sullivan, "Strange Radio Signals Go On; Stir Speculation of a Moon Shot," *New York Times*, November 7, 1957, pp. 1, 20, https://timesmachine.nytimes.com/timesmachine/1957/11/07/issue.html.

51 Multiple names for canine cosmonaut: "Husky Safe, Reds Say As Protests Pour In," *Washington Post*, November 4, 1957, pp. A1, A2. Nate Haseltine, "Dog Reported Alive and Well," *Washington Post*, November 5, 1957, p. A1. Protests at U.N. and Soviet Embassy: "Dogs on Picket Line," *New York Times*, November 5, 1957, p. 12, https://timesmachine.nytimes.com/timesmachine/1957/11/05/issue.html. Reuters, "Moscow Denounces Dog-Lover Protests," *Washington Post*, November 6, 1957, p. A3.

52 Associated Press, "Khrushchev Asks a Satellite Race," *New York Times*, November 6, 1957, pp. 1, 2, https://timesmachine.nytimes.com/timesmachine/1957/11/06/issue.html.

53 United Press International, "Greatest Show Off Earth Is a Soviet Circus Joke," *New York Times*, November 6, 1957, p. 12, https://timesmachine.nytimes.com/timesmachine/1957/11/06/issue.html.

54 The job of presidential science advisor has never had anyone as prominent as its first occupant, although the third science advisor, Kennedy's Jerome Wiesner, became president of MIT after working for Kennedy, and the fourth science advisor, Donald Horning, went on to become president of Brown University.

Text of Eisenhower's address: Dwight D. Eisenhower, "Radio and Television Address to the American People on Science in National Security," November 7, 1957, *American Presidency Project*, https://www.presidency.ucsb.edu/documents/radio-and-television-address-the-american-people-science-national-security.

55 W.H. Lawrence, "President Voices Concern On U.S. Missiles Program, But Not On Satellite," *New York Times*, October 10, 1957, pp. 1, 15, https://timesmachine.nytimes.com/timesmachine/1957/10/10/issue.html.

56 Yanek Mieczkowski, *Eisenhower's Sputnik Moment: The Race for Space and World Prestige* (Ithaca, New York: Cornell University Press, 2013), p. 63.

57 George R. Price, "Arguing the Case for Being Panicky," *Life*, November 18, 1957, pp. 125ff., https://books.google.com/books?id=vVYEAAAAMBAJ&printsec=frontcover&source=gbs_ge_summary_r&cad=0#v=onepage&q&f=false.

58 Milton Bracker, "U.S. Ready to Fire First Satellite Early This Week," *New York Times*, December 2, 1957, pp. 1, 21, https://timesmachine.nytimes.com/timesmachine/1957/12/02/.

59 "The Death of TV-3," *Time*, December 16, 1957, pp. 9–10, http://time.com/vault/issue/1957-12-16/page/13/. Harlen Makemson, *Media, NASA, and America's Quest for the Moon* (New York: Peter Lang Publishing, 2009), p. 33–34.

60 Green and Lomask, *Vanguard*, pp. 208–9.

61 "Excerpts from News Conference on Vanguard Rocket Test," *New York Times*, December 7, 1957, p. 8, https://timesmachine.nytimes.com/timesmachine/1957/12/07/issue.html.

The quote from Dorothy Kilgallen appears in several sources, but typically only the second half—"Someone should go out there and kill it." The full quote is from the *Life* magazine account of the failed Vanguard launch, published the week after it happened, but attributed only to "a female reporter."

"Too Much Talk Too Soon Adds Up To Disaster," *Life*, December 16, 1957, pp. 26–30, https://books.google.com/books?id=1FUEAAAAMBAJ&pg=PA25.

62 Sen. Lyndon Johnson reaction: "Ike is Glum, Orders Report on Failure," United Press International, *The Times* (San Manteo, CA), December 6, 1957, pp. 1, 2, https://www.newspapers.com/image/51951625. *Los Angeles Herald & Express* headline: "Vanguard's Aftermath: Jeers and Tears," *Time*, December 16, 1957, p. 12, http://time.com/vault/issue/1957-12-16/page/16/. London newspaper headlines: "Enoughnik of this," *New York Times*, December 8, 1957, p. 36, https://timesmachine.nytimes.com/timesmachine/1957/12/08/issue.html. Swiss newspaper headline: "If Ridicule Could Kill," *New York Times*, December 8, 1957, p. 36, https://timesmachine.nytimes.com/timesmachine/1957/12/08/issue.html. Polish Army reaction: Reuters, "A Moment of Merriment," *New York Times*, December 8, 1957, p. 36, https://timesmachine.nytimes.com/timesmachine/1957/12/08/issue.html.

Soviet offer of technical assistance: "Russians At U.N. Tweak U.S. on (Satellite) Nose," *New York Times*, December 7, 1957, p. 8, https://timesmachine.nytimes.com/timesmachine/1957/12/07/issue.html.

63 Dudley Clendinen, "What's Doing in Boston," *New York Times*, September 12, 1982, section 10, p. 8, https://timesmachine.nytimes.com/timesmachine/1982/09/12/144946.html. Kennedy family as regulars: David Lamb, "Boston's Historic Pub," *Los Angeles Times*, May 17, 1987, p. 24.

64 Charles Murray and Catherine Bly Cox, *Apollo* (Burkittsville, MD: South Mountain Books, 2004), p. 45 (originally published by Simon & Schuster).

65 Logsdon, *John F. Kennedy and the Race to the Moon*, p. 39.

At a press conference after the Apollo 8 mission around the Moon at Christmas 1968, then President Johnson recalled being asked by President

Kennedy to find a good candidate for NASA administrator—and President Johnson said he had interviewed 28 people before he could persuade James Webb to at least talk to President Kennedy in person about the job. "Transcript of President Johnson's News Conference on Foreign and Domestic Affairs," *New York Times*, December 28, 1968, p. 10, https://timesmachine.nytimes.com/timesmachine/1968/12/28/76926292.html?pageNumber=10, https://timesmachine.nytimes.com/timesmachine/1968/12/28/76926292.html.

66 Seamans, *Aiming at Targets*, p. 78.

67 There are multiple, overlapping, and highly entertaining accounts of the effort to bring in James Webb as NASA's second administrator. See W. Henry Lambright, *Powering Apollo: James E. Webb of NASA* (Baltimore: Johns Hopkins University Press, 1995), pp. 82–85; Logsdon, *John F. Kennedy and the Race to the Moon*, pp. 41–45. The quote from Webb is in Logsdon, p. 42.

68 Lambright, *Powering Apollo*, p. 84.

69 Ibid.

70 Logsdon, *John F. Kennedy and the Race to the Moon*, p. 43.

## 3: "The Full Speed of Freedom"

1 Eugene Kranz, *Failure Is Not an Option* (New York: Simon & Schuster, 2009), p. 56.

2 Fulton quote: Ken Hechler, *Toward the Endless Frontier: History of the Committee on Science and Technology, 1959–1979* (Washington, D.C.: U.S. Government Printing Office, 1980), p. 82, https://archive.org/stream/towardendlessfro00hech#page/80/mode/2up/search/tired.

3 Luna 3's mission to photograph the dark side of the Moon: The space probe did a single loop around the Moon and swooped back toward Earth. It took photos of 70% of the far side of the Moon, 29 pictures in all, but was able to successfully transmit digitized versions of only 17 of them. "Luna 3," NASA Space Science Data Coordinated Archive, March 21, 2017, https://nssdc.gsfc.nasa.gov/nmc/spacecraftDisplay.do?id=1959-008A.

4 Anfuso remarks: Associated Press, "Money Cannot Buy Time for Space Projects," *Corsicana (TX) Daily Sun*, April 13, 1961, pp. 1, 6, https://www.newspapers.com/image/13112535/.

Brooks remarks: United Press International, "Congressmen Call for All-Out Space Program," *Los Angeles Times*, April 13, 1961, p. 2, https://

www.newspapers.com/image/386211719/. Fulton remarks: Logsdon, *John F. Kennedy and the Race to the Moon*, p. 74.

5 There are multiple accounts of this meeting on space policy on Friday, April 14, and the afternoon meeting, without Kennedy, that preceded it. This account is based largely on Sidey, *John F. Kennedy, President*, pp. 117–21; Ted Sorensen, *Counselor* (New York: Harper, 2008), pp. 333–35; Logsdon, *John F. Kennedy and the Race to the Moon*, pp. 75, 77–78.

6 Oddly, Sidey did not report that Sorensen told him, "We're going to the Moon" either in his *Life* magazine story about that evening in the White House (April 21, 1961, p. 26) or in his you-are-there book published in 1963. It is Sorensen who recounts that moment, in *Counselor*, p. 336.

7 The failed Bay of Pigs invasion is well documented. This summary account is largely drawn from these sources: Sidey, *John F. Kennedy, President*, pp. 122–40. Ted Sorensen, *Kennedy: The Classic Biography* (New York: Harper Collins, 2009), pp. 294–309. "The Bay of Pigs Invasion," Central Intelligence Agency: News & Information, April 18, 2016, https://www.cia.gov/news-information/featured-story-archive/2016-featured-story-archive/the-bay-of-pigs-invasion.html.

8 W.H. Lawrence, "Eisenhower Urges Nation To Back Kennedy on Cuba; Khrushchev Chides U.S.," *New York Times*, April 23, 1961, pp. 1, 24, https://timesmachine.nytimes.com/timesmachine/1961/04/23/issue.html.

9 The note from Kennedy to Johnson is signed "John F. Kennedy" on the right corner, which gives the conversational language a certain immediacy. Scans of it are widely available online; it's a vivid document. "Memorandum for Vice President," April 20, 1961, John F. Kennedy Presidential Library and Museum, https://www.jfklibrary.org/Asset-Viewer/6XnAYXEkkkSMLfp7ic_o-Q.aspx.

10 "President Kennedy: Address before the American Society of Newspaper Editors," April 20, 1961, *American Presidency Project*, https://www.presidency.ucsb.edu/documents/address-before-the-american-society-newspaper-editors.

11 "President Kennedy: News Conference, President Kennedy's News Conferences," April 21, 1961, John F. Kennedy Presidential Library and Museum, https://www.jfklibrary.org/Research/Research-Aids/Ready-Reference/Press-Conferences/News-Conference-10.aspx. The Associated Press reported in its story on the press conference that the reporter asking the question about whether the U.S. should beat the Russians to the Moon was William McGaffin of the *Chicago Daily News*: Associated Press, "Kennedy

Clams Up on Cuba," *Cincinnati (OH) Enquirer*, April 22, 1961, p. 1, https://www.newspapers.com/image/101565479/.

12 There are several overlapping accounts of Johnson's quick study of all the urgent questions related to space. The most authoritative is from Logsdon, *John F. Kennedy and the Race to the Moon*, pp. 84–89. Logsdon not only uses memos about the meetings, but he interviewed many of the participants. Separately, Von Braun at first meeting and Welsh observation of Johnson's approach at the meetings: "Ed Welsh: Oral History with Edward Welsh, Executive Secretary of the National Aeronautics and Space Council," July 18, 1969, Lyndon Baines Johnson Library Oral History Collection, via The Miller Center at the University of Virginia, pp. 12–13, 14, http://web1.millercenter.org/poh/transcripts/welsh_edward_1969_0718.pdf.

Webb's reaction to Johnson's full-speed approach, Wiesner's description of April 24, 1961, meeting: Murray and Cox, *Apollo*, pp. 64–65.

13 Webb biographical details: Lambright, *Powering Apollo*, pp. 13 (horse and buggy), 22–27 (Sperry). Webb quote on being able to do what you promise: Lambright, *Powering Apollo*, p. 95. Webb on Johnson's approach to the space question memos: Murray and Cox, *Apollo*, pp. 64–65. Johnson consults von Braun without asking Webb: Murray and Cox, *Apollo*, p. 64. Johnson consults Defense officials without asking McNamara: Logsdon, *John F. Kennedy and the Race to the Moon*, p. 86.

14 "Memorandum for the President," April 28, 1961, John F. Kennedy Presidential Library and Museum, https://www.jfklibrary.org/Asset-Viewer/DjiWpQJegkuIlX7WZAUCtQ.aspx.

15 Logsdon, *John F. Kennedy and the Race to the Moon*, p. 86; Welsh, "Oral History," p. 15

16 Details of Shepard's launch-day preparations: John W. Finney, "Shepard Had Periscope: 'What a Beautiful View,'" *New York Times*, May 6, 1961, pp. 1, 10, https://timesmachine.nytimes.com/timesmachine/1961/05/06/101461355.html.

Rectal thermometer rides to space: "Freedom's Flight," *Time*, May 12, 1961, pp. 52–58.

17 The time of launch of Freedom 7 is confused; some sources report it as 9:34 a.m. Eastern time; some as 10:34 a.m. The confusion arises from the fact that the launch occurred during daylight savings time, which Florida didn't start observing until 1966. So at the launchpad in Cape Canaveral the time was 9:34. In Washington, D.C., at the White House and at NASA headquarters, the launch was at 10:34 a.m.

Delays and their causes: Loyd S. Swenson, Jr., James M. Grimwood, Charles C. Alexander, *This New Ocean: A History of Project Mercury*, NASA, SP-4201, Washington, D.C., 1966, pp. 351–52.

18 Robert Conley, "National Exults Over Space Feat; City Plans to Honor Astronaut," *New York Times*, May 6, 1961, pp. 1, 11, https://timesmachine.nytimes.com/timesmachine/1961/05/06/issue.html.

Photos of President Kennedy, Jacqueline Kennedy, and National Security Council members watching the launch—11 in all: "President Kennedy Views the Lift-Off of Astronaut Cmdr. Alan B. Shepard, Jr., on the 1st U.S. Manned Sub-Orbital Flight," May 5, 1961, John F. Kennedy Presidential Library and Museum, https://www.jfklibrary.org/asset-viewer/archives/JFKWHP/1961/Month%2005/Day%2005/JFKWHP-1961-05-05-A.

19 CBS TV coverage of Shepard flight: "Freedom 7 Flight CBS News Coverage," zellco321, posted December 8, 2016, YouTube, 1:08:05, https://www.youtube.com/watch?v=2OYUykje27g.

350 Mercury reporters, five witnesses to Wright flight: Swenson, Grimwood, Alexander, *This New Ocean*, p. 350.

20 BBC broadcast of NBC coverage: "London Is Elated By Achievement," *New York Times*, May 6, 1961, p. 11, https://timesmachine.nytimes.com/timesmachine/1961/05/06/issue.html.

Voice of America broadcast: "'Voice' Tells the World Of U.S. Space Flight," *New York Times*, May 6, 1961, p. 8, https://timesmachine.nytimes.com/timesmachine/1961/05/06/issue.html.

New York City Hall: Conley, "National Exults Over Space Feat; City Plans to Honor Astronaut," *New York Times*.

21 "Transcript of Space Flight Messages," *New York Times*, May 6, 1961, p. 8, https://timesmachine.nytimes.com/timesmachine/1961/05/06/issue.html.

Cronkite comment: "Freedom 7 Flight CBS News Coverage," YouTube, at 22:30.

22 Recovery details, Shepard comment, call from President Kennedy: "Freedom's Flight," *Time*, May 12, 1961, pp. 52–58.

23 "Roused the country": Conley, "National Exults Over Space Feat; City Plans to Honor Astronaut," *New York Times*.

ABC News special report: "Alan Shepard, First American in Space, ABC news coverage, May 5, 1961," Dan Beaumont Space Museum, October 11, 2011, YouTube, 7:44, Shadell comment at 00:25, https://www.youtube.com/watch?v=6Qjjk6THiJ4.

NBC News special report: "Frank McGee Re-cap Freedom 7, NBC (1961)," Old Movies Reborn, January 18, 2018, YouTube, 9:42, https://www.youtube.com/watch?v=3if-8UbjfXo.

Leonard J. Carter comment: "Freedom's Flight," *Time*.

24 Jean Sprain Wilson, "'Just a Baby Step to What We Shall See,' Says Beaming, Happy Wife of Astronaut," *Express and News (San Antonio, TX)*, May 6, 1961, p. 10-A, https://www.newspapers.com/image/29608961/.

25 Wiesner Committee, *Report to the President-Elect of the Ad Hoc Committee on Space*, January 10, 1961, pp. 15–17, https://www.hq.nasa.gov/office/pao/History/report61.html.

26 Wiesner, who died in 1994, was interviewed extensively by Murray and Cox for their book, *Apollo.* Both his crisp analysis of Kennedy's dilemma and the story about the exchange with President Habib Bourguiba come from *Apollo,* pp. 66–67.

27 Swenson, Grimwood, Alexander, *This New Ocean*, p. 361: "President Kennedy's shore-to-ship radio telephone call to the astronaut was spontaneous, though difficult to link, and symbolic of the American mood that day."

28 John F. Kennedy, "News Conference 11, May 5, 1961," John F. Kennedy Presidential Library and Museum, https://www.jfklibrary.org/archives/other-resources/john-f-kennedy-press-conferences/news-conference-11.

29 Chalmers M. Roberts, "President to Address Congress: To Discuss Urgent National Needs in Person Today," *Washington Post,* May 25, 1961, pp. A1, A9.

30 Quotes from the final Johnson report: "James E. Webb, NASA Administrator, and Robert S. McNamara, Secretary of Defense, to the Vice President, May 8, 1961, with attached: 'Recommendations for Our National Space Program: Changes, Policies, Goals,'" in John M. Logsdon, ed., *Exploring the Unknown: Selected Documents in the History of the U.S. Civil Space Program* (Washington, D.C.: National Aeronautics and Space Administration, 1995). "Disappointing results": p. 448; "Soviets get there first": p. 446; "National vigor": p. 447; "Lunar exploration before the end of this decade": p. 446.

31 Logsdon, *John F. Kennedy and the Race to the Moon*, p. 107.

32 Orloff, *Apollo by the Numbers*, p. 281.

33 John G. Norris, "U.S. to Race Russians to Moon," *Washington Post,* May 20, 1961, pp. A1, A4; Jerry T. Baulch, "JFK to Ask Space, Aid Fund Boost," *Washington Post,* May 24, 1961, p. A1; W. H. Lawrence, "President to Ask Urgent Effort to Land on Moon," *New York Times,* May 24, 1961, p. 1, https://

timesmachine.nytimes.com/timesmachine/1961/05/24/101464643.html ?pageNumber=1.

34 Chalmers M. Roberts, "President to Address Congress," *The Washington Post*, May 25, 1961, p. A1.

35 Text of Kennedy's "Second State of the Union": John F. Kennedy, "Special Message to the Congress on Urgent National Needs," May 25, 1961, *American Presidency Project*, https://www.presidency.ucsb.edu/documents /special-message-the-congress-urgent-national-needs. The full audio of Kennedy's address is widely available. White House Audio Recordings, 1961–63, https://www.jfklibrary.org/Asset-Viewer/Archives/JFKWHA-032 .aspx. There is no video of the full address available on the internet. The eight-minute space section is available at the JFK Library: Video excerpt, John F. Kennedy, "Special Message to Congress on Urgent National Needs," via CBS News, https://www.jfklibrary.org/Asset-Viewer/Archives/TNC-200 -2.aspx.

36 Seamans, *Aiming at Targets*, p. 91; Lambright, *Powering Apollo*, p. 101.

37 Kennedy, "Address at Rice University, Houston, on the Nation's Space Effort."

38 Johnson's "peace through space" quote: Alvin Shuster, "Congress Wary on Cost, but Likes Kennedy Goals," *New York Times*, May 26, 1961, pp. 1, 13, https://timesmachine.nytimes.com/timesmachine/1961/05/26/issue.html. Republican lawmaker quote on Kennedy's deficits: Robert C. Albright, "President's Arms, Space Aims Get Full Backing by Congress," *Washington Post*, May 27, 1961, p. A6.

39 Christian Davenport, "Pence Vows America Will Return to the Moon: The History of Such Promises Suggests Otherwise," *Washington Post*, October 11, 2017, https://www.washingtonpost.com/news/retropolis/wp/2017/10 /10/presidents-love-evoking-jfks-iconic-moon-speech-now-its-the-trump -administrations-turn/?noredirect=on&utm_term=.48264c55df86; John H. Logsdon, "Ten Presidents and NASA," *NASA.gov*, accessed August 20, 2018, https://www.nasa.gov/50th/50th_magazine/10presidents.html; David E. Sanger and Richard W. Stevenson, "Bush Backs Goal of Flight to Moon to Establish Base," *New York Times*, January 15, 2004, https:// www.nytimes.com/2004/01/15/us/bush-backs-goal-of-flight-to-moon -to-establish-base.html; Tariq Malik, "Obama Aims to Send Astronauts to an Asteroid, Then to Mars," *Space.com*, April 15, 2010, https://www.space .com/8222-obama-aims-send-astronauts-asteroid-mars.html.

40 John W. Finney, "Washington Cheers Shepard as Hero," *New York Times*,

May 9, 1961, p. 1, https://timesmachine.nytimes.com/timesmachine/1961/05/09/issue.html.

41 Fifty-fifty chance of beating the Russians: Seamans, *Aiming at Targets*, p. 88. Kranz on Kennedy speech: Kranz, *Failure Is Not an Option*, p. 56. Kraft on Kennedy speech: "Man on the Moon: The Kennedy Challenge," 172HoursontheMoon, posted March 23, 2012, YouTube, 3:38 (Kraft at 1:17), https://www.youtube.com/watch?v=cvHldoh5JF0.

42 Papers of John F. Kennedy, "Special Message to Congress on Urgent National Needs."

43 As of November 1958, with 13 years of increasing passenger flights on commercial airlines during the post–World War II era, 70 percent of Americans still hadn't taken an airplane flight. At that point, just 250,000 U.S. travelers accounted for 40% of all passengers. "Jets across America," *Time*, November 17, 1958, p. 89.

## 4: The Fourth Crew Member

1 Robert G. Chilton, oral history, transcript, NASA Johnson Space Center Oral History Project, interviewed by Summer Chick Bergen, Houston, Texas, April 5, 1999, https://historycollection.jsc.nasa.gov/JSCHistoryPortal/history/oral_histories/ChiltonRG/ChiltonRG_4-5-99.htm.

2 NASA has a set of detailed websites called the *Apollo Lunar Surface Journal* and the *Apollo Flight Journal* about each of the Apollo Moon flights and landings. They have been assembled and meticulously curated—the surface journals by Eric M. Jones, a Ph.D. space scientist who spent most of his career at the Los Alamos National Laboratory; the *Flight Journals* by M. David Woods, a video editor at the BBC and a space historian and author of several authoritative books on the Apollo missions.

The main index pages for each:

*Apollo Flight Journal:* https://history.nasa.gov/afj/index.html
*Apollo Surface Journal:* https://history.nasa.gov/alsj/

One element of the *Journal* compilations is an edited transcript of recorded communications—between Mission Control and the spacecraft and also between the astronauts onboard the spacecraft. Those edited transcripts include commentary—in the style of the Talmud—from the technical debriefings with the astronauts after the missions were finished, and also from interviews. The transcripts are interrupted by explanations from the astronauts of what they were thinking or why they did a particular thing at that moment in the mission. The journals are indispensible and works of scholarship and art. In citing them, I'm including the time reference where the information—including the thinking of the astronauts—is located.

Armstrong, thinking about low-fuel: Eric M. Jones, ed., "The First Lunar Landing," *Apollo 11: Lunar Surface Journal*, 102:44:45 and 102:45:32, https://www.hq.nasa.gov/alsj/a11/a11.landing.html.

3 Gene W. Harms, an engineer for Grumman, the aerospace company that designed and built the lunar modules, remembers showing John Glenn an early mock-up design of the LM's cockpit. "We were very naive about space travel," said Harms. "We knew nothing about zero gravity at the time." An early mock-up had swiveling crew seats, so the astronauts could turn to look at instruments and controls, and out the windows. Glenn came to see that early version. "He said, 'Can I come in? My name is John Glenn.' He was a famous man at that point, people were dying to see him."

Inside the cockpit, Glenn turned to Harms. "Let me ask you a question: Why do you have seats?"

"Everything we have designed at Grumman always had a seat," said Harms.

"We're at zero-g," said Glenn. "We weigh nothing. When we land on the Moon, we're at one-sixth gravity. Maybe we don't need the seat at all." That exchange planted the idea to take the seats out completely. From a telephone interview with Harms by the author, May 5, 2016.

4 Constant weight scrubs on the lunar module: Kelly, *Moon Lander*; story of eliminating LM seats: p. 63; story of thinning out the crew compartment skin: p. 122.

The weight of each lunar module at launch varied depending on the mission. Basic LM dimensions: "Lunar Module: Quick Reference Data," Grumman Corporation, p. LV-1, https://www.hq.nasa.gov/alsj/LM04_Lunar_Module_ppLV1-17.pdf. Launch weights for Apollo 11, *Eagle*: Lunar module, total weight, including propellant: 33,278 pounds; LM Ascent stage, dry: 4,804 pounds; LM Descent stage, dry: 4,483 pounds. Orloff, "Launch Vehicle / Spacecraft Key Facts," in *Apollo by the Numbers*, pp. 276–77.

5 Armstrong discusses running out fuel before touchdown: Jones, ed., *Apollo 11: Lunar Surface Journal*, "The First Lunar Landing," 102:45:32.

6 Armstrong and Aldrin, determined to land rather than abort: Jones, ed.,

*Apollo 11: Lunar Surface Journal*, "The First Lunar Landing," 102:38:42; 102:42:35, and 102:45:32.

7 Armstrong quote on the computer going blank: Don Eyles, *Sunburst and Luminary* (Boston: Four Point Press, 2018), p. 157.

8 "Apollo 11 Lunar Module Powered Descent and Landing on Moon," Smithsonian National Air and Space Museum, posted July 16, 2009, YouTube, 15:59, https://www.youtube.com/watch?v=Kdp5bfcrHME.

9 Jones, ed., *Apollo 11: Lunar Surface Journal*, "The First Lunar Landing," 102:45:31 through 102:46:16.

10 Eyles's reaction to the computer alarms: Eyles, *Sunburst and Luminary*, p. 151. Garman in the staff support room at Mission Control: John R. (Jack) Garman, NASA Johnson Space Center Oral History Project, Edited Oral History Transcript, interviewed by Kevin M. Rusnak, Houston, Texas, March 27, 2001, https://historycollection.jsc.nasa.gov/JSCHistoryPortal/history/oral_histories/GarmanJR/GarmanJR_3-27-01.htm.

Transcript citation on CapCom telling Apollo 11, "Go!": Jones, ed., *Apollo 11: Lunar Surface Journal*, "The First Lunar Landing," 102:42:19.

Garman's handwritten cheat sheet of Apollo computer codes can be seen here: "Interview: Jack Garman, Apollo Guidance Computer," https://www.honeysucklecreek.net/interviews/jack_garman.html.

11 The Apollo command module had two computer keyboards and displays (DSKYs), one on the control panel and one on the side of the capsule alongside the star-sighting telescope and sextant, for easy entry of positioning data.

12 The Apollo Guidance Computer that flew the Moon missions had 2,048 words of erasable memory, what we call random-access memory. It had 36,864 words of fixed memory, what we call read-only memory. In the AGC, a "word" was 15 bits of data. To go from words to bytes, you multiply by 15 bits, then divide by 8 bits per byte. So 2,048 words of RAM translates to 3,840 bytes of data (3.8 KB). And 36,864 words of ROM translates to 69,120 bytes (69.1 KB). Eldon Hall, *Journey to the Moon: The History of the Apollo Guidance Computer* (Reston, VA: American Institute of Aeronautics and Astronautics, 1996), p. 120.

13 James Vincent, "Apple Calls A12 Bionic Chip 'The Smartest and Most Powerful Chip Ever in a Smartphone,'" *The Verge*, September 12, 2018: https://www.theverge.com/circuitbreaker/2018/9/12/17826338/apple-iphone-a12-processor-chip-bionic-specs-speed. Wccf Tech: https://wccftech

.com/apple-a12-specifications-features-cores/. Another comparative measure: The AGC as flown had 5,600 integrated circuits, each of which had the equivalent of six transistors, for a total of 33,600 transistors. The 2018 iPhone Xs has 6.9 billion transistors (Wccf tech, above); iPhone Xs has 200,000 more transistors. iPhone Xs has, roughly, the ability to do 60 million times the transactions per second.

14 Regular at Locke-Ober: Multiple interviews with MIT staff. White suit: pictured at Don Eyles, *Sunburst & Luminary* website, "Picture Gallery," accessed September 10, 2018, https://www.sunburstandluminary.com/SL gallery.html. Ballroom dancing: from a recollection by Apollo astronaut Michael Collins: David L. Chandler, "Michael Collins: 'I could have been the last person to walk on the moon,'" *MIT News*, April 2, 2015, http://news .mit.edu/2015/michael-collins-speaks-about-first-moon-landing-0402. Several of Draper's dancing awards are stored in the Draper collection of papers with the Smithsonian National Air and Space Museum, accessed September 18, 2018, https://sova.si.edu/details/NASM.2001.0019?s= 0&n=10&t=C&q=&i=0#ref39.

15 Given how fascinating Charles Stark Draper's life was, and how much he contributed in technological terms and in terms of the people he helped train, the biographical material on him is startlingly thin. This account comes from two principal sources: Robert A. Duffy, *Charles Stark Draper, 1901–1987: A Biographical Memoir* (Washington, D.C.: National Academy of Sciences, 1994), pp. 121–58, http://www.nasonline.org/publications /biographical-memoirs/memoir-pdfs/draper-charles.pdf; Philip D. Hattis, "How Doc Draper Became the Father of Inertial Guidance," paper presented at the 2018 AAS Guidance and Control Conference, Breckenridge, Colorado, February 2018, https://www.researchgate.net/publication/322963658 _How_Doc_Draper_Became_the_Father_of_Inertia_Guidance.

16 Mark 14 gun sight development and deployment: Duffy, *Charles Stark Draper*, pp. 139–40; Harris, "How Doc Draper Became the Father of Inertial Guidance," pp. 5–7; "Automatic Sight on Our Navy Guns Helps to Win Brilliant Victories," *New York Times*, April 9, 1945, p. 11, https:// timesmachine.nytimes.com/timesmachine/1945/04/09/88212201.html; Associated Press, "Japs Foiled by New Gyro Sight: Makes American Warships Almost Invulnerable to Planes' Attacks," *Arizona Daily Star*, April 9, 1945, p. 10, https://www.newspapers.com/image/162451025.

There is a wonderful and revealing introductory manual for navy gunners learning how to use and maintain their Mark 14 gun sights, available online: "Gun Sight, Mark 14, Gunner's Operating Bulletin No. 2," United States Fleet, Headquarters of the Commander in Chief, https://www.ibiblio.org/hyperwar/USN/ref/Ordnance/GS-Mk14/index.html.

17 Lambright, *Powering Apollo*, p. 26.

18 *Moon Machines*, Episode 3: "The Navigation Computer," directed by Nick Davidson and Christopher Riley, 2008, Dox Productions for The Science Channel, Discovery Communications, clean room material at 10:15. Duffy, *Charles Stark Draper*, pp. 139–40; Norman Sears, telephone interview with the author, January 22, 2016.

19 Details of the flight: Arthur A. Riley, "Pilot Tells How Nature Aided 'Flying Lab' Test," *The Boston Globe*, April 25, 1957, p. 2, https://www.newspapers.com/image/433410673.

Quote from pilot Collins: Bruce Brandon, "Gone West: Col. Charles L. 'Chip' Collins," *Aero-News Network*, Feb. 25, 2015, http://www.aero-news.net/bannertransfer.cfm?do=main.textpost&id=b3bf1d53-d5ec-4afe-9f40-22b20fbd62cf.

20 Thomas Wildenberg, "Origins of Inertial Navigation," *Air Power History*, Winter 2016, pp. 23–24; Don Murray, "'Doc' Draper's Wonderful Tops," *Reader's Digest*, September 1957, p. 64.

21 Richard Witkin, "Device Guides Plane Across U.S. Without Help of Outside Objects," *New York Times*, April 18, 1957, pp. 1, 15, https://timesmachine.nytimes.com/timesmachine/1957/04/18/84959308.html. John B. Knox, "New Jamming-Proof Gyro Pilot Guides Any Kind of Craft Anywhere On Earth," Associated Press, *The Washington Post,* April 18, 1957, pp. A1, A19.

Eric Sevareid flight: John P. Shanley, "Conquest Explains Weather Research," *New York Times*, April 14, 1958, p. 47, https://timesmachine.nytimes.com/timesmachine/1958/04/14/81919069.html.

22 Christopher Morgan, *Draper at 25: Innovations for the 21st Century* (Cambridge, MA: Charles Stark Draper Laboratory Inc., 1998), p. 12 (pages unnumbered), http://citeseerx.ist.psu.edu/viewdoc/download?doi=10.1.1.371.1536&rep=rep1&type=pdf.

23 Ibid., pp. 13–14.

24 "Countdown for Polaris," 1960, posted by Periscope Films, August 4, 2014, YouTube, 10:25, https://www.youtube.com/watch?v=6whcf9R_lds; Hanson W. Baldwin, "Seagoing Missile Base," *New York Times*, January 3, 1960, p. 33, https://timesmachine.nytimes.com/timesmachine/1960/01/03/105172035.html; Mike Billington, "Silent Runners," *Sun-Sentinel*

(Fort Lauderdale, FL), August 1, 1993, p. G1, https://www.newspapers.com/image/238299602/.

25 Harris, "How Doc Draper Became the Father of Inertial Guidance," pp. 10–12; Donald MacKenzie and Graham Spinardi, "The Shaping of Nuclear Weapon System Technology: U.S. Fleet Ballistic Missile Guidance and Navigation: From Polaris to Poseidon," *Social Studies of Science* 18, no. 3 (August 1988): pp. 419–63, https://www.jstor.org/stable/285232; *On Two Fronts: A War-Time Story of the AC Spark Plug Division*, General Motors Corporation, 1943, 64-page booklet, scanned copy via David D. Jackson, "The American Automobile Industry in World War Two," accessed October 1, 2018, http://usautoindustryworldwartwo.com/General%20Motors/ac-sparkplug-twofronts.htm.

26 Hanson W. Baldwin, "2 Polaris Missiles Fired by Submerged Submarine; Hit Mark 1,150 Miles Off," *New York Times*, July 21, 1960, pp. 1, 3 https://timesmachine.nytimes.com/timesmachine/1960/07/21/issue.html.

27 Harris, "How Doc Draper Became the Father of Inertial Guidance," pp. 10–12.

28 Richard H. Battin, transcript of an oral history conducted by Rebecca Wright, April 18, 2000, Lexington, MA, Johnson Space Center Oral History Project, https://historycollection.jsc.nasa.gov/JSCHistoryPortal/history/oral_histories/BattinRH/BattinRH_4-18-00.htm.

29 There are several accounts of Hal Laning, Walt Trageser, and Richard Battin working on the MIT Mars probe, the navigation it would require and the computer that would fly it. Each of those accounts has some unique details. David G. Hoag, "The History of Apollo Onboard Guidance, Navigation and Control," *Journal of Guidance, Control, and Dynamics* 6, no. 1 (January–February 1983): 4; David A. Mindell, *Digital Apollo* (Cambridge, MA: MIT Press, 2011), pp. 99–101; Richard H. Battin, "Second Breakwell Memorial Lecture: 1961 and All That," *Acta Astronautica* 39, no. 6 (1996): 409–11, https://ac.els-cdn.com/S0094576596001531/1-s2.0-S0094576596001531-main.pdf.

30 "List of Lunar Probes," Wikipedia, updated January 4, 2019, https://en.wikipedia.org/wiki/List_of_lunar_probes#1958%E2%80%931960. Elizabeth Howell, "A Brief History of Mars Missions," March 17, 2015, Space.com, https://www.space.com/13558-historic-mars-missions.html.

31 MIT staff member aboard a Polaris submarine: Robert G. Chilton, transcript of oral history conducted by Summer Chick Berger, April 5, 1999, Johnson Space Center Oral History Project, p. 30, https://historycollection

.jsc.nasa.gov/JSCHistoryPortal/history/oral_histories/ChiltonRG/Chilton RG_4-5-99.htm.

The MIT Mars computer, built and working, August 1961: R. L. Alonso, A. I. Green, H. E. Maurer, and R. E. Oleksiak, "R-358: A Digital Control Computer: Developmental Model 1B," Cambridge, MA: Instrumentation Laboratory, Massachusetts Institute of Technology, for the U.S. Air Force, April 1962, pp. 9, 35, https://www.ibiblio.org/apollo/hrst/archive/1712.pdf.

32 Accounts of the selection of the MIT Instrumentation Lab include Mindell, *Digital Apollo*, pp. 104–8; Morgan, *Draper at 25*, p. 31; Courtney G. Brooks, James M. Grimwood, and Loyd S. Swenson Jr., *Chariots for Apollo: A History of Manned Lunar Spacecraft* (Washington, D.C.: NASA, 1979), pp. 38–41.

Webb had confidence in Draper: Lambright, *Powering Apollo*, p. 106.

Seamans called Draper his "mentor": Seamans, *Aiming at Targets*, pp. 26-28, 45-49.

An image of the yellowed telegram to MIT has been posted online by MIT, via Ian A. Waltz, "Perspective Is Everything," *AeroAstro Magazine*, 2008–9, http://web.mit.edu/aeroastro/news/magazine/aeroastro6/intro.html. A larger version of the image itself is at http://web.mit.edu/aeroastro/news/magazine/aeroastro6/img/apollotelegram-lrg.jpg.

33 Hugh L. Dryden was brilliant and an important figure in the history of rocketry and flight in the U.S. He earned a PhD in physics and math from Johns Hopkins University at age 20, in 1919, the youngest person ever to receive a PhD from Hopkins. He supervised development of the X-15 rocket plane and was director of the National Advisory Committee on Aeronautics, the agency that preceded NASA. He was acting administrator of NASA until Webb was appointed by Kennedy, and Webb insisted to Kennedy that Dryden be allowed to stay on as deputy administrator if Webb were to take the top job, to provide both continuity and technical expertise in the administrator's office. Lambright, *Powering Apollo*, p. 84.

34 "In June": the time of the meeting is offered differently in different accounts, but NASA's own history says specifically that Draper went to Washington in June and met with Webb, Seamans, and Dryden, Brooks, Grimwood, and Swenson in *Chariots for Apollo,* p. 41. The dialogue is reproduced to varying degrees of detail in Brooks, Grimwood, and Swenson, *Chariots for Apollo*, p. 41; Eyles, *Sunburst and Luminary*, p. 35; Morgan, *Draper at 25*, p. 31. Sometimes when Draper told the story of assuring

Webb that MIT could do the Apollo navigation, Draper said the software would be ready "when you need it," and sometimes he said "before you need it." In the account from the Smithsonian's Paul Ceruzzi, Draper said "before you need it." Paul E. Ceruzzi, *Beyond the Limits: Flight Enters the Computer Age* (Cambridge, MA: MIT Press, 1989), p. 97. In an account from the annual magazine of MIT's Department of Aeronautics and Astronautics, which includes quotes from Draper recalling the meeting, he also said "before you need it." John Tylko, "MIT and Navigating the Path to the Moon," *AeroAstro*, no. 6 (2008–9), p. 1, http://web.mit.edu/aeroastro/news/magazine/aeroastro6/aeroastro6.pdf.

Given Draper's enthusiasm for the project and his style of salesmanship, "before you need it" would seem to fit his manner.

35 Draper's letter to Seamans: "MIT and Project Apollo," *MIT Institute Archives and Special Collections*, accessed October 4, 2018, https://libraries.mit.edu/archives/exhibits/apollo/. An image of the letter itself in the MIT archives: "Letter from Charles Stark Draper to Robert C. Seamans, Jr., November 21, 1961," *MIT Institute Archives and Special Collections*, https://libraries.mit.edu/archives/exhibits/apollo/letter.html.

Brigadier General Donald Flickinger's role: Tom Wolfe, *The Right Stuff* (New York: Picador, 1979), p. 57.

The reaction at NASA headquarters, via Robert Seamans: Robert Seamans, transcript of an oral history, #3, conducted by Carol Butler, June 22, 1999, Johnson Space Center Oral History Project, pp. 13–14, https://historycollection.jsc.nasa.gov/JSCHistoryPortal/history/oral_histories/SeamansRC/SeamansRC_6-22-99.htm.

36 The aerospace heritage was deep all around. AC Spark Plug had provided spark plugs for both Charles Lindbergh and Amelia Earhart; Kollsman Instrument had been founded by Paul Kollsman, who invented the altimeter, and so, before Doc Draper, had first enabled "instrument flying" of airplanes.

37 Hoag, "The History of Apollo Onboard Guidance, Navigation and Control," p. 6.

38 W. David Woods provides a detailed account of the return of the Apollo command and service modules from the Moon, starting with a section titled, "The Long Fall to Earth," right through reentry. W. David Woods, *How Apollo Flew to the Moon* (Chichester, UK: Praxis Publishing, 2009), pp. 444–503; discussion of velocity on the way home, p. 468.

The reentry corridor is 40 miles wide, from this NASA video: "Spacecraft Communications: 'The Vital Link,' 1967 NASA Goddard Space Flight Center," posted by Jeff Quitney, March 15, 2017, YouTube, 28:41, 26:00, https://www.youtube.com/watch?v=4WP9dCvsMUY&feature=youtu.be.

39 Mindell, *Digital Apollo*, p. 158.

40 David Scott, "Speech at the Opening of the Computer Museum," Marlboro, MA, June 10, 1982, *KLabs.org*, http://www.klabs.org/history/history_docs/ech/agc_scott.pdf.

41 Hall, *Journey to the Moon*, p. 139.

42 IBM computers in Mission Control: James E. Tomayko, "Computers in Spaceflight," pp. 245, 248, 249–50.

43 Reparability as a strategy: Brooks, Grimwood, and Swenson, *Chariots for Apollo*, p. 135. Soldering iron: Hall, *Journey to the Moon*, p. 92.

44 Lunar module weight and repair issues: Kelly, *Moon Lander*, pp. 72–73, 117–20. Alan Shepard on in-flight maintenance: Jim Miller, transcript of conference proceedings, "Apollo Guidance Computer History Project: Second Conference: Trust in Automatic Systems," Cambridge, MA, September 14, 2001, https://authors.library.caltech.edu/5456/1/hrst.mit.edu/hrs/apollo/public/conference2/automatic.htm.

45 Hoag, "The History of the Apollo Onboard Guidance, Navigation, and Control," p. 9.

46 Eyles, *Sunburst and Luminary*, pp. 52–53.

47 *Moon Machines*, Episode 3: "The Navigation Computer," Battin "one-cubic-foot" story at 13:35; Battin, oral history, https://historycollection.jsc.nasa.gov/JSCHistoryPortal/history/oral_histories/BattinRH/BattinRH_4-18-00.htm.

48 Hall, *Journey to the Moon*, pp. 87–88.

49 "The Apollo Guidance Computer, Part One: Eldon Hall," recorded June 10, 1982, posted by Computer History Museum, October 2, 2014, YouTube, quote at 19:00, https://www.youtube.com/watch?v=PbX8OtPe3eY, 56:09.

50 Hall, *Journey to the Moon*, pp. 79–82; Eldon C. Hall, "Autobiography," *IEEE Annals of the History of Computing*, April–June 2000, p. 27.

51 Cline Fraser, transcript of conference proceedings, "Apollo Guidance Computer History Project: First Conference: Joe Gavin's Introduction," Cambridge, MA, July 27, 2001, https://authors.library.caltech.edu/5456/1/hrst.mit.edu/hrs/apollo/public/conference1/gavin-intro.htm.

52 “The Apollo Guidance Computer, Part One: Eldon Hall,” quote at 27:10. The prices MIT paid for early microchips are listed in a chart in Eldon Hall’s meticulous book—every individual chip purchase, by date, with the amount purchased, the vendor, and the price paid per chip: 13,966 chips purchased in 1962 and 1963, ranging from an order of 4 to an order of 4,100. Significantly, Hall also notes whether the order was delivered on time or was late. Hall, *Journey to the Moon*, p. 80.

53 Eldon Hall, transcript of conference proceedings, “Apollo Guidance Computer History Project: First Conference: Joe Gavin’s Introduction.”

54 Most of the story of the MIT IL’s leap to integrated circuits is well told by Hall in *Journey to the Moon*, pp. 79–85. Go-ahead letter from NASA: Hall, *Journey to the Moon*, p. 188. Hall quote, “heated debates”: “The Apollo Guidance Computer, Part One: Eldon Hall,” quote at 28.50.

55 “Manufacturers’ Sales and Retail Value of Home Appliances, 1950 to 1962,” p. 814, in *Statistical Abstract of the United States 1963*.

56 Hall, “Autobiography,” p. 30.

The first commercial computer to use integrated circuits was the RCA Spectrum 70, introduced in 1965. The first IBM model to use integrated circuits was the System/370 introduced in 1970. Paul. E. Ceruzzi, *A History of Modern Computing* (Cambridge, MA: MIT Press, 2003), p. 163.

57 The MIT IL’s standards for quality, testing program, and results are reported in Jayne Partridge and L. David Hanley, “The Impact of the Flight Specifications on Semiconductor Failure Rates,” *Proceedings of the 6th Annual Reliability Physics Symposium*, Los Angeles, November 6–8, 1967, pp. 20–30, https://ieeexplore.ieee.org/document/4207753. The Freon “bath” test is described in Mindell, *Digital Apollo*, p. 133, and described in technical terms in M. D. Holley, W. L. Swingle, S. L. Bachman, et al., “Apollo Experience Report: Guidance and Control Systems: Primary Guidance, Navigation, and Control System Development,” NASA Technical Note D-8277, Lyndon B. Johnson Space Center, May 1976, https://ntrs.nasa.gov/archive/nasa/casi.ntrs.nasa.gov/19760016247.pdf.

58 Patridge and Hanley, “The Impact of Flight Specifications on Semiconductor Failure Rates,” pp. 22, 24, 26.

59 Hall, *Journey to the Moon*, p. 1.

60 Ibid., focused purchased, p. 84; Fairchild succeeded by Philco, p. 23.

61 Raymond D. Speer, “Strict Control Kept Out Semiconductor Flaws,” *Electronic Design* 17 (August 16, 1969), reproduced in full in Hall, *Journey to the Moon*, pp. 147–48.

62 Mindell, *Digital Apollo*, p. 158.

63 Ibid., pp. 96, 108.

64 Tomayko, "Computers in Spaceflight," p. 29.

65 Details of the Manned Space Flight Network from Sunny Tsiao, *"Read You Loud and Clear": The Story of NASA's Spaceflight Tracking and Data Network*, NASA SP-2007-4232 (Washington, D.C.: NASA, 2008), accessed 15 October 2018, http://www.rvtreasure.com/NASA/STADAN%20Network%20history.pdf.

Cost information for the tracking network from William R. Corliss, "Histories of the Space Tracking and Data Acquisition Network (STADAN), the Manned Space Flight Network (MSFN) and the NASA Communications Network (NASCOM)," CR-140390 (Washington, D.C.: NASA, June 1974), p. 353.

66 Mindell, *Digital Apollo*, p. 138.

67 Ibid., p. 165. The comparison of the displays to a slide rule is Mindell's.

68 The story of the verb-noun structure, and its creation, is told several places, but at least one of them gets it wrong. The correct versions include: Mindell, *Digital Apollo*, p. 166. Ramon Alonso, transcript of conference proceedings, "Apollo Guidance Computer History Project: First Conference: Human-Machine Interface," Cambridge, MA, July 27, 2001, https://authors.library.caltech.edu/5456/1/hrst.mit.edu/hrs/apollo/public/conference1/interface.htm.

Another MIT staff member, Hugh Blair-Smith, attributes the verb-noun idea, incorrectly, to other colleagues at MIT in an online essay from 20 years ago, "Hugh Blair-Smith's Annotations to Eldon Hall's *Journey to the Moon*," https://www.ibiblio.org/apollo/hrst/HughBlairSmithsAnnotations.html.

In an interview, Blair-Smith acknowledges that "indeed Ramon Alonso is inventor" of the verb-noun syntax. Hugh Blair-Smith, telephone interview with the author, October 16, 2018.

69 If you want to know what using a personal computer with Microsoft's early operating system, MS-DOS, was like, or you want to be reminded what it was like, see Chris Hoffman, "PCs before Windows: What Using MS-DOS Was Actually Like," *How-To Geek*, May 11, 2014, https://www.howtogeek.com/188980/pcs-before-windows-what-using-ms-dos-was-actually-like/.

70 Scott, "Speech at the Opening of the Computer Museum."

71 Tomayko, "Computers in Spaceflight," p. 123 (13,000 keystrokes), p. 58 ("like playing a piano").

## 5: The Man Who Saved Apollo

1 Ed Copps, quoting Bill Tindall, in "Bill Tindall and Conflicts within Apollo," conference proceedings, "Apollo Guidance Computer History Project: Fourth Conference," Cambridge, MA, September 6, 2002, https://authors.library.caltech.edu/5456/1/hrst.mit.edu/hrs/apollo/public/conference4/tindall.htm.

Howard W. (Bill) Tindall Jr. was a legendary expert on rendezvous and orbital mechanics, a senior NASA engineer in Houston, and a man with several misleading and opaque job titles. During the critical late planning years of Apollo, 1967 to 1970, his official title was chief of Apollo data priority coordination, which was often shortened to chief of Apollo mission techniques. He was also deputy chief, and then chief of the Mission Planning and Analysis Division in Houston. In practice, it was Tindall's job to figure out the navigation techniques necessary to fly to the Moon and back and make sure everything from the computer software to Mission Control was lined up to do that. See Murray and Cox, *Apollo*, p. 284.

2 Associated Press, "Venus Shot Fails as Rocket Strays," *New York Times*, July 23, 1962, pp. 1, 2, https://timesmachine.nytimes.com/timesmachine/1962/07/23/issue.html.

Continued to send signals until it hit the Atlantic Ocean at 357 seconds: N. A. Renzetti, "Tracking and Data Acquisition Support for the Mariner Venus 1962 Mission: Technical Memorandum No. 33-212," NASA, Jet Propulsion Laboratory, July 1, 1965, p. 9, https://ntrs.nasa.gov/archive/nasa/casi.ntrs.nasa.gov/19650023548.pdf.

3 "NASA Science: Solar System Exploration: Missions: Mariner 2," accessed October 13, 2018, https://solarsystem.nasa.gov/missions/mariner-02/in-depth/.

4 There are many explanations for what happened. From the mainstream media at the time, the best is Marvin Miles, "'Hyphen' Blows Up Rocket," *Los Angeles Times*, August 5, 1962, p. H1, https://www.newspapers.com/image/381425643/.

NASA has no Mariner 1 "failure panel" report available online. The most thorough and authoritative scholarly explanation is from Paul Ceruzzi of the Smithsonian in *Beyond the Limits*, pp. 202–3. The relevant passage can be read online: https://arstechnica.com/civis/viewtopic.php?t=862715.

5 Gladwin Hill, "For Want of Hyphen Venus Rocket is Lost," *New York Times*, July 28, 1962, pp. 1, 2, https://timesmachine.nytimes.com/timesmachine

/1962/07/28/87314569.html. Marvin Miles, "'Hyphen' Blows Up Rocket," *Los Angeles Times*, August 5, 1962, sec. H, p. 1. Arthur C. Clarke, *The Promise of Space* (New York: Harper & Row, 1968), p. 225.

6 George Mueller, at Space Technology Laboratories (STL), from Yvette Smith, ed., "NASA History: Remembering George Mueller, Leader of Early Human Spaceflight," NASA, October 15, 2015, https://www.nasa.gov/feature/remembering-george-mueller-leader-of-early-human-spaceflight. STL responsibility for Mariner 1 guidance software: Hill, "For Want of a Hyphen Venus Rocket Is Lost." Battin, oral history, https://historycollection.jsc.nasa.gov/JSCHistoryPortal/history/oral_histories/BattinRH/BattinRH_4-18-00.htm. Battin tells the same story in Richard H. Battin, "Some Funny Things Happened on the Way to the Moon," *Journal of Guidance, Control, and Dynamics* 25, no. 1 (January–February 2002), p. 4.

7 The mission rules for each of the Apollo missions are posted publicly: "Mission Rules," *Apollo Lunar Surface Journal*, accessed October 20, 2018, https://www.hq.nasa.gov/alsj/alsj-MissionRules.html.

8 Shirley Hinson, who went to work for NASA calculating and then hand-plotting flight trajectories in 1959 at about age 22, describes this early era well in her oral history. Hinson went on to be part of mission planning for Gemini, Apollo, and Skylab, and then was a section chief at the Johnson Space Center. Shirley H. Hinson, oral history transcript, NASA Johnson Space Center Oral History Project, interviewed by Rebecca Wright, Louisburg, NC, May 2, 2000, https://historycollection.jsc.nasa.gov/JSCHistoryPortal/history/oral_histories/HinsonSH/HinsonSH_5-2-00.htm.

9 Kranz, *Failure Is Not An Option*, p. 156.

10 In Chris Kraft's memoir *Flight* (New York: Plume, 2002), p. 276, the timing of his conversation with George Low is vague. It appears Kraft thinks he first sent Tindall to MIT in spring 1967. But memos and other contemporaneous accounts make it clear Tindall started going to MIT almost weekly in spring 1966. Malcolm Johnston, telephone interview #1, December 14, 2015.

11 Tindall does orbital calculations of Echo 1: "Rendezvous Planner: Howard Wilson Tindall Jr.," *New York Times*, December 16, 1965, p. 29, https://timesmachine.nytimes.com/timesmachine/1965/12/16/95920363.html. Details of Echo 1: Charles Q. Choi, "1st Communications Satellite: A Giant Space Balloon 50 Years Ago," *Space.com*, August 18, 2010, https://www.space.com/8973-1st-communication-satellite-giant-space-balloon-50-years.html.

The quote from Tindall's mother comes from the reprint of a profile of Tindall from the *Cape Codder* newspaper, December 23, 1965, reprinted in Brown University's alumni magazine. Tindall's mother's name, beyond "Mrs. Tindall Sr.," is not provided. "Something, Indeed, to Tell His Children," *Brown Alumni Monthly*, January 1966, p. 6, https://archive.org/details/brownalumnimonth664brow/page/n7.

12 Tindall's strategic effort to always be second-in-command: Catherine T. Osgood, oral history transcript, NASA Johnson Space Center Oral History Project, interviewed by Rebecca Wright, Houston, TX, November 15, 1999, https://historycollection.jsc.nasa.gov/JSCHistoryPortal/history/oral_histories/OsgoodCT/OsgoodCT_11-15-99.htm; Shirley H. Hinson, oral history transcript, NASA Johnson Space Center Oral History Project, interviewed by Rebecca Wright, Louisburg, NC, May 2, 2000, https://historycollection.jsc.nasa.gov/JSCHistoryPortal/history/oral_histories/HinsonSH/HinsonSH_5-2-00.htm; Jane Tindall, telephone interview with the author, January 13, 2016.

13 Bill Tindall's work, problems and thinking are documented in detail in memos he wrote, day by day during Apollo, and that were so popular and appealing they acquired the nickname Tindallgrams. There were hundreds of these memos, written with such clarity and immediacy that people across NASA looked forward to them, and they were invaluable, helping to shape important decisions about the Apollo missions. There is no comprehensive collection of the Tindallgrams on the internet. There are, rather, three separate major sets, which overlap, and which are of varying quality, in terms of the legibility of their scanning.

For purposes of citation in these Notes, they are labelled Tindallgrams #1, Tindallgrams #2, and Tindallgrams #3. In the references, the memos will be cited by title, date, NASA memo number, and page number in the collection, with all page numbers refering to the pdf page numbers, for consistency. Tindall titled the memos mostly without capitalization, and in the references, we follow his style.

Tindallgrams #1 is a collection assembled by Malcolm Johnston, an engineer at the Instrumentation Lab who worked closely with Tindall, and who assembled this partial collection as a tribute to him after his death. It includes 186 Tindallgrams on 380 pages, and the scan quality is good. https://www.hq.nasa.gov/alsj/tindallgrams02.pdf.

Tindallgrams #2 is a collection scanned in by a NASA historian, Glen Swanson. It is more complete, at 525 pages, but suffers from uneven scan quality, and from the fact that the memos are not in date order. https://www.hq.nasa.gov/alsj/tindallgrams01.pdf.

Tindallgrams #3 is a collection labelled "KSC" (Kennedy Space Center), broken into four files by year, 1967 to 1970. The scan quality and organization are uneven, but the total number of Tindallgrams is large—the four years total 889 pages.

1967: https://www.hq.nasa.gov/alsj/1967_tindallgrams.pdf
1968: https://www.hq.nasa.gov/alsj/1968_tindallgrams.pdf
1969: https://www.hq.nasa.gov/alsj/1969_tindallgrams.pdf
1970: https://www.hq.nasa.gov/alsj/1970_tindallgrams.pdf

A few Tindallgrams are available only in other collections outside these, and those references will be called Additional Tindallgrams, and a URL will be provided.

The data about how much over-capacity MIT's program were from: "Spacecraft computer requirements for AS-207/208, AS-503, and AS-504, May 12, 1966, #66-FM1-59, Tindallgrams #1, pdf, p. 14. "Apollo spacecraft computer program development newsletter," May 31, 1966, #66-FM1-68, Additional Tindallgrams, pdf, p. 2, http://web.mit.edu/digitalapollo/Documents/Chapter7/tindallgrams.pdf.

Statistics on increase in AGC computer memory from 1962 to 1966: Hoag, *The History of Apollo Onboard Guidance, Navigation and Control*, p. 8.

14 Fred Martin, in "Meetings with Bill Tindall," conference proceedings, "Apollo Guidance Computer History Project: First Conference," Cambridge, MA, July 17, 2001, https://authors.library.caltech.edu/5456/1/hrst.mit.edu/hrs/apollo/public/conference1/tindall.htm.

15 The following newspaper stories and advertisements use the word "softwear" instead of "software" in the context of computer programming, all published between 1962 and 1971, including a story in the *New York Times* about whether software can be patented, which spells the word both ways in the course of the story. Harold Chucker, "Courage, Cash Needed in Making Computers," *Minneapolis Star*, March 7, 1962, p. 5D, https://www.newspapers.com/image/187998041; UPI, "Softwear's Her Forte," *Quad City Times* (Davenport, IA), May 24, 1963, p. 10, https://www.newspapers.com/image/302705284; "Computers: What Has Made Systems Programming Corporation the Nation's Fastest Growing Computer Softwear Firm?," advertisement, *Los Angeles Times*, January 26, 1964, p. 12, https://

www.newspapers.com/image/381551029/?terms=softwear%2Band%2Bcomputer; "Computer Softwear Professionals, Control Data Corporation, Data Products Group," advertisement, *St. Louis Post-Dispatch*, November 4, 1965, p. 6C, https://www.newspapers.com/image/139123536; Cornelia K. Wyatt, "Japan Agrees to Let the U.S. Enter Market for Computers," *New York Times*, February 19, 1967, p. F7, https://timesmachine.nytimes.com/timesmachine/1967/02/19/83576259.html; "Capital Commerce: Heads Bethesda Firm," *Washington Post*, July 10, 1968, p. F8; Stacy V. Jones, "Computer Software Unpatentable," *New York Times*, October 23, 1968, p. 59, https://timesmachine.nytimes.com/timesmachine/1967/02/19/83576259.html; Dan Morgan, "Polish Passion for Technology May Breed Political Change," *Washington Post*, October 15, 1970, p. A16; Maurice Corina, "British Computer Chief Calm in Crisis," *New York Times*, September 5, 1971, section 5, p. 2, https://timesmachine.nytimes.com/timesmachine/1971/09/05/79152297.html.

16 Lambright, *Powering Apollo*, pp. 5–10.

17 It's unclear when exactly Margaret Hamilton started using the phrase "software engineering" at MIT, and she may well have coined it herself without hearing or seeing it elsewhere. Her comments are from this interview: Jaime Rubio Hancock, "Margaret Hamilton, the Engineer Who Took the Apollo to the Moon," *Verne*, December 25, 2014, https://medium.com/@verne/margaret-hamilton-the-engineer-who-took-the-apollo-to-the-moon-7d550c73d3fa.

It is worth tracking the use of the phrase "software engineering," Hamilton's experience aside. A comprehensive search of newspaper databases reveals the first use of "software engineering" that can be discovered in the popular press was in an advertisement, "AASLI, systems management, data processing, software engineering R&D, Associated Aero Science Laboratories, Inc.," in the *Los Angeles Times*, January 9, 1966, p. I-7, https://www.newspapers.com/image/382387460. It was used commonly in computer firm employment ads starting in 1966.

The first academic or professional use of "software engineering" in a public context was apparently by the Harvard professor and information scientist Anthony Oettinger, who used it in a wry and funny scientific lecture on March 13, 1967, where he argued passionately both for the maturation of software as part of computer science and for the interconnections between computing, software, and traditional engineering. "The notion of software engineering is, thank goodness, beginning to be heard of more and more. . . . By being cut off from engineering departments, computer science departments lose sight of the fact that their symbol systems have a mission, which is to make machines work, to make them work efficiently and economically as well as elegantly." Anthony Oettinger, "The Hardware-Software Complementarity," lecture, "Academic Role of Computers," Annual Meeting of the Division of Mathematical Sciences, National Academy of Sciences, March 13, 1967, reprinted in *Communications of the Association for Computing Machinery* 10, no. 10 (October 1967), pp. 604–6.

In October 1968 the North Atlantic Treaty Organization held a conference in Munich on software engineering. A report on that conference is online: http://homepages.cs.ncl.ac.uk/brian.randell/NATO/nato1968.PDF.

The first reference to the phrase "software engineering" in the editorial content of newspapers that I could discover was in a mention that an IBM employee, Richard C. Hastings, would be attending that conference: "U. of R. Names 2 Top Aides," *Democrat and Chronicle* (Rochester, NY), October 8, 1968, p. 1C, https://www.newspapers.com/image/136854311. The phrase does not appear in the pages of the *New York Times* until October 1975.

18 Hoag, "The History of Apollo Onboard Guidance, Navigation and Control," p. 10.

19 Malcolm Johnston, telephone interview #2, December 18, 2015.

20 Ed Copps, in "Bill Tindall and Conflicts within Apollo," conference proceedings, "Apollo Guidance Computer History Project: Fourth Conference," Cambridge, MA, September 6, 2002, https://authors.library.caltech.edu/5456/1/hrst.mit.edu/hrs/apollo/public/conference4/tindall.htm.

21 There are two accounts of this meeting to air unhappiness at the Instrumentation Lab during Tindall's early days: Martin, quoted in *Moon Machines*, Episode 3: "The Navigation Computer," at 26:00; Martin, in "Meetings with Bill Tindall," conference proceedings, "Apollo Guidance Computer History Project: First Conference," https://authors.library.caltech.edu/5456/1/hrst.mit.edu/hrs/apollo/public/conference1/tindall.htm.

22 The quotes are from the Tindallgram cited below, but in the collections it is dated May 12, 1966, which must be a mislabelling, because it recounts events that happened on May 13 and 14: "Spacecraft computer requirements for AS-207/208, AS-503, and AS-504, May 12, 1966, #66-FM1-59, Tindallgrams #1, PDF p. 14.

23 "Apollo spacecraft computer program development newsletter," May 31, 1966, #66-FM1-68, Additional Tindallgrams, PDF p. 2, http://web.mit.edu/digitalapollo/Documents/Chapter7/tindallgrams.pdf.

24 "Spacecraft Computer Program Development Newsletter," February 27, 1967, #67-FM1-18, Tindallgrams #3 (1967), pdf, pp. 206–08., http://www.collectspace.com/resources/tindallgrams/1967_tindallgrams.pdf.

25 "Spacecraft computer program status report," June 2, 1966, #66-FM1-70, Tindallgrams #1, PDF, pp. 24–25.

26 Johnston, telephone interview #1.

27 "Lunar orbit revolution counter for 'C'," October 2, 1968, #68-PA-T-213A, Tindallgrams #1, pdf, p. 163.

28 "Let's have no unscheduled water dumps on the F mission," February 24, 1968, #69-PA-T-31A, Tindallgrams #1, PDF, p. 220.

29 "How to land next to Surveyor—a short novel for do-it-yourselfers," August 1, 1969, #69-PA-T-114A, Tindallgrams #1, pdf, pp. 311–15.

"Astronauts Pay a Visit to Surveyor 3," *Apollo 12*, April 17, 2014, https://www.nasa.gov/content/astronauts-pay-a-visit-to-surveyor-3.

The precise figure that Apollo 12 landed 535 feet from Surveyor 3 comes from Orloff, *Apollo by the Numbers*, p. 116.

For the record, Tindall wrote a fresh Tindallgram three months later—and just 10 days before Apollo 12 blasted off—revising his prediction: "Based on things that have happened since [writing that memo] . . . my feeling now is that as long as the systems work as well as they have in the past, we have a pretty good chance of landing near the Surveyor. And I would rather be on record as predicting that, than predicting a miss." "Apollo 12 Descent—Final comments," November 4, 1969, #69-PA-T-142A, Tindallgrams #1, pdf, p. 353.

30 This calculation is explained in Chapter 1, Note 58, p. 350–52.

31 "Tindallgrams," *Apollo Lunar Surface Journal*, https://www.hq.nasa.gov/alsj/alsj-Tindallgrams.html.

32 "Apollo spacecraft computer programs—or, a bucket of worms," June 13, 1966, #66-FM1-75, Additional Tindallgrams, pdf, pp 4–7. http://web.mit.edu/digitalapollo/Documents/Chapter7/tindallgrams.pdf.

33 "Another Apollo spacecraft computer program status report," July 1, 1966, 66-FM1-78, Additional Tindallgrams, pdf, pp. 8–11, http://web.mit.edu/digitalapollo/Documents/Chapter7/tindallgrams.pdf.

34 "Program Development Plans are coming!!," October 11, 1966, #66-FM1-124, Tindallgrams #1, pdf, pp. 55–57.

35 "AGC program for AS-501/502—Final status report," November 3, 1966, #66-FM1-141, Tindallgrams #1, pdf, p. 74.

36 "Mission rules needed for use with AGC self-check," September 21, 1966, #66-FM1-109, Tindallgrams #1, pdf, p. 44.

37 "More about computer self check," January 25, 1967, #67-FM1-11, Tindallgrams #3 (1967), pdf, p. 218.

38 "LM DPS low level light fixing," November 25, 1968, #68-PA-T-257A, Tindallgrams #3 (1968), p. 304.

39 "AS-206 Spacecraft Computer Program Newsletter," January 31, 1967, Tindallgrams #1, pdf, pp. 102–3.

40 Murray and Cox, *Apollo*, p. 288.

41 "Bucket of worms": "Apollo spacecraft computer programs—or, a bucket of worms," June 13, 1966, #66-FM1-75, Additional Tindallgrams, pdf, pp. 4–7. http://web.mit.edu/digitalapollo/Documents/Chapter7/tindallgrams.pdf.

"Holy waste of time, Batman!": "AGS Program Status for AS-278," November 14, 1966, #66-FM1-148, Tindallgrams #1, pdf, p. 78.

"Some Things Ed Copps Is Worried About": "Some Things Ed Copps Is Worried About," May 6, 1967, #67-FM1-36, Tindallgrams #3 (1967), pdf, p. 154.

"A rather unbelievable proposal. . .": "LM Rendezvous Radar Is Essential," August 1, 1968, #68-PA-T-183A, Tindallgrams #3 (1968), pdf, p. 427.

42 Murray and Cox, *Apollo*, p. 288.

43 "Descent engine gimbal polarity error," April 21, 1967, #67-FM1-32, Tindallgrams #3 (1967), pdf, p. 164.

44 Howard W. Tindall in "Managing the Moon Program: Lessons Learned from Project Apollo, Proceedings of an Oral History Workshop," Johnson Spaceflight Center, Houston, TX, July 21, 1989, p. 23, https://ntrs.nasa.gov/archive/nasa/casi.ntrs.nasa.gov/19990053708.pdf.

45 Early discussions at MIT about using the lunar module as a "lifeboat" to bring the astronauts home: "LGC computer requirements to provide DPS backup of SPS," September 21, 1966, #66-FM1-110, Tindallgrams #1, pdf, p. 45. "Use of DPS for lunar mission aborts," August 24, 1966, 67-FM-T-63, Tindallgrams #3 (1967), pdf, p. 117.

As the missions to the Moon grew closer, the topic of using the lunar module and its engine to get home in an emergency came up repeatedly. See, for instance: "F and G mission cis-lunar and abort plan," January 21, 1969, 69-PA-T-10A, Tindallgrams #1, pdf, p. 198.

Tindall passing on Faget's analysis: "LM propulsion of the LM/CSM configuration as an SPS backup technique," July 31, 1968, #68-PA-T-175A, Tindallgrams #3 (1968), pdf, p. 431-434.

46 Tomayko, "Computers in Spaceflight," p. 52.

47 Norm Sears, telephone interview #4 with the author, February 3, 2016 ("commanded enough respect) and interview #5, February 10, 2016 ("it was a gift").

48 "In which is described the Apollo spacecraft computer programs currently being developed," March 24, 1967, #67-FM1-24, Tindallgrams #3 (1968), pdf, p. 188.

49 Hoag comment: David Hoag, "Meetings with Bill Tindall," conference proceedings, "Apollo Guidance Computer History Project: First Conference," Cambridge, MA, July 27, 2001, https://authors.library.caltech.edu/5456/1/hrst.mit.edu/hrs/apollo/public/conference1/tindall.htm. Copps comment: Ed Copps, "Bill Tindall and Conflicts within Apollo," conference proceedings, "Apollo Guidance Computer History Project: Fourth Conference," Cambridge, MA, September 6, 2002, https://authors.library.caltech.edu/5456/1/hrst.mit.edu/hrs/apollo/public/conference4/tindall.htm. Johnston comment: Malcolm Johnston, telephone interview #3, December 29, 2015.

50 "In which is described the Apollo spacecraft computer programs currently being developed," March 24, 1967, #67-FM1-24, Tindallgrams #3 (1968), pdf, p. 188.

51 Battin, oral history, April 18, 2000.

52 Hoag, "The History of Apollo Onboard Guidance, Navigation and Control," p. 10.

53 James A. Hand, ed., "MIT's Role in Project Apollo: Volume 1: Project Management, Systems Development, Abstracts and Bibliography," October 1971, MIT, Charles Stark Draper Laboratory, p. 18.

54 Kipp Teague, "Project Mercury Drawings and Technical Diagrams," *NASA*, accessed November 10, 2018, https://history.nasa.gov/diagrams/mercury.html; *Project Gemini Familiarization Manual: Supplement*, MAC Control No. C-119162, MacDonnell, September 30, 1965, pp. 8-212 to 8-219, https://www.ibiblio.org/apollo/Documents/GeminiManualVol1Sec2.pdf.

55 Eyles, *Sunburst and Luminary*, p. 137.

56 Scott, "Speech at the Opening of the Computer Museum."

57 Margaret Hamilton, "The Language as a Software Engineer," speech, 40th International Conference on Software Engineering, May 31, 2018, Gothenburg, Sweden, YouTube, https://www.youtube.com/watch?v=ZbVOF0Uk5lU, at 11:50.

58 The most detailed and charming account of the creation of George is from Don Eyles's memoir, *Sunburst and Luminary* (pp. 121–22), which includes both the mathematical backstory and an explanation of why the compiler is named George: "Laning said the name came from the expression, 'Let George do it'—the title of a radio drama featuring a detective who boasted, 'If the job's too tough for you to handle, you've got a job for me, George Valentine.'" Laning was born on February 14, 1920. Eyles says Laning wrote George in 1952. An academic paper by Laning's friend and colleague Richard Battin says it was "finished in March 1953": Richard H. Battin, "On Algebraic Compilers and Planetary Fly-By Orbits," *Acta Astronautica* 38, no. 12 (1996), pp. 895–96.

59 Larry Hardesty, "Apollo's Rocket Scientists," *MIT Technology Review*, November–December 2009, https://www.technologyreview.com/s/415796/apollos-rocket-scientists/.

60 Lickly quote and Battin quote: Ibid.

61 Ed Copps, "Documents and Restarts," conference proceedings, "Apollo Guidance Computer History Project: Fourth Conference," Cambridge, MA, September 6, 2002, https://authors.library.caltech.edu/5456/1/hrst.mit.edu/hrs/apollo/public/conference4/restarts.htm.

62 Eyles, *Sunburst and Luminary*, pp. 78–83; Mindell, *Digital Apollo*, pp. 150–51.

63 Copps quotes: Ed Copps, "Documents and Restarts." Copps paper: Edward M. Copps Jr., "Recovery from Transient Failures of the Apollo Guidance Computer," MIT Instrumentation Laboratory, August 1968, pp. 1, 7, 8, 9.

64 In its final configuration, each Apollo guidance and navigation computer had fixed memory (core-rope memory) of 36,864 words. Each word was 16 bits, for a total core-rope memory of 589,824 bits. Hall, *Journey to the Moon*, p. 120.

65 There are two good video accounts of how the Apollo flight computer was constructed, including the weaving work of the women at Raytheon in Waltham: *Moon Machines*, Episode 3: "The Navigation Computer," Raytheon portion at 21:00; "MIT Science Reporter: Computer for Apollo (1965)," material about manufacturing starts at 14:00.

66 Mindell, *Digital Apollo*, p. 171, points out that "rope mother" was an ironic designation when there were almost no women shepherding specific versions of mission software.

67 Ramon Alonso, "Software Issues," conference proceedings, "Apollo Guidance Computer History Project: First Conference," Cambridge, Mass.,

July 27, 2001, https://authors.library.caltech.edu/5456/1/hrst.mit.edu/hrs/apollo/public/conference1/software.htm.

68 David Bates, Eldon Hall, and Ed Blondin, "Ed Blondin's Introduction," conference proceedings, "Apollo Guidance Computer History Project: Third Conference," Cambridge, MA, November 30, 2001, https://authors.library.caltech.edu/5456/1/hrst.mit.edu/hrs/apollo/public/conference3/blondin.htm.

69 Jack Poundstone, "On the Factory Floor," conference proceedings, "Apollo Guidance Computer History Project: Third Conference," Cambridge, MA, November 30, 2001, https://authors.library.caltech.edu/5456/1/hrst.mit.edu/hrs/apollo/public/conference3/floor.htm.

70 Alonso, "Software Issues."

71 "Small program change needed in the AS-501/502 AGC program," November 25, 1966, #66-FM1-168, Tindallgrams #1, pdf, pp. 81-82.

72 Ramon Alonso and J. H. Laning Jr., "R-276: Design Principles for a General Control Computer," MIT Instrumentation Laboratory, April 1960, p. 8.

73 Richard Witkin, "Electrical Difficulty Aboard the Spacecraft Is Brief, Unnerving and Mysterious," *New York Times*, November 15, 1969, p. 22, https://timesmachine.nytimes.com/timesmachine/1969/11/15/issue.html.

74 David Bates and Herb Briss, "Formalization of Project Management," conference proceedings, "Apollo Guidance Computer History Project: Third Conference," Cambridge, MA, November 30, 2001, https://authors.library.caltech.edu/5456/1/hrst.mit.edu/hrs/apollo/public/conference3/formalization.htm.

75 United Press International, "Call Apollo a Triumph of the Squares," *Chicago Tribune*, December 28, 1968, p. 4A, https://chicagotribune.newspapers.com/image/376562797.

For an assessment of U.S. intelligence understanding of the state of the Soviet space program in late 1968, see, among others: Dwayne A. Day, "Spooky Apollo: Apollo 8 and the CIA," *The Space Review*, December 3, 2018, http://www.thespacereview.com/article/3617/1. *Time* cover story: "Race for the Moon," *Time*, December 6, 1968, http://content.time.com/time/subscriber/article/0,33009,844661,00.html.

76 Christopher C. Kraft Jr., oral history transcript, NASA Johnson Space Center Oral History Project, interviewed by Rebecca Wright, Louisburg, NC, May 23, 2008, accessed November 4, 2018, https://historycollection.jsc.nasa.gov/JSCHistoryPortal/history/oral_histories/KraftCC/KraftCC_5-23-08.htm.

77 W. David Woods and Frank O'Brien, "Apollo 8: Day 3: Lunar Encounter," *Apollo Flight Journal*, August 8, 2018, https://history.nasa.gov/afj/ap08fj/12day3_lunar_encounter.html, 069:32:35. Comparison of calculated lunar orbit to actual orbit: Mindell, *Digital Apollo*, p. 178.

78 *Moon Machines*, "Episode 5: Navigation," Kosmala comment at 34:20, Martin comment at 33:40. Johnston, interview #2 with the author, December 18, 2015.

79 Frank Borman, who was commander of Apollo 8, is widely credited with saying that the Moon is not made of green cheese, but of American cheese, a line that made front pages across the country, including in the *Washington Post*, and in some newspapers was given its own headline and brief story. The first citation is to the Associated Press account, crediting Borman. The confusion arises because the NASA transcript of the exchange, which paraphrases the pilot and the reply, credits the answer to Bill Anders. Credibility for the news accounts attributing the line to Borman the day after the splashdown comes from the fact that they include much more detail than the NASA transcript, including the name of the helicopter commander. Associated Press, "Astronauts Land Safely: Pinpoint Splashdown in Pacific Climaxes Historic Moon Flight," *Evening Sun* (Baltimore), December 27, 1968, p. A1; David Woods and Frank O'Brien, eds., *Apollo Flight Journal*, "Apollo 8: Day 6: The Maroon Team—Splashdown," https://history.nasa.gov/afj/ap08fj/28day6_maroon_splash.html, 147:02:03 and following exchanges.

80 There are multiple excellent accounts of the Apollo 14 "abort switch" problem, from different points of view but that do not conflict with each other. Backroom account, and use of flashlight for first tap: Kranz, *Failure Is Not an Option*, pp. 345–49; Eric M. Jones, "Apollo 14 Flight Journal: Landing at Fra Mauro," *Apollo Lunar Surface Journal*, https://history.nasa.gov/alsj/a14/a14.landing.html, 105:46:23 and following exchanges.

81 Don Eyles account: Don Eyles, telephone interview #1, January 27, 2016. Don Eyles, book talk, *Sunburst and Luminary*, MIT Museum, March 14, 2018, posted April 10, 2018, YouTube, https://www.youtube.com/watch?v=eotnk1wVSB8; Eyles, *Sunburst and Luminary*, pp. 17–18 (hiring at MIT); 255–67 (complete account of events of Apollo 14).

82 "Apollo 14 Flight Journal: Landing at Fra Mauro," 106:25:29.

83 Garman, transcript, oral history, March 27, 2001.

84 Eyles, *Sunburst and Luminary*, pp. 257–58.

85 Garman, transcript, oral history, March 27, 2001.

86 Edgar Dean Mitchell, "Guidance of Low-Thrust Interplanetary Vehicles," SciD dissertation, Massachusetts Institute of Technology, June 1964.

87 Eric M. Jones, "Apollo 14 Flight Journal: Post-Landing Activities," *Apollo Lunar Surface Journal*, accessed November 5, 2018, https://history.nasa.gov/alsj/a14/a14.postland.html, 108:21:21 and following exchanges.

88 Richard Witkin, "Defective Switch Posed a Problem," *New York Times*, February 6, 1971, p. 1, https://timesmachine.nytimes.com/timesmachine/1971/02/06/issue.html; Victor K. McElheny, "Faulty Switch Made Landing a Heart-in-Mouth Thriller: MIT Experts Raced to Change Program," *Boston Evening Globe,* February 5, 1971, pp. 1, 3, https://www.newspapers.com/image/435188447; Gerard Weidmann, "Apollo Team at MIT Lab Saves Day," *Boston Evening Globe*, February 5, 1971, pp. 1, 2, https://www.newspapers.com/image/435188447.

89 The text of the *Rolling Stone* story on Don Eyles is available on the *Rolling Stone* website. In two separate locations there are images—one of the opening page of that issue, showing the headline and original layout; one showing the portrait of Eyles that ran with the original story, which is not part of the story posted online. Text: Timothy Crouse, "Don Eyles: EXTRA! Weird-Looking Freak Saves Apollo 14!," *Rolling Stone*, March 18, 1971, https://www.rollingstone.com/politics/politics-news/don-eyles-extra-weird-looking-freak-saves-apollo-14-40737/. Opening page from 1971: http://www.vasulka.org/archive/Publications/RollingStone/EverybodyOnTV.pdf. Portrait of Eyles from 1971: https://78.media.tumblr.com/tumblr_lnl0g6OE1q1qz5s16o1_1280.jpg.

90 The metal tags are visible on some pieces of Apollo guidance and navigation equipment posted online. The easiest to read is at the website of Heritage Auctions, which sometimes auctions Apollo memorabilia: "Apollo Guidance Computer: Original Display and Keyboard (DSKY)," Lot #41178, Heritage Auctions, https://historical.ha.com/itm/explorers/apollo-guidance-computer-original-display-and-keyboard-dsky-unit/a/6033-41178.s?ic4.

NASA still uses the original insignia. It was retired in 1975 for a more modern logo, which just used the agency's letters; the Apollo-era logo was brought back by Administrator Dan Goldin in 1992 to convey the idea that "the magic is back at NASA." Steve Garber, "NASA 'Meatball' Logo," *NASA History Program Office*, October 2, 2018, https://history.nasa.gov/meatball.htm.

91 Hoag writes, "No program errors were ever uncovered during the missions." Hoag, "The History of Apollo Onboard Guidance, Navigation, and Control," p. 9 (hardware), p. 10 (software).

92 Tylko, "MIT and Navigating the Path to the Moon."

93 Hardesty, "Apollo's Rocket Scientists."

94 Ed Mitchell, insert at 108:18:23, in Jones, "Apollo 14 Flight Journal: Post-Landing Activities."

95 Brooks, Grimwood, and Swenson, *Chariots for Apollo*, p. 283.

96 Orloff, *Apollo by the Numbers,* p. 305.

97 Malcolm Johnston, ed., *Collected Tindallgrams: Memos by Howard W. Tindall Jr.*, May 1996. The Chris Kraft quote is in the one-page foreword to this informal collection, which was assembled by Johnston and distributed to Tindall's acquaintances about seven months after his death, November 20, 1995. The collection is now available online: https://www.hq.nasa.gov/alsj/tindallgrams02.pdf.

98 Kranz, *Failure Is Not an Option*, p. 261.

99 Eyles, *Sunburst and Luminary*, pp. 66–69.

## 6: JFK's Secret Space Tapes

1 Eisenhower: United Press International, "Ike Raps Kennedy's Fiscal Plans, Says 'Nuts' to Moon Race," *Tampa Tribune,* June 13, 1963, p. 1, https://www.newspapers.com/image/330184888/.

Kennedy: "Transcript of Presidential Meeting in the Cabinet Room of the White House," November 21, 1962, NASA: History, p. 17, https://history.nasa.gov/JFK-Webbconv/pages/transcript.pdf.

2 Victor Cohn, "Experts See Race for Lunar Sample," *Washington Post,* July 14, 1969, p. A1.

3 "First visit by a U.S. astronaut to the Soviet Union": Siddiqi, *Challenge to Apollo*, p. 693; Cohn, "Experts See Race for Lunar Sample"; Associated Press, "Apollo 11 Crewmen Given Day Off; Outlook Is Good," *Burlington (VT) Free Press,* July 14, 1969, p. 1, https://www.newspapers.com/image/200521120.

4 John Noble Wilford, "Moscow Says That Luna 15 Won't Be in Apollo's Way; Americans Check Module," *New York Times*, July 19, 1969, pp. 1, 10, https://timesmachine.nytimes.com/timesmachine/1969/07/19/issue.html.

The other *New York Times* stories on Luna 15 that day were: "U.S. Space Aides Cautiously Pleased by Russian Amity" (Richard D. Lyons, p. 1), "Cooperative Scientist: Mstislav Vsevolodovich Keldysh" (p.10), and "Soviet Hints Luna Will Stay in Orbit" (Lawrence Van Gelder, p. 10).

5 Asif Siddiqi provides a detailed account of the mission of Luna 15 in July 1969 in *Challenge to Apollo*, pp. 693–96. Additional details in "Report from Jodrell Bank," *New York Times*, July 22, 1969, p. 29, https://timesmachine

.nytimes.com/timesmachine/1969/07/22/issue.html. Distance from Sea of Tranquility to Sea of Crises, on the surface of the Moon, via Wolfram Alpha: http://www.wolframalpha.com/input/?i=what+is+the+distance+between+Mare+Tranquillitatis+and+Mare+Crisium+on+the+Moon%3F; Bernard Gwertzman, "Luna Mission Ends: Soviet Craft Down on Moon—Tass Says Work Is Finished," *New York Times,* July 22, 1969, pp. 1, 29, https://timesmachine.nytimes.com/timesmachine/1969/07/22/issue.html.

6 Siddiqi, *Challenge to Apollo*, p. 696.

7 W. David Woods, Kenneth D. MacTaggart, and Frank O'Brien, "Apollo 11: Day 7, Part 1, Leaving the Lunar Sphere of Influence," *Apollo Flight Journal*, February 10, 2017, 148:23:13, https://history.nasa.gov/afj/ap11fj/22day7-leave-lsi.html.

8 The samples Apollo 11 carried home included 50 individual rocks; some samples of regolith, lunar dust and dirt; and two core samples drilled out of the soil. "Apollo 11 Mission: Lunar Sample Overview," Lunar and Planetary Institute, accessed November 8, 2018, https://www.lpi.usra.edu/lunar/missions/apollo/apollo_11/samples/.

9 John W. Finney, "Grissom Receives Medal for Flight," *New York Times*, July 23, 1961, p. 35, https://timesmachine.nytimes.com/timesmachine/1961/07/23/issue.html; "Hoosier Fete for Gus Gets 'No Go' from NASA," *Indianapolis Star*, July 25, 1961, p. 19, https://www.newspapers.com/image/106282185/; Tom Wicker, "Kennedy Gives Shepard Medal; Quiet Throngs Hail Astronaut," *New York Times*, May 9, 1961, pp. 1, 35, https://timesmachine.nytimes.com/timesmachine/1961/05/09/issue.html.

10 Associated Press, "Astronomers Poll Cool to Manned Moon Trip," *New York Times,* August 1, 1961, p. 3, https://timesmachine.nytimes.com/timesmachine/1961/08/01/118045731.html. United Press International, "Douglas Questions Spending Billions to Land Man on Moon," *San Bernardino County (CA) Sun*, January 27, 1962, p. 1, https://www.newspapers.com/image/51567474/.

11 *New York Times* editorials: "To the Moon and Back," *New York Times*, January 15, 1962, p. 26, https://timesmachine.nytimes.com/timesmachine/1962/01/15/issue.html; "To the Moon," *New York Times*, August 4, 1962, p. 18, https://timesmachine.nytimes.com/timesmachine/1962/08/04/issue.html; "The Effort in Space," *New York Times,* September 14, 1962, p. 30, https://timesmachine.nytimes.com/timesmachine/1962/09/14/issue.html.

12 Ward Cannel, "Computer Expert Is Wary about Letting Machines Do the Thinking," *Wisconsin Rapids Daily Tribune*, November 1, 1961, p. 4,

https://www.newspapers.com/image/244555446/; Richard Witkin, "How Far, How Fast?," *New York Times Book Review*, September 27, 1964, p. 50, https://timesmachine.nytimes.com/timesmachine/1964/09/27/issue.html.

13 Audiotapes and transcripts of Kennedy's press conferences are archived at the John F. Kennedy Library. "President Kennedy: News Conference, President Kennedy's News Conferences," February 7, 1962, U.S. State Department, John F. Kennedy Presidential Library and Museum, https://www.jfklibrary.org/archives/other-resources/john-f-kennedy-press-conferences/news-conference-23.

"President Kennedy: News Conference, President Kennedy's News Conferences," June 14, 1962, U.S. State Department, John F. Kennedy Presidential Library and Museum, https://www.jfklibrary.org/archives/other-resources/john-f-kennedy-press-conferences/news-conference-36.

"President Kennedy: News Conference, President Kennedy's News Conferences," August 22, 1962, John F. Kennedy Presidential Library and Museum, https://www.jfklibrary.org/archives/other-resources/john-f-kennedy-press-conferences/news-conference-41.

14 Rudy Abramson, "Kennedy Vows U.S. to Be First in Space," *Nashville Tennessean*, September 12, 1962, pp. 1, 2, https://www.newspapers.com/image/112923225/.

15 E. W. Kenworthy, "Kennedy Is Assured on Moon Program in Space-Base Tour," *New York Times*, September 12, 1962, pp. 1, 13, https://timesmachine.nytimes.com/timesmachine/1962/09/12/issue.html; E.W. Kenworthy, "Kennedy Asserts Nation Must Lead in Probing Space," *New York Times*, September 13, 1962, pp. 1, 16, https://timesmachine.nytimes.com/timesmachine/1962/09/13/issue.html

16 Full video of Kennedy's speech at Rice University, September 12, 1962, is available online: "President Kennedy's Speech at Rice University," posted by NASA STI Program, September 13, 2011, https://www.youtube.com/watch?v=VaFTVR-hZqg&t=13m28s.

Quotes from the speech that follow are from the text posted at the American Presidency Project, UCSB: "Address at Rice University in Houston on the Nation's Space Effort," September 12, 1962, https://www.presidency.ucsb.edu/documents/address-rice-university-houston-the-nations-space-effort.

17 Bob Woodward and Patrick E. Tyler, "JFK Secretly Taped White House Talks," *Washington Post*, February 4, 1982, pp. A1, A22, https://www.washingtonpost.com/archive/politics/1982/02/04/jfk-secretly-taped-white-house-talks/87d23567-3ae3-4e0c-841c-5ebbe3f8e470/?utm_term

=.aaa792af18fe; Edward A. Gargan, "Kennedy Secretly Taped Sessions in White House," *New York Times*, February 4, 1982, p. A16, https://timesmachine.nytimes.com/timesmachine/1982/02/04/199324.html?pageNumber=16. The Kennedy Library also provides an account of the taping system: "The JFK White House Tape Recordings," The John F. Kennedy Presidential Library and Museum, https://www.jfklibrary.org/learn/about-jfk/jfk-in-history/white-house-tape-recordings.

18 A record of the November 21, 1962, meeting on NASA's budget in the White House Cabinet Room is available as both an audio recording and a transcript. All quotes that follow are directly from the audio recording, which is sometimes hard to make out because 10 people attended, and often talked over each other. But the transcript, although posted on NASA's website, is not completely accurate and should not be relied on without checking passages against the audiotape. The Kennedy Presidential Library never posted the entire tape or a transcript, but simply "made it available" on request. The full audio is from the University of Virginia's Miller Center.

Transcript: "Transcript of Presidential Meeting in the Cabinet Room of the White House," November 21, 1962, NASA, History, https://history.nasa.gov/JFK-Webbconv/pages/transcript.pdf.

Audio: "Meeting on the NASA Budget: John F. Kennedy Presidency, Secret White House Tapes," The Miller Center of the University of Virginia, November 21, 1962, https://millercenter.org/the-presidency/secret-white-house-tapes/meeting-nasa-budget-0.

19 This central exchange from the November 21, 1962, meeting starts at 32:25 on the audio recording and p. 14 in the transcript.

20 NASA's spending when Kennedy became president: $0.7 billion. NASA's spending in FY64 (1963): $2.6 billion, an increase of three times in two years. NASA's spending in FY65 (1964): $5.1 billion. NASA's spending in FY66 (1965): $5.9 billion. *Statistical Abstract of the United States*, 1967, Table No. 544, Federal Administrative Budget Expenditures, by Organization Unit: 1961 to 1968, p. 389, https://www2.census.gov/library/publications/1967/compendia/statab/88ed/1967-06.pdf.

21 This exchange: audio recording, 38:00, transcript p. 17.

22 This exchange, in which Kennedy says, "I'm not that interested in space": audio recording, 39:30, transcript, p. 17.

23 John F. Kennedy, "Excerpts from a Speech by Sen. John F. Kennedy at the Valley Forge Country Club, Valley Forge, PA" (advance release text), October 29, 1960, *American Presidency Project*, https://www.presidency.ucsb.edu

/documents/excerpts-from-speech-senator-john-f-kennedy-valley-forge-country-club-valley-forge-pa.

24 Cabinet Room meeting, audio recording 40:55, transcript, p. 18.

25 Philip H. Abelson, "Manned Lunar Landing," *Science*, vol. 140, no. 3564 (April 19, 1963), p. 267, http://science.sciencemag.org/content/140/3564; *Today* show appearance: Logsdon, *John F. Kennedy and the Race to the Moon*, p. 199; "Science Editor Skeptical of Manned Lunar Shot," *New York Times*, April 19, 1963, p. 86, https://timesmachine.nytimes.com/timesmachine/1963/04/19/82058441.html; Howard Simons, "Webb Defends U.S. Men-on-Moon Plan," *Washington Post*, April 21, 1963, p. A10; Howard Simons, "Moon Madness? Scientists Divided on Apollo," *Washington Post*, May 12, 1963, p. A9.

26 Howard Simons, "Scientist Calls Project Apollo Drain of Talent," *Washington Post*, June 11, 1963, p. A3; "Space: Some Opinions on Reaching the Moon," *Los Angeles Times*, June 16, 1963, p. G4, https://www.newspapers.com/image/381588856.

27 No reporters were permitted into the breakfast where former president Eisenhower spoke, so the press accounts of what he said came from interviewing attendees afterward. The United Press International story had the line from Eisenhower as "Anybody who would spend $40 billion in a race to the Moon for national prestige is nuts." The Associated Press account had a slightly different version: "To spend $40 billion to be the first to reach the Moon is just nuts." Interestingly, neither the *New York Times* nor the *Washington Post* did their own story on the Eisenhower comments, and neither paper used a full quote; they both just used the word "nuts," on which the two accounts agree. The only reason to prefer the UPI account is that in his own Oval Office conversations about space policy, Eisenhower specifically balked at doing an expensive space program for reasons of prestige, giving that line a certain credibility. If the UPI line is not correct, it was never corrected in subsequent days. UPI, "Eisenhower Calls Moon Race 'Nuts,'" *Times Record* (Troy, NY), June 12, 1963, p. 1, https://www.newspapers.com/image/56745599/; "Eisenhower Meets Kennedy on Rights," *New York Times*, June 13, 1963, pp. 1, 13, https://timesmachine.nytimes.com/timesmachine/1963/06/13/issue.html.

28 Tom Wicker, "President in Plea," *New York Times*, June 12, 1963, pp. 1, 20, https://timesmachine.nytimes.com/timesmachine/1963/06/12/issue.html; Claude Sitton, "Alabama Admits Negro Students; Wallace Bows to Federal Force; Kennedy Sees 'Moral Crisis' in U.S.," *New York Times*, June 12,

1963, pp. 1, 20, https://timesmachine.nytimes.com/timesmachine/1963/06/12/issue.html; "Transcript of President's Address," *New York Times*, June 12, 1963, p. 20, https://timesmachine.nytimes.com/timesmachine/1963/06/12/issue.html.

29 Claude Sitton, "N.A.A.C.P. Leader Slain in Jackson; Protests Mount," *New York Times*, June 13, 1963, pp. 1, 12, https://timesmachine.nytimes.com/timesmachine/1963/06/13/issue.html.

30 Tad Szulc, "Kennedy Asks Break in Cold War; New Atom Parley Set in Moscow; U.S. to Forgo Atmospheric Tests," *New York Times*, June 11, 1963, pp. 1, 16, https://timesmachine.nytimes.com/timesmachine/1963/06/11/issue.html; Seymour Topping, "Russians Stirred by Kennedy Talk about Cold War," *New York Times*, June 13, 1963, pp. 1, 4, https://timesmachine.nytimes.com/timesmachine/1963/06/13/issue.html.

31 Stuart H. Loory, "Are We Wasting Billions in Space?," *Saturday Evening Post*, September 14, 1963, pp. 13ff.

32 The complete recording of this meeting between Kennedy and Webb is available through the John F. Kennedy Presidential Library. The approximate time of the quotes used will be noted in the notes to follow. The Kennedy Library provided only a partial transcript of sections of the tape in the news release announcing the recording being made available. That partial transcript (which is not in every case correct) is also noted below. No complete transcript could be found publicly available on the internet. John F. Kennedy and James Webb, sound recording of meeting in the Oval Office, White House, September 18, 1963, John F. Kennedy Presidential Library and Museum, "Meetings: Tape #111: Lunar Program (James Webb)," 46 minutes of audio, https://www.jfklibrary.org/asset-viewer/archives/JFKPOF/MTG/JFKPOF-MTG-111-004/JFKPOF-MTG-111-004; "JFK Library Releases Recording of President Kennedy Discussing Race to the Moon," press release, John F. Kennedy Presidential Library and Museum, May 25, 2011, https://www.jfklibrary.org/about-us/news-and-press/press-releases/jfk-library-releases-recording-of-president-kennedy-discussing-race-to-the-moon.

33 Kennedy-Webb Oval Office meeting, 5:45.

34 Ibid., 38:50.

35 Ibid., 39:55.

36 Ibid., 34:20.

37 Ibid., 33:10.

38 Ibid., 21:30.

39 The phrase "Kennedy's Folly" is suggested in this essay analyzing the rhetoric of Kennedy's speech at Rice University: John W. Jordan, "Kennedy's Romantic Moon and Its Rhetorical Legacy for Space Exploration," *Rhetoric and Public Affairs* 6, no. 2 (Summer 2003): 209–31.

40 Kennedy-Webb Oval Office meeting, 25:25.

41 Ibid., 28:00.

42 Ibid., 42:15 ("younger folks"); 42:55 ("staggering things"); 43:05 ("I predict").

43 Ibid., 28:50.

44 John F. Kennedy, "Address before the 18th General Assembly of the United Nations," September 20, 1963, John F. Kennedy Presidential Library and Museum, text: https://www.jfklibrary.org/archives/other-resources/john-f-kennedy-speeches/united-nations-19630920; audio recording: https://jfkl.prod.acquia-sites.com/asset-viewer/archives/JFKWHA/1963/JFKWHA-218/JFKWHA-218.

45 Thomas J. Hamilton, "Kennedy Asks Joint Moon Flight by U.S. and Soviet as Peace Step," *New York Times*, September 21, 1963, pp. 1, 6; John W. Finney, "Washington Is Surprised by President's Proposal," *New York Times*, September 21, 1963, pp. 1, 7, https://timesmachine.nytimes.com/timesmachine/1963/09/21/issue.html; Carol Kirkpatrick, "President Urges Joint U.S.-Soviet Moon Trip," *Washington Post*, September 21, 1963, pp. A1, A10.

46 Kennedy-Webb Oval Office meeting, 37:10.

47 Finney, "Washington Is Surprised by President's Proposal."

48 "President Kennedy: News Conference, President Kennedy's News Conferences," July 17, 1963, U.S. State Department, John F. Kennedy Presidential Library and Museum, https://www.jfklibrary.org/archives/other-resources/john-f-kennedy-press-conferences/news-conference-58.

49 Historian John M. Logsdon has researched deeply the details of the Kennedy administration's repeated overtures to the Russians on space cooperation; there were several, at various levels, starting even before Kennedy's May 25, 1961, "go to the Moon" speech. He devotes two chapters of his book on Kennedy and space just to the question of U.S.-U.S.S.R. cooperation. See Logsdon, *John F. Kennedy and the Race to the Moon*, pp. 159–96. "President's 1961 Proposal on Moon Flight Revealed," *New York Times*, September 22, 1963, pp. 1, 35, https://timesmachine.nytimes.com/timesmachine/1963/09/22/issue.html.

50 Richard Witkin, "Joint Moon Trip by 1970 Doubted," *New York Times*, September 23, 1963, pp. 1, 13, https://timesmachine.nytimes.com/timesmachine/1963/09/23/issue.html.

51 House cuts $250 million and "boomerang effect": John W. Finney, "Funds for Space in Peril in House," *New York Times*, September 25, 1963, pp. 1, 15, https://timesmachine.nytimes.com/timesmachine/1963/09/25/issue.html. House defeats cut by $1.25 billion: "House Unit Split on Space Budget," *New York Times*, September 28, 1963, p. 3, https://timesmachine.nytimes.com/timesmachine/1963/09/28/issue.html. Thomas comment on cuts: Robert C. Toth, "Outlay for Space Cut to $5.1 Billion," *New York Times*, October 8, 1963, pp. 1, 26, https://timesmachine.nytimes.com/timesmachine/1963/10/08/issue.html.

52 Accounts of the House passage of the NASA budget, and the ban on use of money for a joint U.S.-U.S.S.R. mission: Robert C. Toth, "House Opposes Joint Moon Trip; Votes NASA Fund," *New York Times*, October 11, 1963, pp. 1, 19, https://timesmachine.nytimes.com/timesmachine/1963/10/11/issue.html; UPI, "House Votes to Ban Joint Shot to Moon; Also Cuts Funds for Project," *Tampa (FL) Tribune*, October 11, 1963, p. 1. Comment from Teague: Robert Sherrod, "Let's Go to the Moon Together," *New York Times*, June 17, 1972, p. 29, https://timesmachine.nytimes.com/timesmachine/1972/06/17/issue.html.

53 "National Security Action Memorandum #271," November 12, 1963, John F. Kennedy Presidential Library and Museum, https://www.jfklibrary.org/asset-viewer/national-security-action-memorandum-number-271; Louis Harris, "Public Puts Limits on Russian Dealings," *Iowa City Press-Citizen*, December 16, 1963, p. 8, https://www.newspapers.com/image/363501064/.

54 Associated Press, "NASA Chief Sees No Early Prospect of U.S.-Soviet Team Flights to Moon," *Washington Post*, October 14, 1963, p. A1.

55 Richard Witkin, "Saturn Is Orbited; Its 10-Ton Payload Tops Any of Soviet," *New York Times*, January 30, 1964, pp. 1, 12, https://timesmachine.nytimes.com/timesmachine/1964/01/30/issue.html.

56 Kennedy's visit to Cape Canaveral, November 16, 1963: Marjorie Hunter, "President, Touring Canaveral, Sees Polaris Fired," *New York Times*, November 17, 1962, pp. 1, 44, https://timesmachine.nytimes.com/timesmachine/1963/11/17/167964182.html; Rex Newman and Walter Mack, "Polaris Firing Pleases Kennedy," *Orlando (FL) Sentinel*, November 17, 1963, pp. 1A, 77A, 9D (pictures), https://www.newspapers.com/image/223695429; Associated Press, "Kennedy Goes to Sea for Polaris Launch," *Tampa (FL) Tribune*, November 17, 1963, https://www.newspapers.com/image/330589527/.

57 John F. Kennedy, "Remarks in San Antonio at the Dedication of the Aerospace Medical Health Center," San Antonio, Texas, November 21, 1963, *American Presidency Project*, https://www.presidency.ucsb.edu/documents/remarks-san-antonio-the-dedication-the-aerospace-medical-health-center.

58 John F. Kennedy, "Remarks Prepared for Delivery at the Trade Mart in Dallas," Dallas, Texas, November 22, 1963, *American Presidency Project*, https://www.presidency.ucsb.edu/documents/remarks-prepared-for-delivery-the-trade-mart-dallas.

59 LBJ renames Cape Canaveral; painters hang new sign: Lyndon B. Johnson, "The President's Thanksgiving Day Address to the Nation," White House, November 28, 1963, *American Presidency Project*, https://www.presidency.ucsb.edu/documents/the-presidents-thanksgiving-day-address-the-nation; Associated Press, "Mrs. Kennedy Made Request," *New York Times*, November 30, 1963, p. 8, https://timesmachine.nytimes.com/timesmachine/1963/11/30/89979688.html; "Cape Redesignation Quickly Put into Effect," *Orlando (FL) Sentinel*, November 30, 1963, p. 1-B, https://www.newspapers.com/image/223701704/.

60 Overall LBJ budget figures: Associated Press, "Budget Calls Halt to Spending," *Press and Sun-Bulletin* (Binghamton, NY), January 21, 1964, pp. 1, 8, https://www.newspapers.com/image/254178859/; Edwin L. Dale, Jr., "$97.9 Billion Budget Puts Stress on Poverty Fight; Reduces Cost of Defense," *New York Times*, January 22, 1964, pp. 1, 18, https://timesmachine.nytimes.com/timesmachine/1964/01/22/issue.html.

"No second-class ticket": United Press International, "Space Program Hits New High," *Record* (Hackensack, NJ), January 21, 1964, p. 5, https://www.newspapers.com/image/491120987/.

## 7: How Do You Fly to the Moon?

1 Kelly, *Moon Lander*, p. 205.

2 "F6F-3 Hellcat," National Naval Aviation Museum, accessed December 10, 2018, https://www.navalaviationmuseum.org/attractions/aircraft-exhibits/item/?item=f6f-3_hellcat; "2 Grumman Records Hailed by Navy Aide," *New York Times*, April 10, 1945, p. 36, https://timesmachine.nytimes.com/timesmachine/1945/04/10/88213008.html.

3 Richard D. Lyons, "Lunar Module to Burn Up in Air," *New York Times*, March 8, 1969, p. 12, https://timesmachine.nytimes.com/timesmachine/1969/03/08/88983827.html.

4 Small parts of the story of the "egress rope" and the March 1964 design review (that mock-up was designated "TM-1") are told in several places: Kelly, *Moon Lander*, pp. 88–90; Brooks, Grimwood, and Swenson, *Chariots for Apollo*, p. 151; Gene Harms, telephone interview, May 30, 2016.

The picture of the TM-1 mock-up, with the rope along its side, appears in Joshua Stoff, *Building Moonships: The Grumman Lunar Module* (Portsmouth, NH: Arcadia, 2004), p. 20. The caption to that picture mentions the worry that a ladder could be damaged and become unusable while the LM was landing on the Moon. Stoff also discusses the rope in the lunar module episode of the documentary series *Moon Machines*, Episode 4: "The Lunar Module," directed by Nick Davidson, June 2008, Dox Productions for The Science Channel, Discovery Communications. Stoff discusses the rope starting at 11:35, followed by video of the test of the rope and the block-and-tackle.

5 Logsdon, *John F. Kennedy and the Race to the Moon*, p. 146.

6 Scott, "Speech at the Opening of the Computer Museum."

7 This complexity of orbital mechanics is why rockets on Earth, headed for orbit, can't simply be launched when it's convenient: they have "launch windows." Every rocket—every spaceship, every satellite—launched from Earth is following a plan to get where its payload is going in orbit, whether that's astronauts headed for rendezvous with the ISS or a weather satellite headed for "rendezvous" with a specific spot over the Earth. If something delays a launch 10 minutes outside "the window," you don't just hurry up and launch anyway, like an airliner running late. Because you won't get where you're going. You stand down, and try again at the next launch window that gets you where you need to go.

8 James R. Hansen, *Enchanted Rendezvous*, Monographs in Aerospace History #4 (Washington, D.C.: NASA, 1995), p. 2, https://ntrs.nasa.gov/archive/nasa/casi.ntrs.nasa.gov/19960014824.pdf.

9 Ibid., pp. 5–6.

10 Ibid.

11 Murray and Cox, *Apollo*, p. 100.

12 Hansen, *Enchanted Rendezvous*, p. 8.

13 The online, pdf version of Hansen's *Enchanted Rendezvous* cited above includes, after the published manuscript, a "Key Documents" section, which includes a two-page list of all of Houbolt's LOR presentations through the summer of 1962—he lists 29 presentations over 33 months: pdf pages 75–76.

14 Hansen, *Enchanted Rendezvous*, pp. 9–10.
15 Murray and Cox, *Apollo*, p. 96.
16 Hansen, *Enchanted Rendezvous*, p. 20.
17 Ibid., "Letter to Dr. Robert C. Seamans, Jr. from John C. Houbolt," dated November 15, 1961, Key Documents, pdf, pp. 55–63.
18 Seamans, *Aiming at Targets*, p. 98.
19 Ibid.
20 Murray and Cox, *Apollo*, p. 107.
21 Hansen, *Enchanted Rendezvous*, p. 25.
22 Murray and Cox, *Apollo*, pp. 116–17.
23 Hansen, *Enchanted Rendezvous*, p. 25.
24 The story of Houbolt attending the practice briefing at NASA HQ is in Murray and Cox, *Apollo*, p. 122. The story of von Braun's conversion to LOR is from Murray and Cox, *Apollo*, pp. 121–22; Hansen, *Enchanted Rendezvous*, pp. 25–27. The full text of von Braun's talk endorsing LOR is reprinted (including von Braun's signature on page 11) in Hansen, *Enchanted Rendezvous*, Key Documents, pdf, pp. 64–74.
25 "Lunar ferry": John F. Finney, "$3.8 Billion Voted for Space Plans," *New York Times,* July 12, 1962, pp. 1, 12, https://timesmachine.nytimes.com/timesmachine/1962/07/12/issue.html. Houbolt in Paris: Hansen, *Enchanted Rendezvous*, p. 27.

Although specific references have been cited above, it's worth noting again the value and depth of the two definitive accounts of how NASA chose the way it would fly to the Moon: Hansen's *Enchanted Rendezvous* is about nothing but Houbolt and the decision of how to fly to the Moon. Murray and Cox, in *Apollo*, devote most of three chapters (7, 8, and 9) to the debate around the decision.

26 John W. Finney, "Rendezvous of Satellites in Space by the Mid-Sixties Is Predicted," *New York Times*, May 24, 1961, p. 18, https://timesmachine.nytimes.com/timesmachine/1961/05/24/101464719.html; Howard Simons, "NASA Outlines Plans to Link Craft in Orbit," *Washington Post*, May 24, 1961, p. A2.
27 Barton C. Hacker and James M. Grimwood, *On the Shoulders of Titans: A History of Project Gemini*, SP-4203 (Washington, D.C.: NASA, 1977), p. 246.
28 Gemini 4 ground-to-air mission transcript: https://www.jsc.nasa.gov/history/mission_trans/GT04_TEC.PDF, 01:33:10.
29 "Preliminary GT-4 Flight Crew Debriefing: Part I," NASA: Space

Operations Branch, Flight Crew Support Division, June 16, 1965, pp. 60, 64, 74, https://www.scribd.com/document/57053688/Preliminary-GT-4-Flight-Crew-Debriefing-Transcript-Part-I

30 John Noble Wilford, "First Flight Test of Lunar Landing Craft Expected Tomorrow," *New York Times*, January 21, 1968, p. 78, https://timesmachine.nytimes.com/timesmachine/1968/01/21/76928578.html; David Sheridan, "How an Idea No One Wanted Grew Up to Be the LM," *Life*, March 14, 1969, p. 20.

31 Kelly, *Moon Lander*, pp. 169, 173–74.

32 "The *Eagle* Has Landed: The Lunar Module Story," Grumman Corporation, 1989, posted by Dan Beaumont Space Museum, July 29, 2015, YouTube, https://www.youtube.com/watch?v=vjDdu7WzjQw, "Tumbler" appears at 13:10.

33 Thomas J. Kelly, oral history transcript, NASA Johnson Space Center Oral History Project, interviewed by Kevin M. Rusnak, Cutchogue, NY, September 19, 2000, p. 14, https://historycollection.jsc.nasa.gov/JSCHistoryPortal/history/oral_histories/KellyTJ/KellyTJ_9-19-00.htm.

34 Kelly, *Moon Lander*, p. 107.

35 Fred Haise, interview by filmmaker Mike Marcucci, July 16, 2003, recording provided to the author.

36 Kelly, oral history, p. 40

37 The account of the shattered window comes from: Joe Gavin, oral history transcript, NASA Johnson Space Center Oral History Project, interviewed by Rebecca Wright, Amherst, MA, January 10, 2003, https://historycollection.jsc.nasa.gov/JSCHistoryPortal/history/oral_histories/GavinJG/GavinJG_1-10-03.htm; Gavin, interviewed in *Moon Machines*, Episode 4: "The Lunar Module," starting at 28:00; Orvis E. Pigg and Stanley P. Weiss, "Apollo Experience Report: Spacecraft Structural Windows," Houston: Johnson Space Center, NASA TN D-7493, September 1973, pp. 8, 11–13, https://www.lpi.usra.edu/lunar/documents/apolloSpacecraftWindows.pdf.

38 Kelly, *Moon Lander*, pp. 141–42.

39 Gavin, oral history.

40 Richard D. Lyons, "Lunar Module to Burn Up in Air," *New York Times*, March 8, 1969, p. 12, https://timesmachine.nytimes.com/timesmachine/1969/03/08/88983827.html. Use and disposition of each lunar module manufactured: "Location of Apollo Lunar Modules," Smithsonian National Air and Space Museum: Spacecraft & Vehicles, accessed 10 January 2019,

https://airandspace.si.edu/explore-and-learn/topics/apollo/apollo-program/spacecraft/location/lm.cfm.

41 Kelly, *Moon Lander*, pp. 173–174.

42 The author added up flight times for each of the nine lunar modules piloted by astronauts using data for each mission from Orloff, *Apollo by the Numbers*. "Flight time" for each lunar module was calculated from the moment of undocking to touchdown on the Moon; and then from the moment of liftoff from the Moon to the moment of docking. (There was a tenth lunar module that flew, uncrewed, in Apollo 5.)

An examination of the transcripts of each Apollo flight shows that no astronauts made any kind of farewell comments as the lunar modules were jettisoned back into space.

43 Armstrong comment: Eric M. Jones, ed., "Return to Orbit," *Apollo 11 Lunar Surface Journal,* https://www.hq.nasa.gov/alsj/a11/a11.launch.html, 124:22:11.

Conrad comment: W. David Woods and Lennox J. Waugh, eds., "Apollo 12, Day 6: From the Snowman to Docking," *Apollo Flight Journal*, https://history.nasa.gov/afj/ap12fj/15day6_ftstd.html, 143:21:40.

In his memoir *Moon Lander* (p. 206), Kelly reports that in an after-flight debriefing, the Apollo 9 astronauts Jim McDivitt and Rusty Schweickart, the first to fly the lunar module, said "[it] is a great flying machine. And when it's just the ascent stage alone, it's very quick. It snaps to the controls like a fighter plane, or a sports car. It was super to fly!" Kelly does not say which astronaut made those observations, although McDivitt, as commander, would likely have been at the flight controls.

44 William H. Honan, "Le Mot Juste for the Moon," *Esquire*, July 1969, pp. 53-56, 139-141.

45 Grumbling about *Life* story on Houbolt: Murray and Cox, *Apollo*, pp. 99-100. Low's endorsement of Houbolt's importance: Hansen, *Enchanted Rendezvous*, pp. 36-37 (note 128). Houbolt and von Braun in Mission Control: Hansen, *Enchanted Rendezvous*, p. 28.

## 8: NASA Almost Forgets the Flag

1 United Press International, "U.S. Flag to Fly on Moon," *Panama City News*, June 13, 1969, p. 7, https://www.newspapers.com/image/39141483.

2 Here is a 5-minute video clip showing Armstrong and Aldrin erecting the first American flag on the Moon, as described in this section. The video is

a split screen with two views, in sync: On the left, the video from the TV camera the astronauts set up on the Moon; on the right, the video from the camera mounted in a window of the lunar module, shooting down onto the astronauts. "Apollo 11 Flag," posted by MoonInGoogleEarth, July 13, 2009, 5:01, https://www.youtube.com/watch?v=_H20GUvUfl4.

The flag raising comes about 45 minutes into the Moon walk. It barely shows up in the flight transcripts, because Armstrong and Aldrin didn't discuss it with Mission Control, and exchanged only a few words as they worked together to get the flag in place. At 110:09:05, CapCom McCandless tells Mike Collins, orbiting in the command module, "The EVA is progressing beautifully. I believe they are setting up the flag now." At 110:10:16, Armstrong says to Aldrin, "See if you can pull that end off a little bit. Straighten that end up a little?" Jones, ed., "One Small Step," *Apollo 11 Lunar Surface Journal*, https://www.hq.nasa.gov/alsj/a11/a11.step.html.

3 Richard Nixon, "Inaugural Address," January 20, 1969, *American Presidency Project*, https://www.presidency.ucsb.edu/documents/inaugural-address-1.

4 Memo, George M. Low to Robert Gilruth, January 23, 1969, files of the Johnson Space Center History Office.

5 Jack Kinzler was a great talker and a great storyteller. He has three full-length oral histories on file with NASA, in two of which he discusses the flag and the plaque he designed and fabricated for the Apollo missions. Jack A. Kinzler, oral history transcript, NASA Johnson Space Center Oral History Project, interviewed by Roy Neal, Houston, TX, April 27, 1999, https://historycollection.jsc.nasa.gov/JSCHistoryPortal/history/oral_histories/KinzlerJA/KinzlerJA_4-17-99.htm.

6 There is a definitive history of NASA's effort to get a flag on the Moon, written by Anne Platoff, which is the starting point for the reporting in this chapter. Platoff was generous in sharing her own research and documents.

Anne M. Platoff, "Where No Flag Has Gone Before: Political and Technical Aspects of Placing a Flag on the Moon," NASA Contractor Report 188251, Johnson Space Center, August 1993, https://ntrs.nasa.gov/archive/nasa/casi.ntrs.nasa.gov/19940008327.pdf.

7 The first meeting of the full committee was April 1, 1961, according to Brooks, Grimwood, and Swenson, *Chariots for Apollo,* p. 330, https://history.nasa.gov/SP-4205.pdf. The date of the meeting is not completely clear in Kinzler's accounts in his oral histories with NASA.

8 Kinzler, oral history, April 27, 1999.

9 Jack A. Kinzler, oral history transcript, NASA Johnson Space Center Oral

History Project, interviewed by Paul Rollins, Houston, TX, January 16, 1998, https://historycollection.jsc.nasa.gov/JSCHistoryPortal/history/oral_histories/KinzlerJA/KinzlerJA_1-16-98.htm.

10 Christopher Columbus's flags: Steven Kreis, ed., "The *Journal* of Christopher Columbus (1492)," The History Guide, August 4, 2009, http://www.historyguide.org/earlymod/columbus.html; Lewis and Clark's flags: Joseph Mussulman, "The Expedition's Flags," Discovering Lewis and Clark, http://www.lewis-clark.org/article/665; Roald Amundsun's flag: Evan Andrews, "The Treacherous Race to the South Pole," History.com, January 17, 2017, https://www.history.com/news/the-treacherous-race-to-the-south-pole; Robert Peary's flags: "The Discoverer of the North Pole," U.S. Capitol Visitor Center, https://www.visitthecapitol.gov/exhibitions/artifact/commander-robert-pearys-sledge-party-posing-flags-north-pole-photograph-april; Details of Peary's wife making the flag: "Peary Polar Expedition Flags," CRW Flags, https://www.crwflags.com/fotw/flags/us_peary.html.

11 Kinzler, oral history transcript, April 28, 1999.

12 The hand-drawn diagram for the Apollo flag is reproduced in Platoff, "Where No Flag Has Gone Before," figure 2, p. 2.

13 Kinzler, oral history transcript, April 27, 1999.

14 Kinzler recalls the effort to keep the plaque and the flag a secret until the Moon landing itself in an hour-long interview he did with historian Anne M. Platoff, August 30, 1992, at his home. Platoff provided the author a copy of that video.

Tom Moser described his role in a telephone interview, December 12, 2017.

15 The specific details of the temperature analysis are from Kinzler, interviewed by Platoff.

16 Platoff, "Where No Flag Has Gone Before," p. 3.

17 Neil Armstrong weighed 172 pounds when Apollo 11 was launched (29 pounds on the Moon); Buzz Aldrin weighed 167 pounds (28 pounds on the Moon). From Orloff, *Apollo by the Numbers*, p. 312,

18 Sheridan, "How an Idea No One Wanted Grew Up to Be the LM," p. 20.

19 The lunar module press briefing document describes the ladder as having nine rungs, nine inches apart, with a gap of 18 inches between the front edge of the LM porch to the first rung, and with the lowest rung being 30 inches from the Moon's surface. That's 120 inches from the front edge of the porch to the Moon's surface—10 feet: https://www.hq.nasa.gov/alsj/LM04_Lunar_Module_ppLV1-17.pdf, p. LV-12.

20 Moser, telephone interview, December 12, 2017.

21 Paine's testimony and the Congressional reaction: United Press International, "Congress Furor: Will Moon Crew Plant UN Flag?," *Detroit Free Press*, June 7, 1969, p. 1A, https://www.newspapers.com/image/98919566/; United Press International, "Lawmakers Warn Space Chief Against U.N. Flag on Moon," *Tampa Tribune*, June 7, 1969, p. 2-A, https://www.newspapers.com/image/331251125/.

For context, when Sir Edmund Hillary and the Nepalese Sherpa Tenzing Norgay became the first people to reach the top of Mt. Everest, the tallest mountain on Earth, on June 1, 1953, they brought with them and hoisted three flags: the British Union Jack (for Hillary), the flag of Nepal (for Norgay), and the United Nations flag. Reuters, "2 of British Team Conquer Everest; Queen Gets News as Coronation Gift; Throngs Line Her Procession Route," *New York Times*, June 2, 1953, p. 1, https://timesmachine.nytimes.com/timesmachine/1953/06/02/issue.html.

22 United Press International, "Senator Wants American Flag Put on Moon," *Tennessean*, June 1, 1969, p. 24-D, https://www.newspapers.com/image/113304352/.

23 Leo Rennert, "Solons Moon Over Old Glory," *Fresno Bee*, June 22, 1969, p. 2-C, https://www.newspapers.com/image/25902946.

24 *Muncie Evening Press:* United Press International, "Roudebush Helps Put Old Glory on Moon." *Muncie Evening Press,* June 11, 1969, p. 6, https://www.newspapers.com/image/250301816/; *Orlando Sentinel:* "Apollo 11 Crew Can Plant US Flag Only," *Orlando Sentinel*, June 11, 1969, p. 1-A, https://www.newspapers.com/image/224270703; *Philadelphia Inquirer:* William Hines, Planting a Flag on the Moon Stirs a Foofaraw in Congress," *Philadelphia Inquirer*, July 1, 1969, p. 19, https://www.newspapers.com/image/179992153/.

25 "U.S. Has Tried 5 Times to Send Rocket to Moon," *New York Times*, September 14, 1959, p. 18; Vincent Buist, "860-Pound Red Missile Hits Moon, Plants Soviet Union's Coat of Arms," *Washington Post*, September 14, 1959, p. 1A; Max Frankel, "Soviet Rocket Hits Moon after 35 Hours; Arrival Is Calculated within 84 Seconds; Signals Received till Moment of Impact," *New York Times*, September 14, 1959, p. 1; Harrison E. Salisbury, "Khrushchev Gets Big but Quiet Welcome from 200,000 on Arrival in Washington; Has 'Frank' 2-Hour Talk with Eisenhower," *New York Times*, September 16, 1959, pp. 1, 18.

26 United Press International, "Transcript of the President's News Conference on Foreign and Domestic Matters," *New York Times*, September 18, 1959,

p. 18, question #18, https://timesmachine.nytimes.com/timesmachine/1959/09/18/88821871.html; "President Skeptical on Moon Pennants," *New York Times*, September 18, 1959, p. 2, https://timesmachine.nytimes.com/timesmachine/1959/09/18/issue.html.

27 Rod Pyle, "Fifty Years of Moon Dust: Surveyor 1 Was a Pathfinder for Apollo," NASA, June 2, 2016, via NASA JPL, https://www.nasa.gov/feature/jpl/fifty-years-of-moon-dust-surveyor-1-was-a-pathfinder-for-apollo.

28 "A 23-Cent U.S. Flag Is on the Moon," *Tallahassee (FL) Democrat*, June 2, 1966, p. 1A, https://www.newspapers.com/image/245130969/.; "'Old Glory' Flying High," *Denton (TX) Record Chronicle*, June 2, 1966, p. 1A, https://www.newspapers.com/image/24376494/

29 "NASA Technical Report No. 32-1023: Surveyor 1 Mission Report," Jet Propulsion Laboratory, Pasadena, CA, 1966: Part 1: Mission Description and Performance, https://ntrs.nasa.gov/archive/nasa/casi.ntrs.nasa.gov/19660026658.pdf ("perfect soft-landing," p. XI); Part 2: Scientific Data and Results, https://ntrs.nasa.gov/archive/nasa/casi.ntrs.nasa.gov/19670000738.pdf; Part 3: Television Data, https://ntrs.nasa.gov/archive/nasa/casi.ntrs.nasa.gov/19670007837.pdf.

30 Images of the Hughes Aircraft memos regarding the investigation into "the flag incident" are collected at this website, from a lifelong Hughes Aircraft engineer, Jack Fisher: *Surveyor 1 and the American Flag*, http://www.hughesscgheritage.com/surveyor-i-and-the-american-flag-jack-fisher/. The story of NASA and JPL's irritation at Hughes over the flag is told at "The Story of Surveyor 1," http://www.hughesscgheritage.com/the-story-of-surveyor-i-jack-fisher/.

31 Details of commemorative stamp on Iwo Jima flag-raising: "Iwo Jima 1945, the Photograph and the 3¢ Green Stamp," *Linn's Stamp News*, February 6, 2015, https://www.linns.com/news/us-stamps-postal-history/2015/february/iwo-jima-1945-the-photograph-and-the-3-green-stamp.html.

32 "Treaty on Principles Governing the Activities of States in the Exploration and Use of Outer Space, including the Moon and Other Celestial Bodies," http://www.unoosa.org/pdf/publications/ST_SPACE_061Rev01E.pdf.

33 Memo, Willis H. Shapley to Thomas O. Paine, undated (May 15, 1969, perhaps), "Report of the Committee on Symbolic Activities for First Lunar Landing," NASA Headquarters History Office, Washington, D.C.; copy provided by Platoff.

34 The most exhaustive account of Armstrong's thinking about the words he said when stepping on to the Moon comes from his authorized biography,

Hansen, *First Man*, pp. 493–96. Hansen even raises and dispenses with the theory that Armstrong's words were inspired by a phrase in J. R. R. Tolkien's *The Hobbit.*

35 Platoff interview with Kinzler.

36 Personal work files of Jack Kinzler, provided to Anne Platoff in the course of her research for *Where No Flag Has Gone Before.* Kinzler provided various work files related to the development of the flag and the plaque to Platoff, who generously provided scans of those documents to the author in 2017.

37 Platoff interview with Kinzler. Kinzler says in both his oral history and his interview with Platoff that he installed the flag and the plaque the morning before liftoff. But he appears to be misremembering this detail. His own "Weekly Activities Report" from the period, which Platoff found in the Johnson Space Center history office, records, "The flag and plaque were installed on the LM of Apollo 11 on Wednesday, July 9, 1969, at 4:00 a.m. under the supervision of Jack A. Kinzler."

38 Michael Collins and Edwin E. Aldrin, "The *Eagle* Has Landed," in Edgar M. Cortwright, ed., *Apollo Expeditions to the Moon*, NASA SP-350 (NASA: Washington, D.C.), 1975, p. 216.

39 "Apollo 11 Moon Walk CBS News Coverage," posted by zellco321, January 17, 2017, https://www.youtube.com/watch?v=ntyPG1xewJ8, flag planting starts at 01:53:00. Total video is 5:15:09.

40 Eric M. Jones, ed., "ALSEP Off-load," *Apollo 17 Lunar Surface Journal*, 118:19:22, https://www.hq.nasa.gov/alsj/a17/a17.alsepoff.html.

41 No warning of call, from Hansen, *First Man,* around pp. 506–7; the U.S. National Archives has video of the call as broadcast that night; a color photograph of President Nixon, in suit and tie, at his desk talking to the astronauts; and a separate picture of the president's green, push-button telephone: "A Historic Phone Call," National Archives, https://www.archives.gov/presidential-libraries/events/centennials/nixon/exhibit/nixon-online-exhibit-calls.html.

42 "Nixoning the Moon," *New York Times*, July 19, 1969, p. 24, https://timesmachine.nytimes.com/timesmachine/1969/07/19/90112986.html.

43 The presidential phone call to Armstrong and Aldrin, in the Apollo 11 transcript: Jones, ed., "One Small Step," starting at 109:52:40; in the video "Apollo 11 Moon Walk CBS News Coverage," starting at 2:01:35.

44 "Plaque and flag for Apollo 12," Memorandum from George M. Low to Robert R. Gilruth, September 6, 1969, Johnson Space Center History Office, copy provided to the author by Platoff.

45 Jones, ed., "ALSEP Off-load," *Apollo 16 Lunar Surface Journal*, 120:23:04, https://www.hq.nasa.gov/alsj/a16/a16.alsepoff.html.

46 Apollo 15 deployment, "million years": Jones, ed., "EVA-2 Closeout," *Apollo 15 Lunar Surface Journal*, 148:53:59, https://www.hq.nasa.gov/alsj/a15/a15.clsout2.html; Apollo 17 deployment, "hate to touch it": Jones, "ALSEP Off-load," *Apollo 17 Lunar Surface Journal*, 118:20:35 to 118:23:09; "hate to touch it" at 118:22:18; Apollo 12 deployment, "the flag is up": Jones, ed., "TV Troubles," *Apollo 12 Lunar Surface Journal*, 116:13:40 to 116:20:02, https://www.hq.nasa.gov/alsj/a12/a12.tvtrbls.html.

47 Eric M. Jones, ed., "Apollo 12 CDR and LMP Cuff Checklist," *Apollo 12 Lunar Surface Journal*, https://www.hq.nasa.gov/office/pao/History/alsj/a12/cuff12.html. For the curious, images of the cuff checklists for all 12 astronauts who walked on the Moon are indexed here: "Available Checklists," *Apollo Lunar Surface Journal*, https://www.hq.nasa.gov/alsj/surcl.html.

48 The quotes from Conrad and Bean about the Playboy playmates in their cuff checklists come from the *Playboy* magazine story about the prank, published 25 years after Apollo 12's mission: D. C. Agle, "Playmates on the Moon," *Playboy*, December 1994, pp. 138–39, 213.

49 The photo of Pete Conrad, taken by Alan Bean and also showing Bean's reflection and Reagan Wilson, is image #AS12-48-7071, available from the Apollo 12 mission photo library: https://www.hq.nasa.gov/alsj/a12/AS12-48-7071HR.jpg. Conrad's discovery that a Playboy playmate was photographed on the Moon: Agle, "Playmates on the Moon."

50 *Washington Post*: "'The *Eagle* Has Landed'—Two Men Walk on the Moon," in Abby Phillip, "The *Eagle* Has Landed": How The Post Covered the Apollo 11 Landing," July 21, 2014, https://www.washingtonpost.com/news/post-nation/wp/2014/07/21/the-eagle-has-landed-how-the-washington-post-covered-the-apollo-11-landing/?utm_term=.befc90b865da. *New York Times*: "Men Walk on the Moon," https://timesmachine.nytimes.com/timesmachine/1969/07/21/issue.html.

51 Fred Seibert, telephone interview, February 20, 2018; Siebert is cartoon producer and was working for MTV at the time the channel debuted. With his colleagues, he came up with the idea for using the NASA images of Aldrin and the flag as MTV's first animated logo. The resulting animation ran at the top and bottom of every hour for the first five years MTV was on the air, and became closely identified with the channel. At the meeting where he came up with the idea, Siebert remembers one of his colleagues saying, "Space is very rock 'n' roll." MTV stopped using the animation,

permanently, on January 28, 1986, the day of the space shuttle *Challenger* disaster.

MTV has a story explaining the origin of the animated logo: Madeline Roth, "Ever Wondered Why the VMA Statue is a Moonman?," MTV news, August 27, 2016, http://www.mtv.com/news/2924701/vma-statue-moonman/.

## 9: How Apollo Really Did Change the World

1 John F. Kennedy, "Special Message to the Congress on Urgent National Needs," May 25, 1961, *American Presidency Project*, https://www.presidency.ucsb.edu/documents/special-message-the-congress-urgent-national-needs.

2 The story of the lunar rovers being developed, canceled, and then revived through the cleverness and determination of GM engineers Sam Romano and Ferenc Pavlics is told well, and in varying overlapping detail, in at least four places. The accounts do not differ or conflict, but they have different emphases and different details. Main citations are here; specific details attributed just below. "To the Moon with Ferenc Pavlics," YouTube, posted March 19, 2015, by the Scottsdale Center for Performing Arts, written by Christine Harthun; produced, directed and edited by Jared White, https://www.youtube.com/watch?v=My4sr87MlhM; David Clow, "The Law of the Stronger: Ferenc Pavlics and the Apollo Lunar Rover," *Quest*, vol. 18, no. 1 (January 2011): 7–19; *Moon Machines*, Episode 6: "The Lunar Rover," directed by Duncan Copp, June 2008, Dox Productions for The Science Channel, Discovery Communications; Anthony Young, *Lunar and Planetary Rovers* (Berlin: Springer Books, 2007), pp. 3–23.

Romano quote, "I decided it can be done, it should be done": *Moon Machines,* 2:20, 10:30. Pie-shaped storage compartment: *Moon Machines,* 11:30. Details of the Pavlics model and Astronaut GI Joe: "To the Moon with Ferenc Pavlics," 14:50. Von Braun quote, "We must do this!": *Moon Machines,* 14:20. Key to GM's victory: *Moon Machines,* 17:10. Wheel design and piano wire: "To the Moon with Ferenc Pavlics," 14:50; *Moon Machines*, 23:50.

3 John Nobel Wilford, "Astronauts Explore Moon 6½ Hours, Drive Electric Car on Rough Terrain," *New York Times*, August 1, 1971, pp. 1, 49, https://timesmachine.nytimes.com/timesmachine/1971/08/01/issue.html.

"rocking-rolling ride": Eric M. Jones, ed., "Driving to Elbow Crater,"

*Apollo 15 Lunar Surface Journal*, 121:55:44, https://www.hq.nasa.gov/alsj/a15/a15.elbowtrv.html.

4 Geologists looking over astronauts' shoulders: Walter, Sullivan, "Expert Observers on the Earth Share Moon Explorers' Glimpses into the Past," *New York Times*, August 1, 1971, p. 49, https://timesmachine.nytimes.com/timesmachine/1971/08/01/issue.html.

Jones, ed., "Driving to Elbow Crater," "keep talking": 121:59:37, 180-degree spinout: 123:28:55.

5 Orloff, *Apollo by the Numbers*, p. 298.

6 Eric M. Jones, "The Genesis Rock," *Apollo 15 Lunar Surface Journal*, 145:42:44 to 145:43:40, https://www.hq.nasa.gov/alsj/a15/a15.spur.html; John Noble Wilford, "Astronauts Take 8-Mile Ride on the Moon, Thrilled by Discovery of Ancient Rocks," *New York Times,* August 2, 1971, pp. 1, 11, https://timesmachine.nytimes.com/timesmachine/1971/08/02/79146080.html.

7 John Noble Wilford, "Moon 'Genesis Rock' 4 Billion Years Old," *New York Times*, September 18, 1971, pp. 2, 57, https://timesmachine.nytimes.com/timesmachine/1971/09/18/issue.html.

8 Liftoff filmed from rover: John Noble Wilford, "Astronauts Leave Moon and Dock Safely; Ascent of the Module Televised to Earth," *New York Times,* August 3, 1971, pp. 1, 14, https://timesmachine.nytimes.com/timesmachine/1971/08/03/issue.html. Cost of rover: Richard Witkin, "Lunar Rover Gives U.S. Space Men More Mobility Than Previous Crews Had," *New York Times,* July 31, 1971, p. 8, https://timesmachine.nytimes.com/timesmachine/1971/07/31/issue.html. Rover stamp: United Press International, "Astronauts to Cancel New Stamp on Moon," *New York Times,* August 2, 1971, p. 11, https://timesmachine.nytimes.com/timesmachine/1971/08/02/issue.html.

9 Kennedy, "Speech to Congress on Urgent National Needs."

10 John M. Logsdon, "Perspective: John F. Kennedy's Space Legacy and Its Lessons for Today," *Issues in Science and Technology*, vol. 27, no. 3 (Spring 2011).

11 Roger D. Launius publication references: *Apollo: A Retrospective Analysis*, Monographs in Space History, No. 3 (Washington, D.C.: NASA, 1994), https://ntrs.nasa.gov/archive/nasa/casi.ntrs.nasa.gov/19940030132.pdf; "Public Opinion Polls and Perceptions of U.S. Human Spaceflight," *Space Policy*, vol. 19, no. 3 (August 2003), pp. 163–75, https://www.sciencedirect.com/science/article/abs/pii/S0265964603000390; "Heroes in a Vacuum: The Apollo Astronaut as Cultural Icon," *Florida Historical Quarterly*, vol. 87, no. 2 (Fall 2008), pp. 174–209, https://repository.si.edu/handle/10088/17609.

12 Earl Warren speech: Associated Press, "Social Studies Should Be before Space Research," *Lebanon (PA) Daily News*, June 2, 1969, p. 12, https://www.newspapers.com/image/14819395/. Edward Kennedy speech: Associated Press, "Kennedy Calls for Space Fund Cut after Moon Goal," *Troy (NY) Record*, May 20, 1969, p. 28, https://www.newspapers.com/image/58847677/.

13 "Questions and Answers at the News Conference Held by the Apollo 11 Astronauts," transcript, *New York Times*, August 13, 1969, p. 28, https://timesmachine.nytimes.com/timesmachine/1969/08/13/issue.html.

14 Amitai Etzioni, "A Critic Finds the Gains Weren't Worth the Efforts," *New York Times*, December 3, 1972, p. 68, https://timesmachine.nytimes.com/timesmachine/1972/12/03/93421883.html.

15 Claude Lévi-Strauss had such complicated, interesting, and entertaining views on the Moon landings, as he shared them with John Noble Wilford of the *New York Times*, that they are worth quoting in full. They are also an echo of the cultural moment of the late 1960s and early 1970s in both the U.S. and Europe: "In this sad century, in this sad world where we live, with the pressure of population, rapidity of communication, the uniformity of culture, we are closed, like a prison. . . . The Apollo shots open a little window. It is the one experience—vicarious, but we can follow it on TV—the one moment when the prison opens on something other than the world in which we are condemned to live." And yet, he said, space exploration also brings on "a tragic and deeply melancholic feeling." John Noble Wilford, "Last Apollo Wednesday; Scholars Assess Program: $25-Billion Space Project Ran 11 Years—Cape Kennedy Presses Countdown," *New York Times*, December 3, 1972, pp. 1, 68, https://timesmachine.nytimes.com/timesmachine/1972/12/03/issue.html.

16 Ibid.

17 John F. Kennedy, "Inaugural Address," January 20, 1962, *American Presidency Project*, https://www.presidency.ucsb.edu/documents/inaugural-address-2.

18 Eliza Griswold, "How 'Silent Spring' Ignited the Environmental Movement," *New York Times*, September 21, 2012, https://www.nytimes.com/2012/09/23/magazine/how-silent-spring-ignited-the-environmental-movement.html. Brief review of the impact of Ralph Nader and *Unsafe at Any Speed*: Casey Williams, "Nader Talks Car Safety on 50th Anniversary of 'Unsafe at Any Speed,'" *Chicago Tribune*, July 3, 2015, https://www.chicagotribune.com/classified/automotive/sc-cons-0702-autocover-unsafe-nader-50-20150626-story.html.

19 Eric Schatzberg, the historian of science from Georgia Tech, is the author of a book about the history of technology—the idea of technology, the use of technology, and the term "technology" itself: *Technology: Critical History of a Concept* (Chicago: University of Chicago Press, 2018). Schatzberg quotes are from telephone interview, April 5, 2018.

20 "The Computer in Society: The Cybernated Generation," *Time*, April 2, 1965, pp. 84–91, http://time.com/vault/issue/1965-04-02/page/88/. Data on Apple iPhone sales: "Q1 2018 Unaudited Summary Data," Apple Inc., undated, https://www.apple.com/newsroom/pdfs/Q1_FY18_Data_Summary.pdf; Tom Alexander, "The Unexpected Payoff of Project Apollo," *Fortune*, July 1969, pp. 114–17, 150–56.

21 Eldon Hall provides records of the cost of chips to MIT for Apollo in *Journey to the Moon*, p. 80. The cost of integrated circuits and computer chips in those early years is well-documented by Richard C. Levin, an economist who went on to be president of Yale University from 1993 to 2013. In 1982 he contributed a chapter to the book *Government and Technical Progress: A Cross Industry Analysis*: Levin, "The Semiconductor Industry," pp. 9–100. Historical integrated circuit prices: Table 2.8: Average Price Per Unit of Transistors and Integrated Circuits, 1954–1972, p. 36.

22 Gordon E. Moore first proposed in a paper in 1965 the idea that the number of transistors on an integrated circuit would double every two years and keep doubling that way for at least a decade; when he wrote that paper in 1965, the major customer for integrated circuits was still the U.S. government, and one of the only users in the country was Apollo. That pace of miniaturization and innovation didn't earn the name "Moore's Law" until 1975; his original paper is remarkably prescient. Gordon E. Moore, "Cramming More Components onto Integrated Circuits," *Proceedings of the IEEE* 86, no. 1 (January 1998)" 82–85, originally published in *Electronics*, April 19, 1965, pp. 114–17.

23 Impact of IBM's 360 series of computers: "IBM System/360," "Mainframe Computers: The First Mainframes," Computer History Museum, accessed November 30, 2018, https://www.computerhistory.org/revolution/mainframe-computers/7/161; Levin, "The Semiconductor Industry," p. 62.

24 Levin, "The Semiconductor Industry," p. 62. Collins on IBM 360: "System 360: From Computers to Computer Systems," in *IBM at 100*, IBM, accessed November 30, 2018, https://www.ibm.com/ibm/history/ibm100/us/en/icons/system360/. IBM 360s used by MIT and NASA: Tomayko, "Computers in Spaceflight," pp. 43, 254–55.

25 Levin, "The Semiconductor Industry," p. 62.

26 Ibid., p. 63, Table 2.17: Government Purchases of Integrated Circuits, 1962–1968.

27 Tomayko, "Computers in Spaceflight," pp. 33–34.

28 Levin, "The Semiconductor Industry," p. 64.

29 Moore, "Cramming More Components onto Integrated Circuits," p. 82.

30 J. A. N. Lee, "Computer Pioneers: Gordon Moore," *IEEE Computer Society*, 1995, https://history.computer.org/pioneers/moore.html.

31 Cost of Apollo, total, in real dollars, simply added up year by year: Orloff, *Apollo by the Numbers*, p. 281. Cost of Apollo, total, inflation-adjusted to 1974 dollars: *Hearings before the Subcommittee on Manned Space Flight of the Committee on Science and Astronautics,* U.S. House of Representatives, 93rd Congress, First Session, H.R. 4567 (superseded by H.R. 7528), part 2, p. 563, https://babel.hathitrust.org/cgi/pt?id=mdp.39015084762718;view=1up;seq=569.

32 R. D. Launius, "Managing the Unmanageable: Apollo, Space Age Management and American Social Problems," *Space Policy*, vol. 24, no. 3 (August 2008): 158–65, https://www.sciencedirect.com/science/article/abs/pii/S0265964608000465?via%3Dihub.

33 Annual spending on Apollo, by year: Orloff, *Apollo by the Numbers*, p. 281. Spending on tobacco products in the United States: Verner N. Grise and Karen F. Griffin, "The U.S. Tobacco Industry," Commodity Economics Division, Economic Research Service, U.S. Department of Agriculture, Agricultural Economic Report No. 589, September 1988, p. 7, https://naldc.nal.usda.gov/download/CAT10407134/PDF.

34 Financial cost of Vietnam: "Vietnam Statistics: War Costs: Complete Picture Impossible," *CQ Almanac 1975*, 31st ed., pp. 301–5, 1976, http://library.cqpress.com/cqalmanac/cqal75-1213988. Number killed in the Vietnam War: Ronald H. Spector, "Vietnam War: 1954–1975," *Encyclopedia Britannica*, November 14, 2018, https://www.britannica.com/event/Vietnam-War.

35 The line on the Apollo 11 plaque, bolted to one of the legs of the lunar module, is in the past tense: "We came in peace for all mankind." Tyson made this observation in a tweet on the occasion of the death of Neil Armstrong, on July 25, 2012: https://twitter.com/neiltyson/status/239474562662793217.

36 Siddiqi, *Challenge to Apollo*, pp. 688–94.

37 T. A. Heppenheimer, *The Space Shuttle Decisions: NASA's Search for a Reusable Space Vehicle*, NASA History Series, SP-4221 (Washington, D.C.:

NASA, 1999), p. 152, https://ntrs.nasa.gov/archive/nasa/casi.ntrs.nasa.gov/19990056590.pdf.

38 Jerome E. Schnee, "Space Program Impacts Revisited," *California Management Review*, vol. 20, no. 1 (Fall 1977), pp. 70–71. Comparative data for the number of astronomers and astrophysicists in the U.S. were provided by Richard Fienberg of the American Astronomical Society. Their membership ranks are a reasonable, if imperfect, proxy for the overall number of astronomers and astrophysicists in the U.S. In 1970 the AAS had 2,600 members; the U.S. population was then 205 million. At the end of 2018 the AAS had 7,600 members, and the U.S. population was 329 million. The AAS was the group whose members Senator Paul Douglas polled in 1961 to see what space scientists thought of the race to the Moon. Schnee's number for astronomers in the U.S. in 1970, 2,500, is very close to the number of AAS members in 1970, 2,600.

39 Eric M. Jones, "One Small Step," *Apollo 11 Lunar Surface Journal*," 1995, https://www.hq.nasa.gov/alsj/a11/a11.step.html, Aldrin at 109:43:24.

40 Logsdon, ed., *Exploring the Unknown: Selected Documents in the History of the U.S. Civil Space Program*, p. 446.

41 Pushinka settled in at the White House so well that she learned to climb the ladder to Caroline and John's slide in the backyard, and slide down. She eventually was bred with the dog the Kennedys already had, a terrier named Charlie, producing four Soviet American puppies. "Kennedys Get Puppy as a Gift from Khrushchev," *New York Times*, June 21, 1961, p. 14, https://timesmachine.nytimes.com/timesmachine/1961/06/21/101467118.html; Associated Press, "Khrushchev Sends Kennedys Daughter of Russ Space Dog," *Baltimore Sun*, June 21, 1961, p. 6, https://www.newspapers.com/image/377431225; Alison Gee, "Pushinka: A Cold War Puppy the Kennedys Loved," BBC, January 6, 2014, https://www.bbc.com/news/magazine-24837199.

42 UPI, "State Agriculture Chief Hits Kennedy's Farm Program as Method of Black Jacking," *Montana Standard* (Butte), May 15, 1962, p. 5, https://www.newspapers.com/image/354818012/.

43 In a wide search of newspaper and magazine databases going back to 1955, Lowell Purdy, the agriculture commissioner of Montana, was found to be the first public official recorded using the phrase "If we can put a man on the Moon." Databases consulted include *Newspapers.com* and the archives of the *Washington Post* and the *New York Times*. The search included three variations on the phrase: "we can put a man on the moon" (the most

common); "we can send a man to the moon" (second most common); and "we can land a man on the moon." The qualifier "if" was left out because people often use the phrase directly: "We can send a man to the moon, so why can't we . . ."

There is one earlier published use of the phrase. A columnist for the *Daily Tar Heel,* the student newspaper at the University of North Carolina in Chapel Hill, used it in a column on November 8, 1958. The column, by a student identified only as J. Harper, appeared routinely on the paper's op-ed page under the heading "Harper's Bizarre." In this one, Harper mused on the power of individual connection with other people, one on one, and why we sometimes can't get along with even those closest to us: "Why is it that we can send a man to the moon, but John Jones can't get along with his next-door neighbor? Or even his wife, all the time?" In November 1958 NASA as an agency was only four months old, and the U.S. was three years from its first launch of a human into space.

44 Ann Waldron, "Here's Ann's Perfect 'House for Children,'" *St. Petersburg (FL) Times*, May 18, 1962, p. D-1, https://www.newspapers.com/image/316734273.

45 Account of the *Nautilus* voyage beneath the Arctic, via press conference transcript from Captain William R. Anderson: Associated Press, "Nautilus Commander Tells of History-Making Voyage," *Baltimore Sun*, August 9, 1958, pp. 1, 3; Felix Belair Jr., "Nautilus Sails under the Pole and 1,830 Miles of Arctic Icecap in Pacific-to-Atlantic Passage," *New York Times*, August 9, 1958, pp. 1, 6, https://timesmachine.nytimes.com/timesmachine/1958/08/09/91401153.html. On the celebration of Anderson and the *Nautilus* crew: Philip Benjamin, "Ticker-Tape Parade and City Hall Ceremony Acclaim Crew of Nautilus," *New York Times,* August 28, 1958, pp. 1, 20, https://timesmachine.nytimes.com/timesmachine/1958/08/28/79459869.html.

46 William Anderson appointed by Kennedy to study, then lead, domestic Peace Corps: William Knighton, "Aide Named for Domestic Peace Corps," *Baltimore Sun,* May 7, 1963, p. 6. The first-reported reference to Anderson's remark while testifying is from the *Baltimore Sun*, from the day after his testimony. But it includes only the first half of the quote. Associated Press, "Peace Corps Asked in U.S.," *Baltimore Sun,* May 28, 1958, p. 5. The full quote was distributed widely to U.S. newspapers as part of the "So They Say" quote feature. First reference: "So They Say," *Kingsport (TN) Times-News,* July 7, 1963, p. 23, https://www.newspapers.com/image

/24520306/. Search of database *Newspapers.com* for Anderson quote finds 29 references in 1963, accessed April 19, 2018.

47 Reference about garbage in Massachusetts (speaker was State Representative Joseph G. Bradley, D-Newton): Associated Press, "On Beacon Hill: Sees Attempt to Railroad Education Report Through," *North Adams (MA) Transcript*, January 21, 1965, p. 8, https://www.newspapers.com/image/54822054/. Reference to missing salmon in Idaho rivers (speaker was John R. Woodworth, director of Idaho's Fish and Game Department): Associated Press, "Idaho Awaits Washington, Oregon Salmon Fish Action," *Idaho State Journal*, April 24, 1966, p. 10, https://www.newspapers.com/image/15845450/.

48 Senator Robert Kennedy Jr. at California Senate hearing: Harry Bernstein, "2 Senators Term Farm Housing in Tulare 'Shameful,'" *Los Angeles Times*, March 16, 1966, p. 3, https://www.newspapers.com/image/382382960/. Governor Ronald Reagan campaigning for Nixon: "Reagan Slashes at Humphrey, Taps Wallace," *Jackson (TN) Sun*, October 13, 1968, p. 1, https://www.newspapers.com/image/283180843/. Vice President Hubert Humphrey campaigning for the presidency: Ernie Hernandez, "Humphrey Replaces 'Happiness' with 'Hope,'" *Pasadena (CA) Independent*, July 1, 1968, p. A1, https://www.newspapers.com/image/64982717/. Humphrey speech to the Westinghouse science awards, March 6, 1967: http://www2.mnhs.org/library/findaids/00442/pdfa/00442-02145.pdf.

49 The story of the father trying to call his stepson in Vietnam: Bill Cryer, "Telephoning Vietnam a Frustrating Task," *Austin (TX) American-Statesman*, December 28, 1968, p. 2, https://www.newspapers.com/image/355846406/. The remark from South Carolina senator James Waddell on his frustration with federal programs was first reported the day after he made it. It was then picked up by the *New York Times* two weeks later in a major series on poverty in the U.S. South, a set of stories that was widely republished across the U.S., including Waddell's comparison. Associated Press, "'We Can Send Man to the Moon but Can't Build an Outhouse,'" *Greenville (SC) News*, February 5, 1969, p. 10, https://www.newspapers.com/image/189189030/; Homer Bigart, "Hunger in America: Stark Deprivation Haunts a Land of Plenty," *New York Times*, February 16, 1969, pp. 1, 56, https://timesmachine.nytimes.com/timesmachine/1969/02/16/90049828.html.

50 Matt Weinstock, "Found at Last—Flexible Cliché for All Occasions," *Los Angeles Times*, September 11, 1967, part II, p. 6, https://www.newspapers.com/image/382422466/.

51 Matt Weinstock, "Lipchitz Sculpture under Public Gaze," *Los Angeles*

*Times*, June 2, 1969, part IV, p. 4, https://www.newspapers.com/image/383058058/.

52 Launius, "Managing the Unmanageable," p. 163; William Greider, "Protesters, VIPs Flood Cape Area," *Washington Post*, July 16, 1969, pp. A1, A7.

53 Voter registration rises dramatically: German Lopez, "How the Voting Rights Act Transformed Black Voting Rights in the South, in One Chart," *Vox*, August 6, 2015, https://www.vox.com/2015/3/6/8163229/voting-rights-act-1965. Black voter turnout, 1964 versus 1960: Alan Flippen, "Black Turnout in 1964, and Beyond," *New York Times*, October 16, 2014, https://www.nytimes.com/2014/10/17/upshot/black-turnout-in-1964-and-beyond.html. Poverty drops dramatically: Ajay Chaudry, et al., "Poverty in the United States: 50-Year Trends in Safety Net Impacts," U.S. Department of Health and Human Services, March 2016, p. 9, https://aspe.hhs.gov/system/files/pdf/154286/50YearTrends.pdf. Social Security cuts poverty: "Social Security and Elderly Poverty," National Bureau of Economic Research, undated, https://www.nber.org/aginghealth/summer04/w10466.html. Median income: Russell Sage Foundation, p. 1, https://www.russellsage.org/sites/all/files/chartbook/Income%20and%20Earnings.pdf.

54 Tim O'Brien, "It's Cape Canaveral Again," *Washington Post*, October 10, 1973, p. A2. The *Post* story reports that Rose Kennedy, the president's mother, wanted his name to stay on the Cape and called Representative Thomas P. (Tip) O'Neill, then just a congressman from Massachusetts, to object to restoring Canaveral. But at the time of the formal renaming back to Canaveral, Senator Ted Kennedy's office said the family had no objection, and the decision lay with the people of Florida.

55 Abelson, "Manned Lunar Landing."

56 Fred Ferretti, "TV Audience for Walk Is Called Disappointing," *New York Times*, February 6, 1971, p. 13, https://timesmachine.nytimes.com/timesmachine/1971/02/06/.

57 Associated Press, Edith M. Lederer, "People as Interested in Apollo Shot 'as Border War in Bolivia,'" *Sheboygan (WI) Press*, January 30, 1971, p. 1, https://www.newspapers.com/image/242825240.

58 "Farewell to the Moon," *New York Times*, December 15, 1972, p. 46, https://timesmachine.nytimes.com/timesmachine/1972/12/15/83452276.html.

59 Richard Witkin, "Substitute 747 Off for London; Engine Trouble Causes Delay," *New York Times*, January 22, 1970, pp. 1, 73, https://timesmachine.nytimes.com/timesmachine/1970/01/22/issue.html.

60 Yes, duct tape went all the way to the surface of the Moon. "Duct Tape Auto Repair on the Moon," Smithsonian National Air and Space Museum, August 31, 2015, https://airandspace.si.edu/stories/editorial/duct-tape-auto-repair-moon; "Duct Tape Saves the Day," NASA, September 24, 2015, https://www.nasa.gov/image-feature/duct-tape-saves-the-day. Details on the damage and repair: John Noble Wilford, "Astronauts, on 2nd Day, Find Orange Lunar Soil; Collect Ancient Rocks," *New York Times*, December 13, 1972, pp. 1, 50, https://timesmachine.nytimes.com/timesmachine/1972/12/13/issue.html.

61 "Body Men Honor Apollo 17," *Asbury Park (NJ) Press*, December 14, 1972, p. 1, https://www.newspapers.com/image/143908268.

62 Interestingly, the online version of the *Wall Street Journal* story has the clever "If we can put a man on the Moon" headline; the print version of the story was headlined "Major Shifts at NASA Urged." Andy Pasztor, "If We Can Put a Man on the Moon, Why Can't We Put a Man on the Moon?" *Wall Street Journal*, January 1, 2018, https://www.wsj.com/articles/if-we-can-put-a-man-on-the-moon-why-cant-we-put-a-man-on-the-moon-1514833480; "Overusages," *New York Times*, April 13, 1986, sec. 4, p. 24, https://timesmachine.nytimes.com/timesmachine/1986/04/13/issue.html.

63 "Newsroom of the Smithsonian: Visitor Stats," Smithsonian, accessed December 4, 2018, https://www.si.edu/newsdesk/about/stats; the Smithsonian museums are closed on Christmas.

64 All quotes from Kennedy's May 25, 1961 speech: John F. Kennedy: "Special Message to the Congress on Urgent National Needs," May 25, 1961, *American Presidency Project*, https://www.presidency.ucsb.edu/documents/special-message-the-congress-urgent-national-needs.

65 Bush quote on shopping: "President Holds Prime Time News Conference," White House, press release, October 11, 2001, https://georgewbush-whitehouse.archives.gov/news/releases/2001/10/20011011-7.html. On Disney World: "At O'Hare, President Says 'Get On Board,'" White House, press release, September 27, 2001, https://georgewbush-whitehouse.archives.gov/news/releases/2001/09/20010927-1.html.

66 The graph of retail spending produced by data from the Federal Reserve Bank of St. Louis is, frankly, astonishing. Spending by U.S. consumers during the first nine months of 2001 was unchanging or trending down, as the U.S. economy slogged through a recession. From September 2001 to October 2001 spending jumped 7.6%—in a single month—and it never again came back down to the levels of 2000 and 2001. You can make your

own graph here: "Advance Retail Sales: Retail (Excluding Food Services) [RSXFS]," U.S. Bureau of the Census, retrieved from Federal Reserve Bank of St. Louis, accessed December 5, 2018, https://fred.stlouisfed.org/series/RSXFS.

67 John F. Kennedy, "Remarks in San Antonio at the Dedication of the Aerospace Medical Health Center," November 21, 1963, *American Presidency Project*, https://www.presidency.ucsb.edu/documents/remarks-san-antonio-the-dedication-the-aerospace-medical-health-center.

# SOURCES / BIBLIOGRAPHY

## A Note on Sources

In early 2018, I was searching for the answer to a question about how the Apollo lunar module landed on the Moon, and I stumbled on a document that is at once remarkable and perfectly ordinary: "Investigation of Lunar Surface Chemical Contamination by LEM Descent Engine and Associated Equipment." We were flying to the Moon, in part, to bring home samples of Moon rock and soil, and to leave instruments to study the Moon. Five years into the effort, a group of scientists confronted the question: How do we possibly get good data—from the return samples, from the instruments left behind—without contaminating everything with material from Earth?

The report considered all kinds of things—engine exhaust as the lunar module landed, more exhaust as it took off, atmospheric venting from the lunar module cabin, out-gassing from the astronauts' spacesuits, including "flatus gases." The Moon, in fact, has a very thin atmosphere. So thin that it contains only slightly more gas than the lunar module would expel from its engine while landing on the Moon. Each time a lunar module landed on the Moon, it would bring enough alien gases with it to very nearly replace the entire lunar atmosphere (p. 6). That's why the scientists were worried.

It is, of course, a brilliant question, and an essential question: No scientist wants to study Moon rocks coated in an unintended glaze of rocket fuel or seasoned with organic molecules from a Moon walker. The report, written by fourteen scientists for Grumman Corporation (the company building the LM), and Arthur D. Little, a Cambridge, Massachusetts, consulting company, has an urgent edge of worry, especially for a scientific and engineering document.

"It is possible," the scientists wrote in the conclusion, "that whole fields of scientific investigation may be forever closed after the first manned mission" (p. 201). Human explorers to the Moon would bring a cloud of Earthly pollution, and in that way, cloud some of the point of going.

The report, which is 206 pages, is dated March 1966. It represents two central facts about researching and writing about the Moon missions: the 10,000 questions that had to be considered to fly to the Moon and back; and the documentation all that analysis and questioning left behind.

Robert Seamans, NASA's associate administrator during much of the 1960s, in the paper he wrote with Frederick Ordway about managing Apollo ("The Apollo Tradition"), tucked this remarkable tidbit into a footnote: "It is estimated that during the course of the Apollo development, some 300,000 tons of documentation were generated. In a single year, the Marshall Space Flight Center alone put out some 22 railway boxcars of data" (pp. 302–03). A single ream of paper—500 sheets—weighs a pound. So 300,000 tons of paper would come to 60 billion sheets. Boxed up, that amount of paper would fill the trailers of 13,636 long-haul trucks. The documentation for Apollo would fill a line of long-haul trucks stretching 200 miles.

The really remarkable thing is how much of that has been scanned and uploaded to computer servers and is accessible to anyone with an internet connection who punches in the right combination of words. NASA and its contractors not only documented how to think about contaminating the Moon, the agency eventually produced an inventory of every object that any nation had left on the Moon, from actual spacecraft to "earplugs" and "golf balls (2)," although the gases left behind are not accounted for ("Catalogue of Manmade Material on the Moon").

For a journalist or a historian, that documentation is both a blessing and a burden. For any particular question, there are answers. For any particular topic about which you might wish to write a few thousand words or a chapter, there is likely a 200-page report, or a book, or several books.

This is, then, necessarily a selective bibliography. Listed below are the sources I used to report and write *One Giant Leap*. They are not all the sources of information on a particular topic, or even all the sources I might have consulted in passing. (Just the twelve astronauts who have walked on the Moon have, between them, written fifteen books.)

With a few exceptions for particularly valuable stories, I have also not included the individual articles from newspapers and magazines that I relied on; there are hundreds. Those are cited individually in the Notes.

I have also not included links to specific YouTube videos in the bibliography; YouTube is an indispensable source of information from the era, and is easily searchable. Specific YouTube videos I relied on are cited in the Notes.

The standard bibliography style is not well adapted to the internet era. In the entries below, I have provided a link for every source that can be found easily on the internet. The bibliography will be posted online at the website of *One Giant Leap*, so those links are easily accessible, at www.onegiantleap.space.

## Online Resources

**American Presidency Project, John T. Woolley and Gerhard Peters, editors**

http://presidency.proxied.lsit.ucsb.edu/

An archive of presidential speeches, campaign speeches, and press conferences.

**Apollo Guidance Computer History Project, David A. Mindell, Slava Gerovitch, Alexander Brown, and Shane Hamilton, editors**

https://authors.library.caltech.edu/5456/1/hrst.mit.edu/hrs/apollo/public/index.html.

An archival-quality compilation of information about the computers that flew the Apollo spacecraft to the Moon. The site includes technical data and links to scientific papers published as the computer was developed, tested, and built, and also to some specialized journalism about the computer from the 1960s.

Just as important, the site contains the transcripts of four conferences held to understand the development of the computer and its impact. Those conference proceedings constitute an oral history of the computer, including stories not told anywhere else.

The site can be puzzling to navigate. Three of the conference transcripts are under the tab "Discussions." The transcript of the fourth conference is listed only under "What's New" on the archive's main home page.

**Apollo Lunar Surface Journals, Eric M. Jones and Ken Glover, editors**
https://www.hq.nasa.gov/alsj/main.html

**Apollo Flight Journals, David Woods, editor**
https://history.nasa.gov/afj/

The Apollo Lunar Surface Journals and Flight Journals are an extraordinary compilation of information about each of the Apollo missions. At the core, the journals include transcripts of all the ground-to-spacecraft communications for each mission. Those transcripts are edited to include commentary about what is happening moment-by-moment in the mission, and audio clips of some significant moments.

The Journals also include compilations of photos from each mission, mission rules as published by NASA, and a range of other items, including, for instance, press kits, guides to equipment, and images of the checklists astronauts wore on their spacesuits as they walked on the Moon.

**Dwight D. Eisenhower Presidential Library, Museum and Boyhood Home,**
https://www.eisenhower.archives.gov/research/online_documents.html

About Eisenhower's approach to Sputnik and the space program, the section, "Sputnik and the Space Race"
https://www.eisenhower.archives.gov/research/online_documents/sputnik.html

**John F. Kennedy Presidential Library and Museum**
https://www.jfklibrary.org

The Kennedy Library has a comprehensive collection of transcripts of President Kennedy's press conferences. It also has original images of many documents, available online, and photographs of events during the Kennedy presidency.

**Newspapers.com**

Newspapers.com is a little-known but extraordinary resource. It is a searchable collection of hundreds of U.S. newspapers, from every state, going back a century or more.

Each page of each issue of each newspaper has been scanned and uploaded—so when you search, you see the print newspaper as readers would have seen it in 1937 or 1967. But you can search the full text of all those newspapers from a single search box, as if the content were simply text. The content that matches your search is highlighted on the newspaper pages in the search results.

The result is not just access to the vast expanse of press coverage of American society going back one hundred years, but also access to how that coverage looked to readers—the size and positioning of headlines and stories, the photos that accompanied them. The content of advertisements also shows up in searches.

Newspapers.com charges a fee for access—but it is small compared to other online news databases, about $80 for six months.

## Tindallgram collections

The "Tindallgrams" are memos written by senior NASA manager and engineer Howard W. "Bill" Tindall during the second half of the 1960s, when Tindall was at the center of guiding Apollo spacecraft to the Moon and back. The memos are indispensable to understanding technical debates around Apollo, and also documenting the challenges that had to be overcome.

There is as yet no central or authoritative archive of Tindallgrams—either online or in a physical library. As of the end of 2018, there were three collections online, of varying quality and comprehensiveness. The designations below—Tindallgrams #1, etc.—are my own.

**Tindallgrams #1**

A collection assembled by MIT engineer and Tindall friend Malcolm Johnston; includes an introduction from Johnston and a table of contents. https://www.hq.nasa.gov/alsj/tindallgrams02.pdf

**Tindallgrams #2**

A collection assembled by NASA historian Glen Swanson; it is larger than the Johnston collection, but the scans are of uneven quality, with some memos unreadable, and the date order is inconsistent. https://www.hq.nasa.gov/alsj/tindallgrams01.pdf

**Tindallgrams #3**

A collection labeled "KSC," for Kennedy Space Center. It is the largest collection, and the memos are sorted by year, but the scan quality is uneven, and within years, the memos are not in consistent date order.
1967: https://www.hq.nasa.gov/alsj/1967_tindallgrams.pdf
1968: https://www.hq.nasa.gov/alsj/1968_tindallgrams.pdf
1969: https://www.hq.nasa.gov/alsj/1969_tindallgrams.pdf
1970: https://www.hq.nasa.gov/alsj/1970_tindallgrams.pdf

## Newspapers and Magazines

*Fortune*
*Life*
*The New York Times*
*Time*
*The Washington Post*

## Video and Audio

"John F. Kennedy and James Webb," sound recording of meeting in the Oval Office, the White House, September 18, 1963." John F. Kennedy Presidential Library and Museum, "Meetings: Tape #111: Lunar Program (James Webb)," 46 minutes of audio, https://www.jfklibrary.org/asset-viewer/archives/JFKPOF/MTG/JFKPOF-MTG-111-004/JFKPOF-MTG-111-004.

"Meeting on the NASA Budget: John F. Kennedy Presidency, Secret White House Tapes." The Miller Center of the University of Virginia, November

21, 1962, ten attendees, 57 minutes of audio, https://millercenter.org/the-presidency/secret-white-house-tapes/meeting-nasa-budget-0.

"MIT Science Reporter: Computer for Apollo (1965)." Hosted by John Fitch, YouTube, posted January 20, 2016, by the Vault of MIT, https://www.youtube.com/watch?v=ndvmFlg1WmE.

*Moon Machines*, Dox Productions for The Science Channel, Discovery Communications, 2008. Six episodes.

"To the Moon with Ferenc Pavlics," YouTube, posted March 19, 2015, by the Scottsdale Center for Performing Arts, written by Christine Harthun; produced, directed and edited by Jared White, https://www.youtube.com/watch?v=My4sr87MlhM.

## Articles and Papers from Magazines, Journals, and Online Periodicals

Abelson, Philip H. "Manned Lunar Landing," *Science*, vol. 140, no. 3564, April 19, 1963, p. 267, http://science.sciencemag.org/content/140/3564.

Agle, D. C. "Playmates on the Moon," *Playboy*, December 1994, pp. 138–39, 213.

Aldrin, Edwin Eugene Jr. "Line-of-Sight Guidance Techniques for Manned Orbit Rendezvous," Thesis, Sci.D., Massachusetts Institute of Technology, January 1963, https://dspace.mit.edu/handle/1721.1/12652.

Alexander, Tom. "The Unexpected Payoff of Project Apollo," *Fortune*, July 1969, p. 156.

Alonso, R. L., A. I. Green, H. E. Maurer, and R. E. Oleksiak. "R-358: A Digital Control Computer: Developmental Model 1B," Cambridge, MA: Instrumentation Laboratory, Massachusetts Institute of Technology, for the U.S. Air Force, April 1962, https://www.ibiblio.org/apollo/hrst/archive/1712.pdf.

Alonso, Ramon, and J. H. Laning Jr. "R-276: Design Principles for a General Control Computer," Instrumentation Laboratory, Massachusetts Institute of Technology, April 1960.

"Apollo Space Suit," International Latex Corporation, company brochure for ASME landmark designation, Septembger 20, 2013, https://www.asme.org/wwwasmeorg/media/ResourceFiles/AboutASME/Who%20We%20Are/Engineering%20History/Landmarks/ApolloBR.pdf.

Battin, Richard H. "On Algebraic Compilers and Planetary Fly-By Orbits," *Acta Astronautica*, vol. 38, no. 12 (June 1996): pp. 895–902.

_________. "Second Breakwell Memorial Lecture: 1961 and 'All That,'" *Acta Astronautica*, vol. 39, no. 6 (September 1996): 407–16, https://ac.els-cdn.com/S0094576596001531/1-s2.0-S0094576596001531-main.pdf.

_________. "Some Funny Things Happened on the Way to the Moon," *Journal of Guidance, Control, and Dynamics*, vol. 25, no. 1 (January–February 2002), pp. 1–7.

Blitz, Matt. "How NASA Made Tang Cool," *Food & Wine*, May 18, 2017, https://www.foodandwine.com/lifestyle/how-nasa-made-tang-cool.

Clow, David. "The Law of the Stronger: Ferenc Pavlics and the Apollo Lunar Rover," *Quest*, vol, 18, no. 1, January 2011, pp. 7–19.

Coplan, B.V., and R.W. King. "Applying the Ablative Heat Shield to the Apollo Spacecraft," Proceedings of the 4th Space Congress, April 3, 1967, https://commons.erau.edu/cgi/viewcontent.cgi?article=3525&context=space-congress-proceedings.

Crouse, Timothy. "Don Eyles: EXTRA! Weird-Looking Freak Saves Apollo 14!," *Rolling Stone*, March 18, 1971, https://www.rollingstone.com/politics/politics-news/don-eyles-extra-weird-looking-freak-saves-apollo-14-40737/.

David, Leonard. "The Moon Smells: Apollo Astronauts Describe Lunar Aroma," *Space.com*, August 25, 2014, https://www.space.com/26932-moon-smell-apollo-lunar-aroma.html.

Day, Dwayne A. "Spooky Apollo: Apollo 8 and the CIA," *The Space Review*, December 3, 2018, http://www.thespacereview.com/article/3617/1.

Draper, Charles Stark. "Origins of Inertial Navigation," *Journal of Guidance, Control and Dynamics*, vol. 4, no. 5, Sept.-Oct. 1981, pp. 449–63.

Duffy, Robert A. *Charles Stark Draper, 1901–1987: A Biographical Memoir* (Washington, D.C.: National Academy of Sciences, 1994), http://www.nasonline.org/publications/biographical-memoirs/memoir-pdfs/draper-charles.pdf.

Fisher, Jack. "Surveyor 1 and the American Flag," *Our Space Heritage: Hughes Aircraft Company*, March 11, 2016, http://www.hughesscgheritage.com/surveyor-i-and-the-american-flag-jack-fisher/.

_________. "The Story of Surveyor 1," *Our Space Heritage: Hughes Aircraft Company*, June 6, 2015, http://www.hughesscgheritage.com/the-story-of-surveyor-i-jack-fisher/.

Hall, Eldon C. "From the Farm to Pioneering with Digital Control Computers: An Autobiography," *IEEE Annals of the History of Computing*, April–June 2000, pp. 22–31.

Hancock, Jaime Rubio. "Margaret Hamilton, the Engineer Who Took the

Apollo to the Moon," *Verne*, December 25, 2014, https://medium.com/@verne/margaret-hamilton-the-engineer-who-took-the-apollo-to-the-moon-7d550c73d3fa.

Hardesty, Larry. "Apollo's Rocket Scientists," *MIT Technology Review*, November–December 2009, https://www.technologyreview.com/s/415796/apollos-rocket-scientists/.

Hattis, Philip D. "How Doc Draper Became the Father of Inertial Guidance," paper presented at the 2018 AAS Guidance and Control Conference, Breckenridge, Colorado, February 2018, https://www.researchgate.net/publication/322963658_How_Doc_Draper_Became_the_Father_of_Inertia_Guidance.

Hoag, David G. "The History of Apollo Onboard Guidance, Navigation and Control," *Journal of Guidance, Control, and Dynamics*, vol. 6, no. 1, (January–February 1983), pp. 4–13.

Honan, William H. "Le Mot Juste for the Moon," *Esquire*, July 1969, pp. 53–56, 139–41.

Jastrow, Robert, and Homer E. Newell. "The Space Program and the National Interest," *Foreign Affairs*, April 1972.

Johnson, Stephen B. "Samuel Phillips and the Taming of Apollo." *Technology and Culture*, vol. 42, no. 4 (Oct. 2001), pp. 685–709.

Jordan, John W. "Kennedy's Romantic Moon and Its Rhetorical Legacy for Space Exploration," *Rhetoric and Public Affairs*, vol. 6, no. 2 (Summer 2003): pp. 209–31.

Krugman, Herbert E. "Public Attitudes Toward the Apollo Space Program, 1965–1975," *Journal of Communication* (Autumn 1977), pp. 87–93.

Launius, Roger D. "Heroes in a Vacuum: The Apollo Astronaut as Cultural Icon," *Florida Historical Quarterly*, vol. 87, no. 2 (Fall 2008): pp. 174–209, https://repository.si.edu/handle/10088/17609.

__________. "Interpreting the Moon Landings: Project Apollo and the Historians," *History and Technology*, vol. 22, no. 3 (September 2006), pp. 225–255.

__________. "Managing the Unmanageable: Apollo, Space Age Management and American Social Problems," *Space Policy*, vol. 24, no. 3 (August 2008): pp. 158–65, https://www.sciencedirect.com/science/article/abs/pii/S0265964608000465?via%3Dihub.

__________. "Project Apollo in American Memory and Myth," Proceedings of Space 2000: Seventh International Conference and Exposition on Engineering, Construction, Operations, and Business in Space, Albuquerque, NM, February 27-March 2, 2000.

———. "Public Opinion Polls and Perceptions of U.S. Human Spaceflight," *Space Policy*, vol. 19, no. 3 (August 2003), pp. 163–75.

Leedham, Charles. "The 'Chip' Revolutionizes Electronics," *New York Times Sunday Magazine*, September 19, 1965, pp. 56ff, https://timesmachine.nytimes.com/timesmachine/1965/09/18/290267062.html.

Levinson, Arlene. "Top 100 News Stories of the Century," *Deseret News* (Salt Lake City, UT), April 15, 1999, https://www.deseretnews.com/article/691495/Top-100-news-stories-of-the-century.html.

Logsdon, John M. "Perspective: John F. Kennedy's Space Legacy and Its Lessons for Today," *Issues*, vol. 27, no. 3 (Spring 2011), http://issues.org/27-3/p_logsdon-3/.

"The Long Shadow from Apollo," *Nature*, vol. 226, April 18, 1970, pp. 197–98.

Loory, Stuart H. "Are We Wasting Billions in Space?" *Saturday Evening Post*, September 14, 1963, pp. 13ff.

Moore, Gordon E. "Cramming More Components onto Integrated Circuits," *Proceedings of the IEEE*, vol. 86, no. 1 (January 1998), pp. 82–85; originally published in *Electronics*, April 19, 1965, pp. 114–17.

Moore, Meg Mitchell. "J. Halcombe Laning Jr.," *MIT Technology Review*, November-December 2009, https://www.technologyreview.com/s/415816/j-halcombe-laning-jr-40-phd-47/.

Morgan, Christopher. *Draper at 25: Innovations for the 21st Century*, Cambridge, MA: Charles Stark Draper Laboratory, 1998, http://citeseerx.ist.psu.edu/viewdoc/download?doi=10.1.1.371.1536&rep=rep1&type=pdf.

Morris, Steven. "How Blacks View Mankind's 'GIANT LEAP'," *Ebony*, September 1970.

Murray, Don. "'Doc' Draper's Wonderful Tops," *Reader's Digest*, September 1957, p. 63–67.

Newport, Frank. "Despite Recent High Visibility, Americans Not Enthusiastic About Spending More Money on Space Program," Gallup, July 28, 1999, http://www.gallup.com/poll/3688/Despite-Recent-High-Visibility-Americans-Enthusiastic-About-Spend.aspx.

Oettinger, Anthony. "The Hardware-Software Complementarity," lecture, "Academic Role of Computers," Annual Meeting of the Division of Mathematical Sciences, National Academy of Sciences, March 13, 1967, reprinted in *Communications of the Association for Computing Machinery*, vol. 10, no. 10 (October 1967): pp. 604–6.

Parker, Martin. "Space Age Management," *Management & Organizational History*, vol. 4, no. 3 (August 2009), pp. 317–32.

Partridge, Jayne, and L. David Hanley. "The Impact of the Flight Specifications on Semiconductor Failure Rates," *Proceedings of the 6th Annual Reliability Physics Symposium*, Los Angeles, November 6–8, 1967, pp. 20–30, https://ieeexplore.ieee.org/document/4207753.

Phillips, Tony. "The Mysterious Smell of Moondust," *Science@NASA: Apollo Chronicles*, January 30, 2006, https://www.nasa.gov/exploration/home/30jan_smellofmoondust.html.

Radnofsky, Caroline. "Putting Man on the Moon," *Al Jazeera*, October 15, 2015, https://www.aljazeera.com/programmes/aljazeeracorrespondent/2015/10/putting-men-moon-151013082436203.html.

"Rendezvous Planner: Howard Wilson Tindall Jr.," *New York Times*, December 16, 1965, p. 29, https://timesmachine.nytimes.com/timesmachine/1965/12/16/95920363.html.

Ring, Paul. "When MetroWest went to the Moon: Area Companies Helped Apollo 11," *MetroWest Daily News* (Framingham, Massachusetts), July 19, 2009, http://www.metrowestdailynews.com/x1730895539/When-MetroWest-went-to-the-moon-Area-companies-helped-Apollo-11.

Roth, Madeline. "Ever Wondered Why the VMA Statue is a Moonman?" *MTV news*, August 27, 2016, http://www.mtv.com/news/2924701/vma-statue-moonman/.

Ruth, Eric. "Man on the Moon," *University of Delaware* Messenger, vol. 25, no. 3 (Dec. 2017), http://www1.udel.edu/udmessenger/vol25no3/stories/alumni-reihm.html.

Schmidt, Stanley F. "The Kalman Filter: It's Recognition and Development for Aerospace Applications," *Journal of Guidance, Control and Dynamics*, v. 4, no. 1 (Jan.–Feb 1981): pp. 4–7.

Schnee, Jerome E. "Space Program Impacts Revisited," *California Management Review*, vol. 20, no. 1 (Fall 1977): pp. 62–73.

Scott, David. "Speech at the Opening of the Computer Museum," Marlboro, MA, June 10, 1982, *KLabs.org*, http://www.klabs.org/history/history_docs/ech/agc_scott.pdf.

Sheridan, David. "How an Idea No One Wanted Grew Up to Be the LM," *Life*, March 14, 1969, p. 20ff.

Sidey, Hugh. "How the News Hit Washington—With Some Reactions Overseas," *Life*, April 21, 1961, pp. 26–27.

"Something, Indeed, to Tell His Children," *Brown Alumni Monthly*, January 1966, p. 6, https://archive.org/details/brownalumnimonth664brow/page/n7.

"Soviet Satellite Sends US Into Tizzy," *Life*, Oct. 14, 1957, pp. 34ff.

Speer, Raymond D. "Strict Control Kept out Semiconductor Flaws," *Electronic Design*, Aug. 16, 1969.

"Suiting the Spaceman," *Flight International*, September 10, 1970, pp. 422–23.

Tennant, Diane. "Forgotten Engineer Was Key To Space Race Success," *The Virginian-Pilot*, November 15, 2009, https://pilotonline.com/news/article_8f205f01-1ec7-54d5-a61f-1708350cbf66.html.

Tylko, John. "MIT and Navigating the Path to the Moon," *AeroAstro*, no. 6 (2008–9): p. 1, http://web.mit.edu/aeroastro/news/magazine/aeroastro6/aeroastro6.pdf.

Wellerstein, Alex. *Restricted Data: The Nuclear Secrecy Blog*, http://blog.nuclearsecrecy.com.

Wildenberg, Thomas. "Origins of Inertial Navigation," *Air Power History*, Winter 2016.

Wilford, John Noble. "Last Apollo Wednesday; Scholars Assess Program: $25-Billion Space Project Ran 11 Years—Cape Kennedy Presses Countdown," *New York Times*, December 3, 1972, pp. 1, 68, https://timesmachine.nytimes.com/timesmachine/1972/12/03/issue.html.

__________. "Race to Space, Through the Lens of Time," *New York Times*, May 23, 2011, https://www.nytimes.com/2011/05/24/science/space/24space.html.

Zak, Anatoly. Russian Space Web, http://www.russianspaceweb.com/.

## Interviews and Oral Histories

Two main oral history collections are referenced below, in addition to interviews conducted by the author and by others.

**The Johnson Space Center Oral History Project from NASA**
https://historycollection.jsc.nasa.gov/JSCHistoryPortal/history/oral_histories/oral_histories.htm.

**The Glennan-Webb-Seamans Project for Research in Space History, from the Smithsonian Air & Space Museum (which includes interviews from many more people than Glennan, Webb and Seamans)**
https://airandspace.si.edu/research/projects/oral-histories/ohp-introduction.html#GWS.

Arthur Applbaum, telephone interview, June 1, 2018

Richard H. Battin, NASA JSC Oral History Project, April 18, 2000

Leslie Berlin, telephone interview, July 13, 2016

Jeff Bezos, in-person interview, August 23, 2016

Willard Bischoff

*Telephone interviews:*

June 21, 2016
June 24, 2016
June 28, 2016
July 1, 2016

*In-person interview:*

July 8, 2016

Hugh Blair-Smith, telephone interview, October 16, 2018

Robert G. Chilton, NASA JSC Oral History Project, April 5, 1999

Aaron Cohen, NASA JSC Oral History Project

September 25, 1998
May 12, 1999

Brian Duff, Smithsonian Air & Space Oral History Project

April 24, 1989
Apriil 26, 1989
May 1, 1989
May 24, 1989

Don Eyles

*Telephone interviews:*

January 27, 2016
February 3, 2016

Maxime A. Faget, NASA JSC Oral History Project

June 18, 1997
August 19, 1998

Bill Famiglietti, telephone interview, August 7, 2017

Saul Ferdman, interview by Mike Marcucci, July 16, 2003

George C. Franklin, NASA JSC Oral History Project, October 3, 2001

John R. (Jack) Garman, NASA JSC Oral History Project, March 27, 2001

Joseph G. Gavin

NASA JSC Oral History Project, January 10, 2003

Interview by Mike Marcucci, August 3, 2003

Robert Gilruth, Smithsonian Air & Space Oral History Project, October 2, 1986

Charles R. Haines, NASA JSC Oral History Project, November 7, 2000

Fred Haise, interview by Mike Marcucci, July 16, 2003

Gene W. Harms, telephone interview, May 5, 2016

James Kirby Hinson, NASA JSC Oral History Project, May 2, 2000

Shirley H. Hinson, NASA JSC Oral History Project, May 2, 2000

Malcolm Johnston

*Telephone interviews:*

December 14, 2015

December 18, 2015

December 29, 2015

January 5, 2016

January 13, 2016

Thomas J. Kelly

NASA HQ Historical Reference Collection, interview by Robert Sherrod, December 13, 1972

NASA JSC Oral History Project, September 19, 2000

Jack A. Kinzler

*Interview by Anne Platoff:*

August 30, 1992

*NASA JSC Oral History Project:*

June 9, 1997

January 16, 1998

April 27, 1999

Joseph Kosmo, telephone interview, January 13, 2019

Christopher C. Kraft Jr., NASA JSC Oral History Project

May 23, 2008

April 14, 2009

February 11, 2010
Agusut 6, 2012

Ain Laats, telephone interview, December 1, 2015

Roger D. Launius, in-person interview, June 9, 2016

Dorothy "Dottie" Lee, NASA JSC Oral History Project, November 10, 1999

Scott A. Manatt
*Telephone interviews:*
July 5, 2016
July 7, 2016

Mike Marcucci, telephone interview, June 8, 2016.

Roger Marino, telephone interview, December 9, 2017

Michael Matonti
*Telephone interviews*
November 21, 2017
November 22, 2017
November 27, 2017
November 28, 2017
December 1, 2017

James A. McDivitt, NASA JSC Oral History Project, June 29, 1999

John Miller
*Telephone interviews:*
December 14, 2015
December 16, 2015
December 30, 2015

David A. Mindell, telephone interview, October 29, 2017

Thomas L. Moser, telephone interview, December 12, 2017

Richard L. Nafzger, NASA JSC Oral History Project, June 12, 2013

Catherine T. Osgood, NASA JSC Oral History Project, November 15, 1999

Michelle Pelersi, telephone interview, June 23, 2016

John William Poduska, telephone interview, December 14, 2017

Sean Redmond, telephone interview, February 24, 2018

William D. Reeves, NASA JSC Oral History Project, March 9, 2009

William Robertson

*Telephone interviews:*

December 12, 2017
December 15, 2017
December 18, 2017

Eric Schatzberg, telephone interview, April 5, 2018

Jerome Schnee, telephone interview, June 15, 2018

Robert C. Seamans Jr.

*Smithsonian Air & Space Oral History Project:*

December 4, 1986
February 25, 1987
April 9, 1987
November 2, 1987
December 8, 1987

*NASA JSC Oral History Project:*

September 30, 1998
November 20, 1998
June 22, 1999

Norman Sears

*Telephone interviews:*

January 15, 2016
January 18, 2016
January 22, 2016
February 3, 2016
February 10, 2016

Fred Seibert, telephone interview, February 20, 2018

Theodore C. Sorensen, JFK Library Oral History Interview, March 26, 1964

Erik Takacs, telephone interview, June 23, 2016

Jane Tindall, telephone interview, January 13, 2016

Brian Troutwine, telephone interview, March 10, 2018

James E. Webb, LBJ Library Oral History Collection, April 29, 1969

Ed Welsh, Lyndon Baines Johnson Library Oral History Collection, July 18, 1969

Claudia "Tiki" Tindall Williams, telephone interview, January 14, 2019

Ed Zander, telephone interview, February 26, 2018

## Books

Aldrin, Buzz, with Ken Abraham. *Magnificent Desolation*. New York: Three Rivers Press, 2009.

Ceruzzi, Paul E. *Beyond the Limits: Flight Enters the Computer Age*. Cambridge, MA: MIT Press, 1989.

_________. *A History of Modern Computing*. Cambridge, MA: MIT Press, 2003.

Clarke, Arthur C. *The Promise of Space*. New York: Harper & Row, 1968.

de Monchaux, Nicholas. *Spacesuit*. Cambridge, MA: The MIT Press, 2011.

Eyles, Don. *Sunburst and Luminary: An Apollo Memoir*. Boston: Four Point Press, 2018.

Gray, Mike. *Angle of Attack: Harrison Storms and the Race to the Moon*. New York: Penguin Books, 1994.

Hall, Eldon. *Journey to the Moon: The History of the Apollo Guidance Computer*. Reston, VA: American Institute of Aeronautics and Astronautics, 1996.

Hansen, James R. *First Man*. New York: Simon & Schuster, 2005.

*Historical Statistics of the United States: Colonial Times to 1957*. Washington, DC: Bureau of the Census, 1960.

Irwin, James B., with William A. Emerson, Jr. *To Rule the Night*, New York: Ballantine Books, 1974.

Johnson, Lyndon Baines. *The Vantage Point*. New York: Popular Library, 1971.

Johnson, Stephen B. *The Secret of Apollo: Systems Management in American and European Space Programs*. Baltimore: Johns Hopkins University Press, 2002.

Kelly, Thomas J. *Moon Lander: How We Developed the Apollo Lunar Module*. Washington, D.C.: Smithsonian Books, 2001.

Kraft, Chris. *Flight: My Life in Mission Control.* New York: Plume, 2002.

Kranz, Eugene. *Failure Is Not an Option.* New York: Simon & Schuster, 2009.

Lambright, W. Henry. *Powering Apollo: James E. Webb of NASA.* Baltimore: Johns Hopkins University Press, 1995.

Launius, Roger D., and Howard E. McCurdy. *Spaceflight and the Myth of Presidential Leadership.* Champaign, IL: University of Illinois Press, 1997.

Levin, Richard C. "The Semiconductor Industry," in *Government and Technical Progress: A Cross Industry Analysis.* Richard R. Nelson, ed. Elmsford, NY: Pergamon Press, 1982, pp. 9–100.

Logsdon, John M. *John F. Kennedy and the Race to the Moon.* New York: Palgrave Macmillan, 2010.

Makemson, Harlen. *Media, NASA, and America's Quest for the Moon.* New York: Peter Lang Publishing, 2009.

McDougall, Walter A. *. . .the Heavens and the Earth: A Political History of the Space Age.* New York: Basic Books, 1985.

Mieczkowski, Yanek. *Eisenhower's Sputnik Moment: The Race for Space and World Prestige.* Ithaca, New York: Cornell University Press, 2013.

Mindell, David A. *Digital Apollo: Human and Machine in Spaceflight.* Cambridge, MA: The MIT Press, 2011.

Murray, Charles, and Catherine Bly Cox. *Apollo.* Burkittsville, MD: South Mountain Books, 2004.

Nelson, Richard R., ed. *Government and Technical Progress.* Elmsford, NY: Pergamon Press, 1982.

Safire, William. *Before the Fall: An Inside View of the Pre-Watergate White House,* New York: Da Capo Press, 1975.

Salinger, Pierre. *With Kennedy.* New York: Doubleday, 1966.

Sayles, Leonard R., and Margaret K. Chandler. *Managing Large Systems: Organizations for the Future.* New Brunswick, NJ: Transaction Publishers, 1993.

Schatzberg, Eric. *Technology: Critical History of a Concept.* Chicago: University of Chicago Press, 2018.

Scott, David Meerman, and Richard Jurek. *Marketing the Moon: The Selling of the Apollo Lunar Program.* Cambridge, MA: The MIT Press, 2014.

Seamans, Robert. *Aiming at Targets: The Autobiography of Robert C. Seamans, Jr.* Honolulu: University Press of the Pacific, 2004.

Sidey, Hugh. *John F. Kennedy, President.* New York: Crest Books, 1964.

Slayton, Deke, with Michael Cassutt. *Deke! US Manned Space: From Mercury to the Shuttle.* New York: Forge Books, 1994.

Slotkin, Arthur L. *Doing the Impossible: George E. Mueller and the Management of NASA's Human Spaceflight Program*. New York: Springer-Praxis Books, 2012.

Sorensen, Ted. *Counselor*. New York: Harper, 2008.

__________. *Kennedy: The Classic Biography*. New York: Harper Perennial, 2009.

*Statistical Abstract of the United States*. Washington, DC: Bureau of the Census. Editions from the following years: 1962 through 1974, 1982-83. All editions of the *Statistical Abstract* are available online at: https://www.census.gov/library/publications/time-series/statistical_abstracts.html.

Stoff, Joshua. *Building Moonships: The Grumman Lunar Module*. Portsmouth, NH: Arcadia, 2004.

Thomson, William Tyrrell. *Introduction to Space Dynamics*. New York: Dover Publications, 1986.

Vladimirov, Leonid. *The Russian Space Bluff: The Inside Story of the Soviet Drive to the Moon*. New York: The Dial Press, 1973.

Watkins, Bill. *Apollo Moon Missions: The Unsung Heroes*. Lincoln, NE: Bison Books, 2007.

Wolfe, Tom. *The Right Stuff*. New York: Picador, 1979.

Woods, W. David. *How Apollo Flew to the Moon*. Chichester, UK: Praxis Publishing, 2009 edition.

Young, Anthony. *Lunar and Planetary Rovers*. Berlin: Springer Books, 2007.

## NASA Books, Reports, and Documents, and Apollo Contractor Reports and Documents

Aronowitz, L., et. al. "Investigation of Lunar Surface Chemical Contamination by LEM Descent Engine and Associated Equipment," Bethpage, NY, Grumman Research Department Report RE-242, January 1966, https://ntrs.nasa.gov/archive/nasa/casi.ntrs.nasa.gov/19660016286.pdf.

"Backgrounder: The Apollo Heat Shield." Press kit, Avco Space Systems Division, Lowell, Mass.

"Catalogue of Manmade Material on the Moon." NASA History Program Office, July 5, 2012, https://history.nasa.gov/FINAL%20Catalogue%20of%20Manmade%20Material%20on%20the%20Moon.pdf.

Brooks, Courtney G., James M. Grimwood, and Loyd S. Swenson Jr. *Chariots for Apollo: A History of Manned Lunar Spacecraft*. Washington, D.C.: NASA, 1979.

Compton, William David. *Where No Man Has Gone Before: A History of Apollo*

*Lunar Exploration Missions*. Washington, DC: NASA, the NASA History Series, 1989.

Copps Jr., Edward M., "Recovery from Transient Failures of the Apollo Guidance Computer," Instrumentation Laboratory, Massachusetts Institute of Technology, August 1968.

Cortright, Edgar M., ed. *Apollo Expeditions to the Moon*. Washington, DC: NASA, SP-350, 1975, https://history.nasa.gov/SP-350/cover.html.

Dick, Steven J., and Keith L. Cowing, eds. "Risk and Exploration: Earth Sea and the Stars," NASA Administrator's Symposium, Monterey, CA: Naval Postgraduate School, September 26-29, 2004, SP-2005-4701, https://history.nasa.gov/SP-4701/riskandexploration.pdf.

Fries, Sylvia Doughty. "NASA Engineers and the Age of Apollo," Washington, DC: NASA, the NASA History Series, SP-4104, 1992, https://history.nasa.gov/SP-4104.pdf.

Ginzberg, Eli, James. W. Kuhn, Jerome Schnee, Boris Yavitz. "Economic Impact of Large Public Programs: The NASA Experience," Salt Lake City, UT: Olympus Publishing, 1976, NASA CR-147952, https://ntrs.nasa.gov/archive/nasa/casi.ntrs.nasa.gov/19760016995.pdf.

Glennan, T. Keith. *The Birth of NASA: The Diary of T. Keith Glennan*. Washington, DC: NASA, 1993, https://history.nasa.gov/SP-4105.pdf.

Green, Constance McLaughlin, and Milton Lomask. *Vanguard: A History*. Washington, D.C.: NASA, NASA Historical Series, 1970.

Hacker, Barton C., and James M. Grimwood. *On the Shoulders of Titans: A History of Project Gemini*. Washington, D.C.: NASA, SP-4203, 1977.

Hand, James A., ed. "MIT's Role in Project Apollo: Volume 1: Project Management, Systems Development, Abstracts and Bibliography," Charles Stark Draper Laboratory, Massachusetts Institute of Technology, October 1971, http://web.mit.edu/digitalapollo/Documents/Chapter5/mitroleapollovi.pdf.

Hansen, James R. "Enchanted Rendezvous: John C. Houbolt and the Genesis of the Lunar-Orbit Rendezvous Concept," Monographs in Aerospace History #4, Washington, D.C.: NASA, December 1995, https://ntrs.nasa.gov/archive/nasa/casi.ntrs.nasa.gov/19960014824.pdf.

Heppenheimer, T. A. *The Space Shuttle Decisions: NASA's Search for a Reusable Space Vehicle*. Washington, D.C.: NASA, NASA History Series, SP-4221, 1999, https://ntrs.nasa.gov/archive/nasa/casi.ntrs.nasa.gov/19990056590.pdf.

Jones, R. L. "Evaluation and Comparison of Three Space Suit Assemblies," Houston: NASA Manned Spacecraft Center, Technical Note D-3482, July 1966, https://ntrs.nasa.gov/archive/nasa/casi.ntrs.nasa.gov/19660022897.pdf.

Launius, Roger D. "Apollo: A Retrospective Analysis," Monographs in Space History, No. 3, Washington, D.C.: NASA, 1994, https://ntrs.nasa.gov/archive/nasa/casi.ntrs.nasa.gov/19940030132.pdf.

Levine, Arnold S. *Managing NASA in the Apollo Era*. Washington, DC: NASA, the NASA History Series, SP-4102, 1982, https://history.nasa.gov/SP-4102.pdf.

Lickly, D. J., H.R. Morth, and B.S. Crawford. "Apollo Reentry Guidance," Cambridge, MA: Instrumentation Laboratory, Massachusetts Institute of Technology, NASA CR-52776, July 1963, https://www.ibiblio.org/apollo/Documents/R-415%20Apollo%20Reentry%20Guidance.pdf.

Logsdon, John M., ed. *Exploring the Unknown: Selected Documents in the History of the U.S. Civil Space Program*. Washington, D.C.: National Aeronautics and Space Administration, 1995.

"Lunar Module: Quick Reference Data." Grumman Corporation, https://www.hq.nasa.gov/alsj/LM04_Lunar_Module_ppLV1-17.pdf.

"Managing the Moon Program: Lessons Learned from Project Apollo, Proceedings of an Oral History Workshop." Johnson Spaceflight Center, Houston, TX, July 21, 1989, https://ntrs.nasa.gov/archive/nasa/casi.ntrs.nasa.gov/19990053708.pdf.

"Mariner-Venus 1962: Final Project Report." Pasadena, CA: Jet Propulsion Lab, NASA, SP-59, 1965, https://history.nasa.gov/SP-59.pdf.

McGee, Leonard A., and Stanley F. Schmidt. "Discovery of the Kalman Filter as a Practical Tool for Aerospace and Industry," Ames Research Center, Moffett Field, CA, NASA TM-86847, November 1985.

Milliken, J. Gordon, and Edward J. Morrison. "Aerospace Management Techniques: Commercial and Governmental Applications," Denver: University of Denver, NASA Contract 06-004-081, November 1971.

*NASA Historical Data Book, 1958–1968*, vol. 1. Washington, DC: NASA, SP-4012, 1976, https://history.nasa.gov/SP-4012v1.pdf.

"NASA Technical Report No. 32-1023: Surveyor 1 Mission Report." Pasadena, CA: Jet Propulsion Laboratory, 1966, volumes 1 to 3.

Orloff, Richard W. *Apollo by the Numbers: A Statistical Reference*. Washington, D.C.: NASA, 2000, https://history.nasa.gov/SP-4029.pdf.

Platoff, Anne M. "Where No Flag Has Gone Before: Political and Technical Aspects of Placing a Flag on the Moon," Houston: Johnson Space Center, NASA Contractor Report 188251, August 1993, https://ntrs.nasa.gov/archive/nasa/casi.ntrs.nasa.gov/19940008327.pdf.

"Proceedings of a Conference on Results of the First U.S. Manned Suborbital

Space Flight" (Mercury-Redstone 3). Washington, DC: NASA, NIH and NAS, June 6, 1961, https://msquair.files.wordpress.com/2011/05/results-of-the-first-manned-sub-orbital-space-flight.pdf.

"Quantifying the Benefits to the National Economy from Secondary Applications of NASA Technology." Princeton, NJ: Mathematica, Inc., June 1975.

Renzetti, N.A. "Tracking and Data Acquisition Support for the Mariner Venus 1962 Mission," Pasadena, CA: Jet Propulsion Lab, NASA Technical Memo 33-212, July 1965.

*Report to the President-Elect of the Ad Hoc Committee on Space* ("The Wiesner Report"). January 10, 1961, https://www.hq.nasa.gov/office/pao/History/report61.html

"Results of the First United States Manned Orbital Space Flight, February 20, 1962." Houston: NASA Manned Spacecraft Center, https://spaceflight.nasa.gov/outreach/SignificantIncidents/assets/ma-6-results.pdf.

Robbins, Martin D., John A. Kelley and Linda Elliott. "Mission-Oriented R&D and the Advancement of Technology: The Impact of NASA Contributions," University of Denver, NASA Contractor Report 126561, May 1972.

Seamans, Jr., Robert C. "Project Apollo: The Tough Decisions," Monographs in Aerospace History, No. 37, Washington, DC: NASA, SP-2007-4537, 2007, https://history.nasa.gov/monograph37.pdf.

Seamans, Jr., Robert C., and Frederick I. Ordway. "The Apollo Tradition: An Object Lesson for the Management of Large-scale Technological Endeavors, *Interdisciplinary Science Reviews*, vol. 2, no. 4 (1977): 270-304.

Siddiqi, Asif A. *Challenge to Apollo: The Soviet Union and the Space Race, 1945–1974*. Washington, DC: NASA, 2000, https://history.nasa.gov/SP-4408pt1.pdf.

Swanson, Glen E. *"Before This Decade Is Out": Personal Reflections on the Apollo Program.* Washington, DC: NASA, the NASA History Series, SP-4223, 1999.

Swenson. Jr., Loyd S., James M. Grimwood, Charles C. Alexander, *This New Ocean: A History of Project Mercury*. Washington, DC: NASA, SP-4201, 1966, https://ntrs.nasa.gov/archive/nasa/casi.ntrs.nasa.gov/19990026158.pdf.

Thompson, Floyd L., et. al. "Report of the Apollo 204 Review Board," Washington, DC: NASA, TM-84105, April 1967, https://history.nasa.gov/Apollo204/summary.pdf.

Tomayko, James E. "Computers in Spaceflight: The NASA Experience," Washington, DC, NASA, Contractor Report 182505, 1988, https://ia600304.us.archive.org/19/items/nasa_techdoc_19880069935/19880069935.pdf.

Tsiao, Sunny. *"Read You Loud and Clear": The Story of NASA's Spaceflight Tracking and Data Network*. Washington, D.C.: NASA, SP-2007-4232, 2008, http://www.rvtreasure.com/NASA/STADAN%20Network%20history.pdf.

## NASA Apollo Experience Reports

Holley, M. D., W. L. Swingle, S. L. Bachman, et al. "Apollo Experience Report: Guidance and Control Systems: Primary Guidance, Navigation, and Control System Development," Houston: Johnson Space Center, NASA Technical Note D-8277, May 1976, https://ntrs.nasa.gov/archive/nasa/casi.ntrs.nasa.gov/19760016247.pdf.

Lee, Dorothy B. "Apollo Experience Report: Aerothermodynamics Evaluation," Houston: Manned Spacecraft Center, NASA Technical Note D-6843, June 1972, https://ntrs.nasa.gov/archive/nasa/casi.ntrs.nasa.gov/19720018273.pdf.

Pigg, Orvis E., and Stanley P. Weiss. "Apollo Experience Report: Spacecraft Structural Windows," Houston: Johnson Space Center, NASA Technical Note D-7493, September 1973, https://www.lpi.usra.edu/lunar/documents/apolloSpacecraftWindows.pdf.

Sperber, K.P. "Apollo Experience Report: Reliability and Quality Assurance," Houston: Johnson Space Center, NASA Technical Note D-7438, September 1973, https://ntrs.nasa.gov/archive/nasa/casi.ntrs.nasa.gov/19730023037.pdf.

West, Robert B. "Apollo Experience Report: Earth Landing System," Houston: Johnson Space Center, NASA Technical Note D-7437, November 1973, https://ntrs.nasa.gov/archive/nasa/casi.ntrs.nasa.gov/19740003586.pdf.

# INDEX

# ABOUT THE AUTHOR

Charles Fishman is an award-winning reporter and *New York Times* bestselling author, and has been writing about space for thirty years.

As a *Washington Post* reporter, Fishman was part of the team investigating the aftermath of the space shuttle *Challenger* disaster. Fishman went on to write about space for *Fast Company*, *The Atlantic*, and *Smithsonian.*

Fishman started his career in newspapers. After the *Washington Post*, he was a writer for, then editor of, *Florida*, the Sunday magazine of the *Orlando Sentinel*; and then assistant managing editor at the *News & Observer* in Raleigh, North Carolina. Fishman was a founding staff member of *Fast Company*, the innovative business magazine, where he worked for more than a decade.

Fishman is the author of *The Wal-Mart Effect*, the first book to pierce Wal-Mart's wall of secrecy and explain how the retailer really works; and of *The Big Thirst,* the bestselling book about water in a generation. He is the coauthor of the #1 *New York Times* bestseller *A Curious Mind*, with Hollywood producer Brian Grazer.

In the wake of his reporting on the space shuttle program, Fishman earned a pilot's license. As part of the reporting for *One Giant Leap*, he flew twice on zero-gravity flights.

Fishman lives in Washington, DC, with his wife, also a journalist, their two children, and two Labradors.